INTERRELATIONSHIP AMONG AGING, CANCER AND DIFFERENTIATION

THE JERUSALEM SYMPOSIA ON QUANTUM CHEMISTRY AND BIOCHEMISTRY

Published by the Israel Academy of Sciences and Humanities, distributed by Academic Press (N.Y.)

1st JERUSALEM SYMPOSIUM: *The Physicochemical Aspects of Carcinogenesis* (October 1968)
2nd JERUSALEM SYMPOSIUM: *Quantum Aspects of Heterocyclic Compounds in Chemistry and Biochemistry* (April 1969)
3rd JERUSALEM SYMPOSIUM: *Aromaticity, Pseudo-Aromaticity, Antiaromaticity* (April 1970)
4th JERUSALEM SYMPOSIUM: *The Purines: Theory and Experiment* (April 1971)
5th JERUSALEM SYMPOSIUM: *The Conformation of Biological Molecules and Polymers* (April 1972)

Published by the Israel Academy of Sciences and Humanities, distributed by D. Reidel Publishing Company (Dordrecht, Boston and Lancaster)

6th JERUSALEM SYMPOSIUM: *Chemical and Biochemical Reactivity* (April 1973)

Published and distributed by D. Reidel Publishing Company (Dordrecht, Boston and Lancaster)

7th JERUSALEM SYMPOSIUM: *Molecular and Quantum Pharmacology* (March/April 1974)
8th JERUSALEM SYMPOSIUM: *Environmental Effects on Molecular Structure and Properties* (April 1975)
9th JERUSALEM SYMPOSIUM: *Metal-Ligand Interactions in Organic Chemistry and Biochemistry* (April 1976)
10th JERUSALEM SYMPOSIUM: *Excited States in Organic Chemistry and Biochemistry* (March 1977)
11th JERUSALEM SYMPOSIUM: *Nuclear Magnetic Resonance Spectroscopy in Molecular Biology* (April 1978)
12th JERUSALEM SYMPOSIUM: *Catalysis in Chemistry and Biochemistry Theory and Experiment* (April 1979)
13th JERUSALEM SYMPOSIUM: *Carcinogenesis: Fundamental Mechanisms and Environmental Effects* (April/May 1980)
14th JERUSALEM SYMPOSIUM: *Intermolecular Forces* (April 1981)
15th JERUSALEM SYMPOSIUM: *Intermolecular Dynamics* (Maart/April 1982)
16th JERUSALEM SYMPOSIUM: *Nucleic Acids: The Vectors of Life* (May 1983)
17th JERUSALEM SYMPOSIUM: *Dynamics on Surfaces* (April/May 1984)

VOLUME 18

INTERRELATIONSHIP AMONG AGING, CANCER AND DIFFERENTIATION

PROCEEDINGS OF THE EIGHTEENTH JERUSALEM SYMPOSIUM ON QUANTUM CHEMISTRY AND BIOCHEMISTRY HELD IN JERUSALEM, ISRAEL, APRIL 29–MAY 2, 1985

Sponsored by the Fondation Edmond de Rothschild, Paris, France
and
The National Foundation for Cancer Research, Bethesda, U.S.A.

Edited by

BERNARD PULLMAN
Institut de Biologie Physico-Chimique
(Fondation Edmond de Rothschild), Paris, France

PAUL O. P. TS'O
The Johns Hopkins University, School of Hygiene and Public Health,
Division of Biophysics, Baltimore, Maryland, U.S.A.

and

EDMOND L. SCHNEIDER
National Institutes of Health, National Institute on Aging,
Bethesda, Maryland, U.S.A.

D. REIDEL PUBLISHING COMPANY

A MEMBER OF THE KLUWER ACADEMIC PUBLISHERS GROUP

DORDRECHT / BOSTON / LANCASTER / TOKYO

Library of Congress Cataloging in Publication Data

Jerusalem Symposium on Quantum Chemistry and Biochemistry (18th : 1985)
Interrelationship among aging, cancer, and differentiation.

(The Jerusalem symposia on quantum chemistry and biochemistry ; v. 18)
Includes index.
1. Carcinogenesis–Congresses. 2. Aging–Physiological aspects–Congresses.
3. Cell differentiation–Congresses. 4. Cancer–Genetic aspects–Congresses.
5. Cancer–Age factors–Congresses. I. Pullman, Bernard, 1919-
II. Ts'o, Paul O.P. (Paul On Pong), 1929- . III. Schneider, Edmond L.
IV. Fondation Edmond de Rothschild. V. National Institute for Cancer Research.
VI. Title. VII. Series.
RC268.5.J47 1985 616.99'4071 85-18330
ISBN 90-277-2117-3

Published by D. Reidel Publishing Company,
P.O. Box 17, 3300 AA Dordrecht, Holland.

Sold and distributed in the U.S.A. and Canada
by Kluwer Academic Publishers,
190 Old Derby Street, Hingham, MA 02043, U.S.A.

In all other countries, sold and distributed
by Kluwer Academic Publishers Group,
P.O. Box 322, 3300 AH Dordrecht, Holland.

Printed in The Netherlands

TABLE OF CONTENTS

PREFACE

In 1980, a distinguished group of scientists gathered in Washington, D.C. for an International Symposium on Aging and Cancer. Among the recommendations of this Symposium was to convene a future meeting to discuss the molecular basis for interrelationships between aging and cancer when the appropriate scientific knowledge was available. That same year, the 13th Jerusalem Symposium on Quantum Chemistry and Biochemistry entitled "Carcinogenesis : Fundamental Mechanisms and Environmental Effects", was held, attended by some 50 international authorities in this field. At this meeting, it became clear that the fundamental process of carcinogenesis is intimately associated with differentiation, which must also be mechanistically related to aging. It was therefore proposed that the next Jerusalem Symposium on Cancer could provide the appropriate forum for the study on the interrelationship among cancer, aging and differentiation.

The impressive advances in our knowledge of the nature of the genome through molecular genetic and physical chemical techniques have now provided the opportunity to examine the interrelationships between these complex biological processes. Through the isolation, cloning and rearranging of genes we are able to dissect and manipulate the genome in a fashion that was unanticipated only a decade ago. At the same time, the increase in longevity and the increased numbers of individuals entering the last decades of life where cancer incidences are highest raise the profound and practical question of whether aging and cancer are linked through common mechanisms. It is equally important to examine the role of differentiation in the processes of aging and cancer, which in turn could provide insight into the mechanisms of age-associated diseases such as cancers.

In this volume, we are fortunate to have the perspectives of the leaders in diverse disciplines each examining the interrelationship of these biological processes. We intentionally chose a mixture of basic scientists, clinical investigators, as well as an epidemiologist to provide the optimum diversity. We also chose to have a wide range of topics from the physical chemical aspects of DNA configurations to the treatment of cancers in older patients. Among the many areas that will be explored are the roles of genetic instability, viruses and proto-oncogenes in cancer and aging.

The close and effective cooperation among the Israel Academy of Sciences and Humanities, the Hebrew University of Jerusalem, the Fondation Edmond de Rothschild of Paris and the National Institute on Aging, National

Institutes of Health, U.S.A., as well as the generous and essential financial support of the National Foundation for Cancer Research, Bethesda, Maryland, U.S.A., are gratefully acknowledged. It is hoped that this volume will provide the useful direction and impetus to future studies focusing on the fundamental relationship among differentiation, aging and cancer.

Bernard Pullman
Edward L. Schneider
Paul O.P. Ts'o

INTERRELATIONSHIPS AMONG AGING, CANCER AND DIFFERENTIATION

Edward L. Schneider, M.D.
National Institute on Aging, Bethesda, MD

Aging, cancer and differentiation represent three complex biological processes. In the past, the complexity of these processes has led many basic researchers to avoid studying these areas. Their argument, which remains valid, is that increased understanding of the nature of molecular and cellular biology will ultimately provide the key to understanding these complex biological processes. Another aspect of that argument was that basic genetic research was far more advanced in prokaryotic organisms where aging, cancer and differentiation are either absent or difficult to detect. However, the recent quantum leaps in technology in cellular and molecular biology have provided the opportunity for investigators to examine complex eukaryotic organisms where aging, cancer and differentiation significantly impact on their studies.

In this introduction, I will focus on aging and its relationship to cancer and differentiation, leaving discussions with primary interest on cancer or differentiation to the experts in these respective fields. Intensive support for aging research has only occurred in the last ten years. It is therefore not suprising that we know little about the fundamental nature of aging processes. We do know that cancer is a disease of our older population. While there are childhood cancers, 50% of malignancies occur in 12% of our population, those over age 65 (1). Some cancers are clearly age-dependent, and deaths from these cancers parallel age-specific mortality (2).

I would like to raise a few questions that might be interesting to explore in relation to aging and cancer. Is the age-associated increase in cancer morbidity and mortality related to the aging process *per se*, to the passage of time, or to a combination of these two factors?

B. Pullman et al. (eds.), Interrelationship Among Aging, Cancer and Differentiation, 1–5.

Experimental malignancies have defined latent periods. Humans exposed to certain carcinogens develop specific tumors after latent periods that can last for decades, e.g. over 20 years for mesotheliomas after asbestos exposure (3). Thus, older humans and older animals may get cancers merely due to the passage of time. Alternatively, intrinsic aging processes such as impaired DNA repair or immunodeficiency might predispose animals and humans to cancers. Several investigators have tried a variety of approaches to discriminate between these two possibilities. Unfortunately, the results of these studies do not provide a clear answer. At one extreme are the studies of Ebbesen where chemical carcinogen exposure of young and old skin transplanted to young hosts resulted in a higher frequency of carcinomas in the older skin tissues (4,5). By contrast, Peto and coworkers (6) found that animals administered carcinogens at different ages had similar time-dependent, but age-independent onset of malignant tumors. Therefore, it is still not clear whether intrinsic aging processes play a role in the age-related increase in cancer observed in man.

What can we learn from cell biology about the interrelationships between aging, cancer and differentiation? Human diploid cells in tissue culture have three choices: they can differentiate or age; often these possibilities are hard to distinguish. Alternatively, they can become transformed and escape from aging. Somatic cell hybridization studies have revealed that senescence appears to be dominant and transformation recessive (7,8). The next step is to discern the molecular basis for cellular transformation and senescence.

What can we learn from the new techniques in molecular genetics about interrelationships between aging, cancer and differentiation? There is increasing evidence that cellular proto-oncogenes are important for normal cell growth and differentiation. The Type II oncogenes, such as myc oncogene, appear to be involved in the first step of cellular transformation converting cells with limited lifespans to the phenotype of unlimited replication (9). This step is important to aging research as well as cancer research. It will be interesting to see if these proto-oncogenes are involved in cellular aging *in vivo* as well as *in vitro*.

What can we learn from new biochemical techniques that will contribute to our knowledge of aging, cancer and differentiation? Of particular interest are new insights into free radical interactions. While these molecules may be vital for the synthesis of prostaglandins and for the action of neutrophiles in killing bacteria, they may also contribute to certain cancers and to aging. In fact, one of the popular theories of aging invokes free radicals as the primary cause of aging and age-associated diseases (10). The work of Cerutti suggests that free radicals can damage cells through both genetic and epigenetic mechanisms (11). Antioxidants have been reported to protect against tumor promotion by phorbol esters, suggesting that these promotors may act through free radical mediated mechanisms (12). It has also been reported that antioxidants increase lifespan and reduce aging processes in rodents (13). However, the results of these studies have been questioned (14). The critical experiments which will clarify the role of free radicals in aging remain to be carried out.

What can we learn from cytogenetics about the interrelationship between aging, cancer and differentiation? Both malignant transformation and aging involve chromosomal abnormalities. Carcinogenesis is almost always accompanied by chromosomal aneuploidy (15). Aging, at least in human peripheral lymphocytes and oocytes, also features increased aneuploidy, albeit in a much less impressive manner (16). Both aging and cancer also produce alterations in sister chromatid exchanges. With aging, there is decreased sister chromatid exchange induction (17,18) while with certain cases of transformation there is a slight increase in these cytogenetic events (19). Finally, those human genetic disorders which feature chromosomal instability also display increased susceptibility to malignancies (20) as well as signs of premature aging (21).

What can we learn from interventions that affect aging, cancer and differentiation? One classic example is dietary restriction. Undernutrtion of calories coupled with appropriate nutrient supplementation in rodents will delay development and extend longevity (22). It will also substantially delay the onset of tumor formation (23). While we have been aware of these findings for almost 50 years, we are still far from an understanding of how undernutrition affects longevity and carcinogenesis.

I have tried to raise a few questions regarding the interactions of three important biological processes. Obviously, there are many more intriguing questions that can be raised. The three fields are at different stages of development, with aging being the least developed of them. It is therefore our hope that research related to cancer and differentiation may provide clues to understanding aging.

ACKNOWLEDGEMENTS

I would like to thank Drs. R. Shmookler Reis and J. Brody for their helpful comments.

REFERENCES

1. Brody, J.: this volume.

2. Brody, J. and Schneider, E.L.: submitted for publication.

3. Selikoff, I.J.: 1977, In, Cancer Risk of Exposure in Origins of Human Cancer. Eds.: Hiatt, H.H., Watson, J.D., Winstein, J.H., (Cold Spring Harbor Labs.), Cold Spring Harbor, New York, pp. 1765-1784.

4. Ebbesen, P.: 1984, Science 183, pp. 217-218.

5. Ebbesen, P.: 1977, J. Natl. Cancer Inst. 58, pp. 1057-1060.

6. Peto, R., Roe, F.J.C., Lee, P.N., Levy, L. and Clack, J.: 1975, Br. J. Cancer 32, pp. 411-426.

7. Pereira-Smith, O.M. and Smith, J.R.: 1982, Somate Cell Genetics 8, pp. 731-742.

8. Pereira-Smith, O.M. and Smith, J.R.: 1983, Science 221, pp. 964-966.

9. Cairns, J. and Logan, J.: 1983, Nature 304, pp. 582-585.

10. Harman, D.J.: 1956, Geront. 11, pp. 298-300.

11. Cerutti, P.: 1985, Science 227, pp. 375-381.

12. McCord, J.M.: 1985, NEJM 312, pp. 159-163.

13. Miquel, J. and Economos, A.C.: 1979, Exp. Geront. 14, pp. 279-285.

14. Schneider, E.L. and Reed, J.R.: 1985, NEJM 312, pp. 1159-1168.

15. Yunis, J.J.: 1983, Science 221, pp. 227-236.

16. Schneider, E.L.: 1985, Cytogenetics of aging. In, Handbook of the Biology of Aging, 2nd Edition. Eds.: Finch, C., Schneider, E.L., Van Nostrand Reinhold, New York (in press).

17. Kram, D., Schneider, E.L., Tice, R.R. and Gianas, P.: 1978, Exp. Cell Res. 114, pp. 471-475.

18. Schneider, E.L. and Gilman, B.: 1979, Human Genet. 46, pp. 57-63.

19. Nichols, W.W., Brandt, C.T., Toji, H.L., Goodley, M. and Segawa, T.N.: 1978, Cancer Res. 38, p. 960.

20. Ray, J.H. and German, J.: 1982, In, Sister Chromatid Exchange. Ed.: Sandberg, A.A., Alan R. Liss, Inc., pp. 553-577.

21. Martin, G.: 1977, Genetics syndromes in man with potential relevance to pathobiology of aging. In, Genetic Effects on Aging, Birth Defects: Original Article Series. Eds.: Bergsma, D., Harrison, D., National Foundation March of Dimes, New York.

22. McCay, C.M., Crowell, M.F. and Maynard, L.A.: 1935, J. Nutr. 10, pp. 63-79.

23. Fernandes, G., West, A. and Good, R.A.: 1979, Clin. Bull. 9, pp. 91-106.

DECELERATION OF CANCER MORTALITY RATES WITH AGE AND TIME

Jacob A. Brody, M.D., Associate Director and
Donald F. Everett, Jr., Statistician
Epidemiology, Demography, and Biometry Program
National Institute on Aging
National Institutes of Health
Bethesda, Maryland 20205

Mortality rates from all causes and from cardiovascular diseases rise exponentially with age. Cancer, on the other hand, first rises rapidly in middle life then plateaus, accounting for less than 12 percent of all deaths among persons over age 80. This pattern does not support a direct relationship between cancer and aging, or a simple association between cancer and cumulative exogenous or endogenous factors. Cancer deaths can be expected to represent an ever decreasing proportion of total mortality although a slight rise in absolute numbers should persist.

In 1979, for the first time the age-adjusted death rates for cancer in the United States declined[1]. This fact has produced surprisingly little comment. Two concepts are discussed in this article. The first is the very modest increase in age-specific mortality rates for cancer after age 65 which brings into question the widely held supposition that cancer pathogenesis is closely related to fundamental aging processes with a relentless increase in host susceptibility. The second is the apparently inevitable decline in cancer mortality as a proportion of total mortality as a result of demographic shifts which markedly increase the proportion of elderly in the total population.

CANCER MORTALITY AFTER AGE 65

Cancer has many causes and in a pure sense should not be aggregated. While death certificates have limitations, Doll and Peto state[2] that they are still useful indications of total cancer mortality, although after age 65 records of site specificity are less reliable. The factors we are discussing, however, are of sufficient magnitude as to minimize the relative impact of these imperfections. Figure 1 shows age-specific death rates for 1976 (a typical year). The well known exponential increase in age-specific mortality is seen with age. Cardiovascular disease (diseases of the heart,

B. Pullman et al. (eds.), Interrelationship Among Aging, Cancer and Differentiation, 7–13.

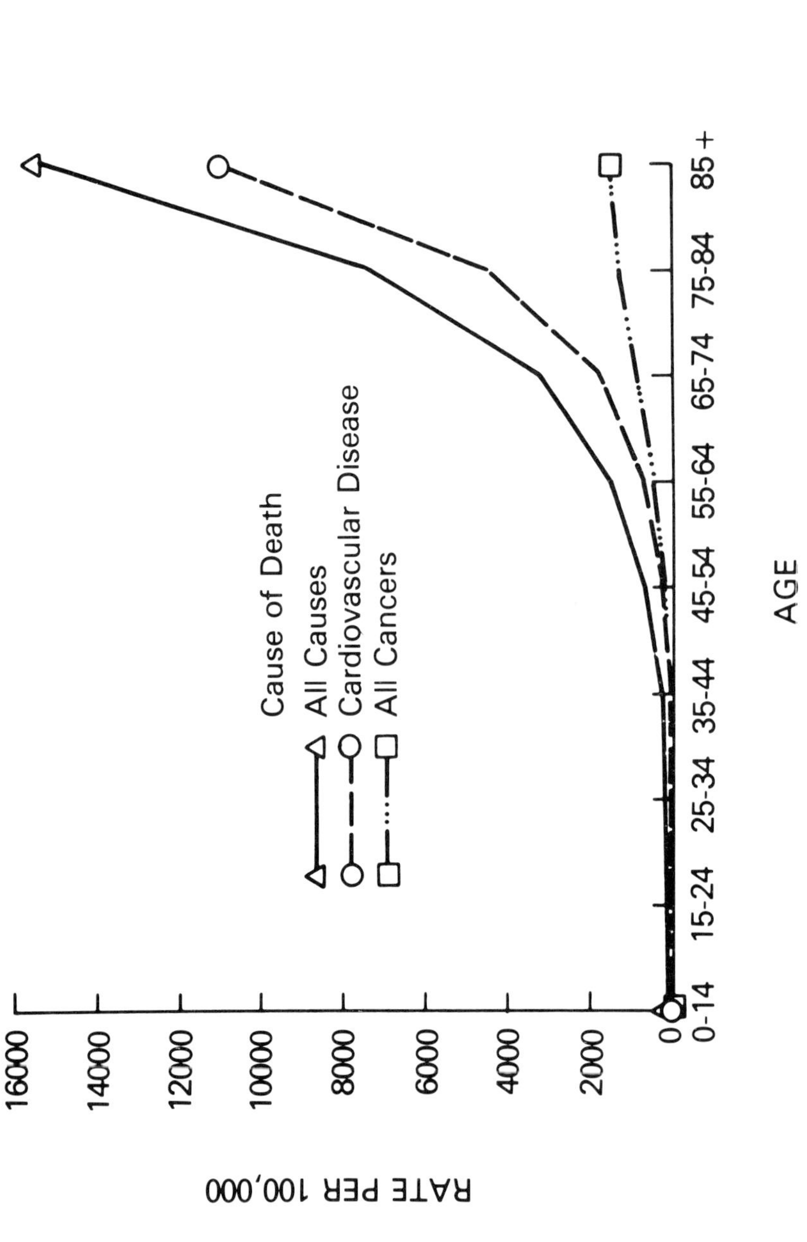

FIGURE 1: DEATH RATES PER 100,000 POPULATION BY CAUSE AND AGE GROUP, 1976

Source: Vital Statistics of the United States, Volume I — Mortality, Part A, National Center for Health Statistics, 1976

TABLE 1. BIOLOGICAL THEORIES OF AGING

1. General Genetic Theories
2. Cellular Genetic Theories--DNA Damage
3. Somatic Mutation--Radiation
4. Error Theory of Aging--Cellular
5. Wear and Tear Theory
6. Diffusion Theories
7. Accumulation Theories
8. Free Radical Theory
9. Cross-linking Theories
10. Cardiovascular System
11. Thyroid Gland
12. Sex Glands
13. Pituitary Gland--Neurohypophysis
14. Stress Theory
15. Immuniological Theories
16. Endocrine Control System
17. Nervous Control Mechanisms

Shock, N.W. Biological Theories of Aging. In: Bi and Schaie, K.W., eds. Handbook of the Psychology New York: Van Nostrand Reinhold Company, 1977:103

cerebrovascular diseases, and arteriosclerosis) also increased exponentially and roughly parallel to total deaths. Cancer, on the other hand, does not rise exponentially and actually tapers off at about age 80. In figure 2, cancer accounts for approximately 30 percent of all deaths for the group age 65 to 69, but by age 80 the proportion declines to under 12 percent. The increasing percentage of deaths caused by cardiovascular disease with aging is noted, rising from 50 percent in the age group 65 to 69, to 65 percent by the age of 80. Although lung cancer comprises a quarter of all cancer deaths its relative impact on these trends is not great (figure 2).

THEORIES OF BIOLOGICAL AGING AND CANCER ETIOLOGY

While no coherent biological theory of aging has gained uniform acceptance, most involve the accumulation of harmful factors or the progressive inability to perform life-sustaining or protective functions. One listing suggested by Shock (table 1) presents theories which have been proposed. The observation (figure 1) that cancer mortality rose and then leveled off at a time in life when deaths from all causes rose exponentially reinforces the knowledge that causes for cancers are complex. Simple theories depending on cumulative effects such as dysfunction at a cellular level, the gradual exhaustion of immunologic capabilities, or the progressive accumulation of harmful products, or radiation damage are not compatible with the observed

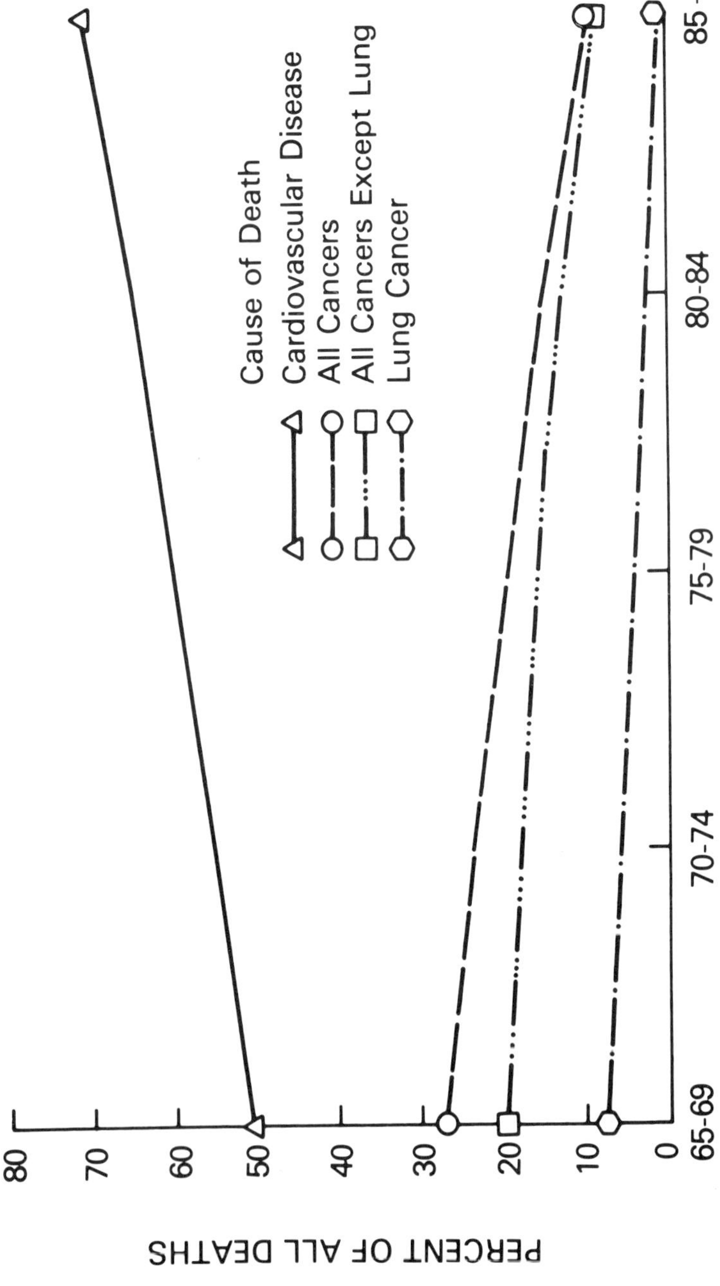

FIGURE 2: MORTALITY BY CAUSE AND AGE AS PERCENT OF TOTAL MORTALITY, 1976

Source: Vital Statistics of the United States, Volume I — Mortality, Part A, National Center for Health Statistics, 1976 decline in the proportional mortality from cancer in later years. Cancer does not occur in a simple relationship with an increasingly aged and vulnerable host.

Early genetic selection of the people most susceptible to carcinogens may occur, but absence of pervasive familial patterns for most cancers suggests that this is unlikely to be a major influence. Also, while the rise in cancer rates with age does persist, albeit at a decelerating rate, it is possible that a small portion of the observed trend could be explained purely by cumulative phenomena. Apparently, most carcinogens become pathogenic under specific and, as yet, poorly understood circumstances. These data could indicate the existence of protective factors which develop with age, perhaps, because cellular mutations are expressed more slowly or are more difficult to induce in cells of older hosts.

POPULATION DYNAMICS AND DECLINING CANCER MORTALITY

The population age 65 and over rose from approximately 3.1 million people in 1900 to 26 million in 1980[3, 4] accompanied by a rise in the percent age 65 and over of the total population from 4.1 to 11.3. Projections by the U.S. Bureau of the Census are that the number of persons over age 65 will reach 33.5 to 36.6 million and constitute about 13 percent of the total population by the year 2000[4]. Within 40 years (2025) as the post World War II baby boom cohort ages, those 65 and over will number approximately 58.5 million and comprise 20 percent of the total, barring major medical advances[4].

Among the 11.3 percent of the population currently over age 65, deaths in this age group comprise 67.2 percent of all deaths[5], compared to 24 percent in 1900. The population 80 years and over is 2.3 percent of the total and accounts for 30.6 percent of all deaths, while at age 85 and over, 1.0 percent of the total population account for 16.8 percent of total mortality.

During the next century, life expectancy will rise to 85 years[6] or more [7, 8, 9, 11, 12]. Given current knowledge, cancer will account for no more than 10 percent of deaths in those age 85 and over, a group experiencing half of all fatalities.

We noted that for the first time the age-adjusted death rates for cancer in the United States declined in 1979[1]. Slight rises occurred in subsequent years. The decline could, of course, be an artifact of mortality statistics, or a true sign that we are in a period in which therapeutic intervention is curing or postponing enough cancer deaths to produce this picture. It is possible, however, that since cancer causes a progressively smaller proportion of deaths in age groups 65 and over, that the shift in deaths to older age groups was responsible for this decline in cancer deaths. Whether artifactual or real, whether medical triumph or demographic shift, cancer mortality statistics bear the closest scrutiny. It has been suggested that planning of prevention trials for heart disease would have been different if we had recognized the importance of early declines in heart disease mortality[10].

DISCUSSION

Medicine is increasingly occupied with the population over age 65 because of the large and increasing number of people in this age range and the disproportionate utilization of health and medical services by older individuals. Several popular beliefs are questioned by the data presented. The death rate from cancer does not accelerate steadily with age. Instead it rises throughout most of adult life, but at about age 80 and over the increase is modest, and cancer assumes a diminishing role in overall mortality. Thus, notions which suggest that cancer emerges because the body ages and progressively loses certain protective functions or because of an accumulation of deleterious substances are oversimplifications. We recommend that thought be given to causal concepts of cancer which relate carcinogens with age-specific factors such as the rate of accumulation, and the status of the host's coping and responding mechanisms.

We cannot ignore the fact that by the year 2025, 20 percent of our population, or some 60 million people will be age 65 and over. All indications are that about half of all deaths will then be occurring in those 80 and over. Less than 12 percent of deaths in the aged population will be from cancer. Unalterable demographic forces will cause increasing declines in cancer deaths relative to total mortality. Health care and financial planning must commence with this realization about the future needs relating to cancer. It is unlikely that we will empty cancer wards as we did tuberculosis sanitoriums. We do, however, believe that meaningful research strategies, policies, and allocations are in order. Basic research into cause, prevention, and cure of cancer remains of prime importance. In terms of clinical practice, there will be an increased demand for oncologists, but an even greater demand for other types of physicians.

REFERENCES

1. U.S. Department of Health and Human Services. National Center for Health Statistics. Advanced Report of Final Mortality Statistics, 1979, Monthly Vital Statistics Report, Vol. 31, No. 6, Supplement, DHHS Publication No.(PHS) 82-1120, September 30, 1982.

2. Doll, R. & Peto, R. The causes of cancer: Quantitative estimates of avoidable risks of cancer in the United States today. Journal of the National Cancer Institute, 66, 1191, 1981.

3. U.S. Bureau of the Census, 1979c. Social and economic characteristics of the older population: 1978, Special studies series p-23, No. 85. U.S. Government Printing Office, Washington, DC, 1979.

4. U.S. Bureau of the Census. America in transition: An aging society, Special studies series p-23, No. 128. U.S. Government Printing Office, Washington, DC, 1983.

5. U.S. Department of Health and Human Services. National Center for Health Statistics. Final Mortality Statistics, 1978, Monthly Vital Statistics Report, Vol. 29, No. 6, Supplement (2), DHHS Publication No. (PHS) 80-1120, September 17, 1980.

6. Fries, J.F. Aging, natural death, and the compression of morbidity. New England Journal of Medicine, 303, 130,1980.

7. Brody, J.A. Life expectancy and the health of older people. Journal of the American Geriactics Society, 30, 681, 1982.

8. Kirscht, J.P. Patient education, blood pressure control, and the long run. American Journal of Public Health, 73, 134, 1983.

9. Manton, K.G. Changing concepts of morbidity and mortality in the elderly population. Milbank Memorial Fund Quarterly/Health and Society, 60, 183, 1982.

10. Multiple Risk Factor Intervention Trial Research Group: Risk factor changes and mortality results. Journal of the American Medical Association, 248, 1465, 1982.

11. Rosenwaike, I., Yaffe, N., & Sagi, P.C. The recent decline in mortality of the extreme aged: An analysis of statistical data. American Journal of Public Health, 70, 1074, 1980.

12. Schatzkin, A. How long can we live? A more optimistic view of potential gains in life expectancy. American Journal of Public Health, 70, 1199, 1980.

CLINICAL ASPECTS OF CANCER IN THE ELDERLY

Harvey Jay Cohen, M.D., Professor of Medicine
Director, Geriatric Research, Education and Clinical Center,
VA Medical Center and Center for the Study of Aging and Human
Development, Duke University, Durham, NC

ABSTRACT

Physicians are encountering increasing numbers of elderly cancer patients. Initial presentation and diagnosis may be complicated by changes of aging and co-morbid disease. There are differences in biology and behavior of cancer in the elderly including stage of presentaton, tumor type and aggressiveness. Patient management must include assessement of general functional and physiologic status and expected survival rates. Surgery, radiation therapy and chemotherapy can be utilized effectively, but individualization of treatment may be necessary. Psychosocial support and measures to maximize quality as well as quantity of life are important.

INTRODUCTION

Other discussions in this volume will address the basic scientific and epidemiologic aspects of the relationship of cancer and aging. From the standpoint of the clinician however, certain of the epidemiologic features point up the importance of this relationship. Thus, in the United States, over one-half of all cancer occurs in the 11% of the population over age 65, and it is the second or third leading cause of death in the elderly (1,2,3). Since this portion of our population is projected to increase dramtically over the next several decades, the physician will be increasingly confronted by decisions relating to the diagnosis and management of the elderly cancer patient. This chapter deal with some of the basic principles involved in the diagnosis and management of elderly patients with cancer, the importance of some of the clinical manifestations of the biologic changes of aging, the important treatment modalities, and then gives two specific contrasting examples of the response of older patients to malignant disease.

PRESENTATION AND DIAGNOSIS

The first problem facing the clinician is the potential for

B. Pullman et al. (eds.), Interrelationship Among Aging, Cancer and Differentiation, 15–21.

variation in the initial presentation and thus diagnosis of cancer in elderly patients. Diagnosis of cancer in an elderly patient may be complicated by the problem of either the physician or patient or both ignoring potential warning signs and assuming that they are simply changes due to age. Though it is quite clear, as will be discussed further below, that physiologic parameters, and thus overall homeostatic reserve, decrease with age, it is also true that these changes in general, do not result in specific diseases or symptom complexes but rather in an increased susceptibility to various environmental and other stresses (4). Failure to recognize this may lead first to delay on the part of the patient, who may be reluctant to admit to himself or to the physician that specific changes have occurred. The physician may further compound this by assuming that some of these changes are simply age-related. Thus, changes in bowel habits might be attributed to decreased motility with age rather than to the possibility of colon cancer. Many of the other potential symptoms relating to neoplasia are rather non-specific and might be similarily dismissed i.e., symptoms of anorexia, weight loss, general decreased performance status, aches and pains relating to potential bone pain or metastatic cancer, should be considered important clues to diagnosis in an elderly patient. This is especially true if any of these symptoms or signs represent changes in previous status. Thus, for example the new occurrence of anemia, even if mild, represents a change in status and should be evaluated in the elderly patient. This should not be assumed to be simply the "anemia of old age", which in reality does not appear to exist.

On the other hand, the fact that elderly patients frequently tend to have multiple medical problems, thus requiring them to interact more frequently with health care providers, creates an increased possibility that otherwise asymptomatic malignancy may be detected because of diagnostic tests done for other purposes. Thus, the chest x-ray obtained because of a unrelated viral illness might reveal a small lung cancer. The subsequent approach to such unsuspected diagnosis will depend on the physiologic status of the patient, not his chronologic age. There is also an increasing incidence of second and third primaries in the elderly. Thus, as patients are seen for longer and longer periods of time following the initial diagnosis of malignancy, the diagnosis of an unrelated malignancy must be considered if new symptoms are presented. In a number of instances, elderly patients may present with disease in more advanced stages than their younger counterparts. The reasons for this are not totally clear. It has been suggested in some cases that it is due to a greater delay in presentation for elderly patients. Though there is some suggestion that this may occur, it has not been directly correlated with the advanced stage of presentation (5).

On the other hand, these variations may be due to biological variations in tumor behavior in the older host. Such examples clearly exist. Thus, malignant melanoma, for example, tends to occur at a greater depth of penetration at presentation in the older individual, and in a somewhat different distribution of anatomic locations than in the younger counterparts (6). This is has not been directly correlated with any difference in time to presentation following the initial notation of a skin lesion. Thyroid carcinoma is noted to have a poor prognosis in older individuals. When observed more closely for histologic type one finds that elderly individuals have a disproportionate occurrence of anaplastic carcinoma compared to follicular and papillary carcinoma which occur to a greater extent in the younger patient (2). It would appear that this variation in initial tumor type may more responsible for the variation in prognosis than is either a delay in presentation or specific host factors.

PATIENT MANAGEMENT

The management of elderly patients with cancer must be undertaken with a firm knowledge of the changes in physiology occurring with age. Thus, an assessment of the patients overall functional status in addition to physiologic decrements in specific organ systems must be carefully undertaken (7). It is well appreciated that most organ systems undergo a physiologic decline with age that begins sometime after age 30. These declines place the elderly patient at potentially greater risk for the changes brought about by the malignant disease itself or by diagnostic and therapeutic interventions of the physician. Certain important features to be taken into account here include: changes in the cardiovascular system resulting in a decrease in cardiac output, in the pulmonary system resulting in decreased flow rates and decreased vital capacity; changes in the renal system resulting in decreased creatinine clearance; alterations in the nervous system with somewhat decreased nerve conduction times; alterations in gastrointestinal system with decreased peristaltic function of the esophagus, and late gastric emptying, and decreased muscle tone and motor function of the colon; alterations in the function of the bone marrow with some limitation of proliferative capacity of aged marrow cells; and major alterations in the immune system with marked decrease in thymic dependent lymphocyte mediated immune function.

These changes then represent potential problems but in an individual patient may not preclude effective diagnostic and therapeutic endeavors. On the other hand, life expectancy for elderly patients is often surprisingly good so that questions of therapeutic intervention can address not only quality of survival but also the issue of length of survival when discussing elderly patients. Thus, for example, a 70 year old woman has an approximately 15 year average life expectancy and a man approximately 11 years. Even an 80 year old woman has an approximately 9 year life expectancy while a man of that age has one of approximately 7 years (2). Thus, in the usual jargon of cancer treatment a "5 year cure" can be a major accomplishment within the life expectancy of such individuals.

The goals of therapy in such elderly patients must however, be clearly outlined since the heterogenity among the patient group is quite extreme. One must decide whether curative attempts are appropriate or whether treatment geared to maintaining the maximal functional status of an individual but not necessarily curing the disease would be more appropriate. In this regard, the social support system and the skills of a multidisciplinary team are frequently required and the patient's wishes must be strongly considered.

The major modalities of management include surgery, radiation therapy and chemotherapy.

Surgery

Surgery plays a major role in both diagnosis and treatment of neoplastic disease. The prospect of a surgical procedure for an elderly cancer patient may create fears and anxieties both in the patient himself as well as in middle aged children who are often assisting their parents through this crisis. The situation can be made extremely complex by the emotional interaction of children, patients and physicians at the time of these important decisions. The family's grief, and even guilt about their relationship with their elderly parent, may color these decisions. From the biomedical standpoint, it is now quite established that age per se is not contraindication to surgery (8). Though mortality does increase in patients of increasing age, this increase in mortality appears to be more correlated to the incidence of co-morbid disease and altered physiologic status than age itself. Thus, older patients who do not have additional complications of chronic pulmonary or cardiovascular disease appear to have surgical mortality not greatly different from younger individuals. Careful preoperative assessment and monitoring can result in successful outcomes of surgery frequently quite comparable to those in the younger patient.

Radiation Therapy

Radiation therapy can also be effectively delivered to the elderly patient. The exact degree to which radiation induced changes in normal tissue can be expected in the elderly, compared with the younger patient is not clear but it has been estimated that such effects may be enhanced by as much as 10 to 15% in the elderly (9). There is no evidence to suggest that a given cancer in an elderly individual varies in its radiosensitivity as a function of age. Radiation is sometimes offered as an alternative to surgery in a compromised elderly individual. However, it should be recognized that the stress of surgery, though intense, can be completed in a relatively short time and under careful monitoring. The stress induced by radiation is generally present for a much more prolonged period of time, from weeks to months, and may at times be more difficult for an elderly patient to tolerate. Certain specific side effects of radiation may be particularly bothersome; for example, therapy involving the head and neck may produce severe dryness of the mouth and loss of taste sensation which may be particularly bothersome, and abdominal radiation may produce nausea and vomiting leading to the possibility of dehydration which may be a particular problem for such elderly patients.

Chemotherapy

The approach to chemotherapy for elderly individuals had been, for a long time, not carefully studied. In fact, many of the large national trials of cancer treatment systematically excluded patients over the age of 65. This restriction has now been lifted. Chemotherapy has often been considered a frightening undertaking for elderly patients. The efficacy of a particular type of chemotherapy for the specific tumor type involved and the potential for major complications must be considered. Thus, for example, the relatively limited potential for response and remission with chemotherapy for metastatic lung cancer in an elderly patient must be weighed against the certainty of treatment side effects (10). In other cases where the chances for response are considerably higher such as in breast cancer or multiple myeloma, the benefits may outweigh the potential risks. The previously alluded to decline in physiologic reserve in elderly patients must be carefully considered in the design of the treatment regimen. Bone marrow toxicity is generally the major limiting factor at all age groups, and on theoretical grounds decreased bone marrow reserve of elderly patients might predispose to an increased toxicity from drug treatment (2). This however appears to vary considerably with treatment type. Thus, for example, in the treatment of certain solid tumors such as lung cancer, breast cancer and colon cancer (11) as well as in the hematologic malignancy of multiple myeloma, (12) in actual treatment trials, relatively little difference in bone marrow toxicity was observed for older compared with younger patients. On the other hand, in the treatment of acute myelogenous leukemia wherein much more intense therapy has been utilized, there is much more striking impact on bone marrow function and other physiologic reserves in the elderly resulting in a increased mortality rate during the initial phases of treatment (13). The approach to these two hematologic malignancies i.e., acute myelogenous leukemia and multiple myeloma can provide a interesting contrast in the potential outcomes of treatment among the elderly patient with neoplastic disease. In patients with multiple myeloma, it has been demonstrated that multi-agent chemotherapy can be effectively delivered to elderly patients and result in remissions and survival potential which is equal to that of younger patients. Moreover, the cost of such treatment in terms of toxicity is no greater (12). On the other hand, in the treatment of acute myelogenous leukemia, as mentioned above, initial toxicity may be quite high and the initial response rates considerably lower for elderly patients. This is complicated somewhat by the fact that many elderly patients with acute myelogenous leukemia have had antecedent hematologic disorders which predispose to poor prognosis regardless of age. Nevertheless, for those patients who survive the initial phase of treatment, survival for the elderly patient is approximately comparable to that for the younger, though in both cases immediate survivals are still quite short (13). The physician, patient and family must then enter into the treatment decision making process with these, often difficult to resolve, pieces of data in mind.

Supportive Care

The other aspect of the management of the elderly cancer patient which is of great importance is that of supportive care. This can take the form of recognition of the psychological and social needs of the elderly patient including the patient's expectations of therapy, the presence of a spouse in good health, proximity and willingness to help of friends and family, economic status and other such issues. The patient's view of death and dying and the meaning of quality of life to that individual may differ considerably from what the (frequently younger) health professional may assume from his own point of view. It must be remembered that it is the patient's point of view that is primary. Specific aspects of supportive care during treatment must also be considered. Thus, in an attempt to achieve the primary goal of the relief of suffering, treatment of pain related to neoplastic disease is of great importance. Elderly patients may have differences in pain perception with decreased pain response to a given stimulus. Moreover, analgesic medication, especially morphine, may achieve better pain control for comparable dose of drug in the elderly patient than in the younger because of changes in drug metabolism (14). Treatment of nausea and vomiting may be of great importance in the ability to achieve effective treatment responses. Drug choices for this purpose must be made carefully for the elderly patient since some of the side effects of these drugs may be particularly troublesome for them. For example, cannabinoids (THC) are frequently poorly tolerated by the elderly. On the other hand, certain drugs may be of even greater effectiveness. For example, metaclopramide may produce significantly fewer episodes of emesis in elderly patients than in a younger group despite similar blood levels for the two groups (15). The choice of these and other ancillary modes of therapy must be individualized for the particular patient.

Thus, the elderly patient with cancer presents a special set of diagnostic and therapeutic challenges for the clinician, which must be approached with careful attention to the patient's functional status and physiologic reserve. In general, considerations of these parameters and the presence of other major co-morbid diseases should take precedence over considerations of chronologic age per se. We still have a great deal to learn about the potential of elderly patients to respond to treatment regimens and the potential toxicities involved but it is encouraging that increasing attention is being paid to the systematic, careful scientific, exploration of these issues.

REFERENCES

1. Butler, R.N., and Gastel, B.: 1979, Cancer 29, pp. 333-340.
2 Crawford, J.C., and Cohen, H.J.: 1984, Annual Review of Gerontology and Geriatrics 4, pp. 3-32.
3. American Cancer Society: 1984, Ca-A Journal for Clinicians 34, pp. 7-23.
4. Fries, J.F.: 1980, New England Journal of Medicine 303, pp. 130-135.
5. Warnecke, R.B., Havlicek, P.L., and Manfredi, C.: 1983, "Perspectives on Prevention and Treatment of Cancer in the Elderly", pp. 275-287.
6. Levine, J.: 1981, Journal Dermatol. Surg. Oncol. 7, pp. 311-316.
7. Becker, P.M., and Cohen, H.J.: 1984, Journal Amer. Geriat. Soc. 32, pp. 923-929, 1984.
8. Johnson, J.C.: 1983, Journal Amer. Geriat. Soc. 31, pp. 621-625.
9. Gunn, W.G.: 1980, Cancer 30, pp. 337-347.
10. Simes, R.J.: 1985, Journal of Clinical Oncology 3, pp. 462-472.
11. Begg, C.B., and Carbone, P.P.: 1983, Cancer 52, pp. 1986-1992.
12. Cohen, H.J., Silberman, H.R., Forman. W., Bartolucci, A., and Liu, C.: 1983, Journal Amer. Geriat. Soc. 31, pp. 272-277.
13. Peterson, B.A.: 1982, "Adult Leukemias", pp. 199-235.
14. Kaiko, R.F.: 1980, Clin. Pharmacol. Ther. 28, pp. 823-826.
15. Meyer, B.R., Lewin, M., Drayer, D.E.: 1984, Annals of Internal Medicine 100, pp. 393-395.

OVERVIEW OF THE PATHOBIOLOGY OF AGING

George M. Martin
Departments of Pathology and Genetics
University of Washington
Seattle, Washington 98195
USA

ABSTRACT. The nature of biological senescence is reviewed with regards to definitions, phenotypes, genetic controls, potential mechanisms and certain experiments relevant to somatic mutational theories. Of special relevance to the theme of this conference is the evidence for a widespread loss of proliferative homeostasis and the evidence for chromosomal instability, both of which may contribute to the pathogenesis of neoplasia and might be related to aspects of differentiation.

Some Definitions

"Pathobiology" is a term popularized by Donald West King in the 1960's to emphasize the intimate relationships between investigations of normal and abnormal structure and function. Nowhere is the overlap greater than in biogerontology. The dissociation of normative aging from "disorders" of aging is a major delemma that resists any facile resolutions. One such attempt is embodied in the sometimes contrasting definitions of aging and senescence (Medawar, 1957; Leopold, 1975 and 1978), in which the term aging is used to describe essentially all biological phenomena that accompany "accruing maturity with the passage of time," and thus essentially includes all of developmental biology. The term senescence is then reserved for deteriorative processes that lead to the natural death of an organism. These deteriorative alterations in structure and function can be assumed to occur predominantly after sexual maturation. They involve very complex sets of interacting phenomena at all levels of organization, from molecules to populations of organisms. Collectively, these might be summarized thermodynamically as leading to a gradual increase in the entropy of the system.

If one accepts both definitions and includes both pre- and post-maturational development and pathobiology under the rubric of gerontology, which I believe one should do on scientific grounds, then one must essentially cover the universe of biological science in a

B. Pullman et al. (eds.), Interrelationship Among Aging, Cancer and Differentiation, 23–34.

"overview" of the field. Clearly, we can only emphasize a few concepts, especially those that are germane to the theme of our conference - the interfaces of aging, differentiation and cancer.

The Senescent Phenotype

A priori, there is no reason to insist that, in order to qualify as a marker of the senescent phenotype, a given trait should be observed in all species or in all members of a group of closely related species. One can imagine nature-nurture interactions such that "private" patterns of aging might emerge for certain species or for particular individuals within the species. Nevertheless, it is instructive to seek evidence for relatively universal phenomena of aging. I shall review two such examples, the first characteristic of aging organisms ranging from fungi to man and the second characteristic of aging in organisms whose somatic cells are composed, in part, of pools of cells subject to varying degrees of semi-conservative DNA synthesis.

Lipofuscins: Putative Markers of Lipid Peroxidation

Lipofuscin is a generic term for what are probably a heterogeneous group of fluorescent compounds that accumulate in the cytoplasm of aging cells, typically as lysosomal residual bodies (Eldred et al, 1982). At least one class of lipofuscins is widely regarded as a biproduct of lipid peroxidation, a principal intermediate being malondialdehyde, which is known to cross-link amine-containing compounds via NN^1-disubstituted 1-amino-3-imino-propene linkages (-NH-CH=CH-CH=N-) (Eldred et al, 1982). Although few age-related lipofuscins have yet been investigated with any degree of chemical rigor, such pigments appear to accumulate in the aging cells of an amazing variety of organisms, some of which are listed in Table 1. These observations have been interpreted as evidence in favor of the free radical theory of aging (Harman, 1956).

Loss of Proliferative Homeostasis

The senescing organism exhibits a progressive loss of the ability to maintain homeostasis for a wide variety of physiologic systems. A particularly striking aspect, especially well documented in the mammalian species, is the loss of proliferative homeostasis. The histopathologist observes, side by side with regions of tissue atrophy, multifocal areas of cellular hyperplasia. Table 2 lists many examples of seemingly inappropriate hyperplasias in the tissues of aging humans and other mammals. Such aberrations in proliferative homeostasis should be considered in any formulation of the evolution of age-related neoplasias. I shall return to this subject later in this review.

Table 1. Examples of organisms in which there is evidencee of age related accumulations of lipofuscin pigments.

Species Name	Common Name	Reference
Neurospora crassa	bread mold	Munkres and Rana, 1978a
Podospora anserina	dung fungus	Munkres and Rana, 1978b
Paramecium aurelia	paramecium	Sundaraman and Cummings, 1976
Caenorhabditis elegans	round worm	Klass, 1977
Bulla gouldiana	snail	Robles, 1978
Drosophila melanogaster	fruit fly	Sheldahl and Tapel, 1974
Musca domestica	house fly	Donato, 1978
Psittacula kameri	parrot	Singh and Munkherjee, 1972
Mus domesticus	house mouse	Miguel et al., 1978
Rattus norvegius	laboratory rat	Shimasaki et al., 1980
Cavia porcellus	guinea pig	Wilcox, 1959
Felus catus	domestic cat	Ives et al., 1975
Canis familiaris	domestic dog	Munnell and Getty, 1968
Sus scrofa	pig	Nanada and Getty, 1971
Macaca mulatta	rhesus monkey	Brizzee et al., 1974
Homo sapiens	man	Strehler et al., 1959

Evidence for Genetic Control of the Phenotype

The most dramatic and universal aspect of the senescent phenotype is the exponential increase in the probability of organismal death over time. The durations of the life phase between organismal maturation and the beginning of the exponential decline in survival varies enormously among species, the range being perhaps 30-40 fold amongst iteroparous mammals. This constitutes the most obvious line

Table 2. Examples of hyperplasia (typically multi-focal) which can develop during the aging of human subjects or other mammals (modified from Martin, 1979).

Cell Type Which Proliferates	Associated Age-Related Disorder	Reference
Adipocyte	Regional obesity	Bertrand et al, 1978
Arterial myointimal cell	Atherosclerosis	Geer and Haust, 1972
Astrocyte	Gliosis	Brizee et al, 1976
Bronchiolar epithelium	Unknown	Tanaka, 1984
Cartilage, osteocytes and synovial cells	Osteoarthritis (Osteoarthritis)	Sokoloff, 1972
Endometrial glandular epithelium	Post-menopausal hyperplasia	Parks et al, 1958
Epidermal basal cell	Verrucca senilis (Seborrheic keratosis) (Basal cell papilloma)	Hookey, 1931
Epidermal melanocyte and basal cell	Senile lentigo ("liver spots")	Hodgson, 1963
Epidermal squamous cell	Senile keratosis	Hookey, 1931
Fibroblast	Interstitial and regional fibrosis (multiple tissues)	Andrew, 1971
Lymphocyte	Ectopic lymphoid tissue	Andrew, 1971

Table 2, Continued

Cell Type Which Proliferates	Associated Age-Related Disorder	Reference
Lymphocyte (suppressor T cell)	Immunologic deficiency	Segre and Segre, 1976
Oral mucosal squamous cell	Leukoplakia	Bhaskar, 1968
Ovarian cortical stromal cell	Cortical stromal hyperplasia	Schneider and Bechtel, 1956
Pancreatic ductal epithelial cell	Ductal epithelial hyperplasia and metaplasia	Andrew, 1944
Sebaceous glandular epithelium	Senile sebaceous hyperplasia (skin)	Braun-Falco and Thianprusit, 1965
	Fordyce disease (oral mucosa)	Bhaskar, 1968

of evidence supporting a genetic determination of lifespan. Evidence that particular aspects of the senescent phenotype are subject to genetic controls comes from an examination of the effects of mutation and allelic variation within the species. For the case of man, I have systematiclly reviewed those predominantly abiotrophic mutations with effects that overlap to some extent with what one observes during ordinary aging (Martin, 1978; Martin, 1982). A surprisingly large proportion of the human genome (as an upper limit about 7%) may be affected by such mutations. Mutations at some genetic loci appear to accelerate discrete subsets of the senscent phenotype; I refer to such genetic diseases as "segmental progeroid syndromes" (Martin, 1978). A cardinal example is the Werner syndrome (Epstein et al, 1966), an autosomal recessive condition characterized by a wide variety of degenerative and proliferative lesions, including several forms of arteriosclerosis (atherosclerosis, medial calcinosis, arteriolosclerosis), osteoporosis, ocular cataracts, diabetes mellitus, skin atrophy and ulceration, calcification of heart valves, and diverse and often multiple benign and malignant neoplasms. The term "progeroid" was chosen quite deliberately, the suffix -"oid" indicating that, although the phenotypic features of such syndromes appear to be "like" what is found in ordinary aging, there may be distinct differences, including, of course, pathogenetic differences. For example, the ratio of mesenchymal to epithelial neoplasms in the Werner syndrome appears to be substantially higher than what is found in ordinary aging. Other examples of segmental progeroid syndromes are given in Table 3. Of special interest is the Down syndrome (trisomy 21), in which the etiology is an abnormality in gene dosage rather than a gene mutation. A particularly striking feature is that the subjects invariably develop the anatomic and biochemical pathology of dementias of the Alzheimer type by ages 35 to 40.

Some genetic disorders appear to affect predominately a single aspect of the senescent phenotype; I refer to those as unimodal

Table 3. Examples of putative segmental progeroid syndromes (after Martin, 1978). The numbers refer to the entries in the 6th (1983) edition of McKusick's catalogue of <u>Mendelian Inheritance in Man.</u>

17667	Progeria (Hutchinson-Gilford Syndrome)
20890	Ataxia-Telangiectasia
21640	Cockayne Syndrome
26970	Seip Syndrome
27770	Werner Syndrome
----	Down Syndrome (Trisomy 21)

progeroid syndromes (Martin, 1982) Table 4. The biochemical defect in many of these syndromes, and of certain of the segmental progeroid syndromes may be expressed in cell culture, thus opening the door to detailed biochemical genetic studies.

My studies of genetic variants in man and the evolutionary studies of Sacher (1975) and of Cutler (1975) leave little doubt that aging is subject to highly polygenic controls. Certainly hundreds of genes are likely to be involved in man. The only caveat is the possibility that a comparatively small subset of genetic loci (perhaps several dozen) may be of major relevance to aging as it ordinarily occurs. The recent evidence of the striking effects of single locus point mutations (heterochronic mutations) on the timing of discrete aspects of development in Caeorhabditis elegans, for example, is consistent with such major effects of a discrete subset of genes (Ambros and Horvitz, 1984). Another subset of great potential relevance are the genes for the metabolism of DNA, but even in E. coli, there are already dozens of loci known to be involved (Clark and Ganesan, 1975).

Classifications of Theories of Aging

There is no single satisfactory classification of subsets of the various extant theories of aging. A list of potential dichotomous classifications is given in Table 5. It would seem intuitively obvious that theories listed under the two columns are not mutually exclusive, although, in any given organism, one category could predominate. For example, senescence of many flowering plants is sharply programmed, the somatic cell death of the plant being tightly coupled to seed dispersion (Leopold, 1975 and 1978). The closest analogy in higher animals is found in those species of small insectivorous marsupial mice (of the genus Antechinus) that undergo "big bang" reproduction (Calaby and Taylor, 1981; Diamond, 1982).

Table 4. Examples of putative unimodal progeriod syndromes (after Martin, 1982). The numbers refer to the entries in the 6th (1983) edition of McKusick's catalogue of Mendelian Inheritance in Man.

10430	Alzheimer Disease (Presenile and Senile Dementia)
10500	Amyloidosis III (Cardiac Type)
14389	Hypercholesterolemia, Familial (LDL Receptor Disorder)
16235	Neuronal Ceroid-Lipofuscinosis (Parry Type)
17510	Polyposis, Intestinal, I (Familial Polyposis of the Colon)
27870	Xeroderma Pigmentosum I (Group A)

Table 5. Approaches to the classification of theories of aging.

• Changes in Gene Structure	vs.	Changes in Gene Expression
• Stochastic Events	vs.	Determinative or Programmed
• Pathological Alterations	vs.	Physiological Alterations
• Cell Injury	vs.	Cell Adaptation or Differentiation
• Uncoupled Peripheral Events	vs.	Primary Central Events (e.g., Neuroendocrine)

After an intense frenzy of copulation, in the late Australian winter, there is a rapid weight loss, regression of sexual organs and death. In most species, this pattern is particularly striking in males, but may occur in females as well (for example, in A. minimus). The synchronous passage from optimal structure and function to senescence and death occurs during the final few weeks of the 11.5 month life span. Thus, there is not much opportunity to evolve neoplasms. Most mammals, of course, are iteroparous and therfore do not exhibit such sharply programmed senescence and death. The hallmark of aging in man is its insidiously slow progression, at least within the first few decades after sexual maturation. This gives abundant opportunities for diverse nature-nurture interactions, especially considering the enormous genetic heterogeneity of our species and its seemingly limitless range of ecological niches. It is a reasonable proposition that no two human beings can ever be expected to age in exactly the same fashion. Each of us presumably has his own special pattern of vulnerability and resistance.

While environmental factors undoubtedly play important roles (do we have now to worry about "gerontogens" as well as mutagens, carcinogens and teratogens?), mechanisms of aging that assign crucial roles to alterations in the genomes of somatic cells (be they stochastic or programmed) would seem to me to have special merit. Table 5 summarizes a variety of such potential mechanisms. While the classification is based upon the first dichomotized pair listed in Table 5 (alterations in gene structure versus alterations in gene expression), it is by now obvious to all geneticists that normal, physiologic or adaptive changes in gene expression or of the differentiated state may be accompanied by alterations in gene structure. This can vary from the comparatively subtle and reversible inversion associated with phase variation in Salmonella (Zieg et al, 1977) to the complete loss of the genome that accompanies red blood cell maturation in mammals. Nevertheless, it is a convenient point of departure and does, in a general way, tend to relate to the other dichotomizations listed in Table 5.

My colleagues and I at the University of Washington have a special interest in exploring those mechanisms that predict the accumulation of various types of somatic mutations in the tissues of aging mammmals. The results to date, limited mainly to research on aging cohorts of Mus domesticus, give no evidence in support of the protein synthesis error catastrophe (Oregel, 1963 and 1970) or intrinsic mutagenesis (Burnet, 1974) theories of aging, which predict the accumulation of point mutations (Horn et al, 1984). On the other hand, there is striking evidence for increase in the frequency of diverse types of chromosomal mutations (Martin et al, 1985) certain of which might be predictable on the basis of the free radical theory of aging. Recall that the most universal marker of cellular aging, lipofuscin (Table 1) is also predicted by that theory, which emphasizes the importance of such gene products as the superoxide dismutases, catalases and various peroxidases in modulating the effects of active oxygen species. These effects almost certainly include clastogenesis.

My colleagues and I have also investigated the so-called "terminal differentiation" theory of aging as it applies to the phenomenon of the limited replicative life span of normal somatic cells. The latter is manifested by the gradual clonal attenutation of replicative potential (Martin et al, 1984). While it certainly would seem reasonable that certain somatic cell types (eg., epidermal basal cells and skeletal muscle myoblasts) differentiate themselves into reproductive death, there is yet no evidence that such mechansisms obtain for the case of mesenchymal cells, with which most of the research on "in-vitro aging" has been carried out (especially human diploid fibroblast-like cells).

In any event, the well documented phenomenon of clonal attenuation of normal somatic cells may have important implications for the pathogenesis of neoplasia. I have speculated elsewhere (Martin, 1979) that age-related asynchronous clonal attenuations among related families of differentiating cell types may set the stage for loss of proliferative homeostasis that characterizes senescent mammalian organisms. Thus, aging itself could be the major "agent" of tumor promotion. A number of theories of intercellular regulatory circuits emphasizing in particular negative feedback systems have been proposed. Unfortunately, molecular biologists have been mainly preoccupied with positive regulators of growth, mainly because of the ease of the bioassays.

Guidelines for Future Research

From the point of view of research on the interrelationships between differentiation, cancer and aging, it is unfortunate that the most elegant models for investigating the genetic control of aging processes (eg., Caenorhabditis elegans and Drosophila melanogaster) are organisms, the adult forms of which, consist almost entirely of obligate post replicative cells. Since there is no dynamic homeostasis of proliferative populations, one cannot investigate how these can go wrong and produce cancers. Mus domesticus would seem to be the best available experimental model. While there is currently considerable excitement over our ability to manufacture transgenic mice (Gordon, 1983) and there is still some hope for the utility of the mouse teratocarcinoma system (Bradley et al, 1984), old fashioned germ line mutagenesis can still be valuable and has not been sufficiently exploited. An example of a recently characterized germ line mutation of great potential interest to gerontologists is the "wasted" mutation (Shultz et al, 1982). It may be the analogue of the human ataxia telangiectasia mutation, heterozygotes for which exist in human populations at frequences of at least one per hundred (Swift, 1976; Chen et al, 1978). Such heterozygotes may be at high risk for developing a variety of age-related neoplasms (Swift, 1976).

As indicated earlier, there are also great opportunities presented by our ability to culture many different cell types, often in nearly chemically defined media, including cultures from individual human donors with relevant genetic disorders. Consider, for example, what this has done for the field of atherogenesis (Goldstein and Brown, 1983). The methods of cell and tissue culture also offer major opportunities for research on cell-cell interactions and how such interactions are altered with aging. The cells of the immune system would appear to be ideal models for such investigations.

REFERENCES

Ambros, V., and Horvitz, H.R.: 1984, Science **226**, pp. 409-416.
Andrew, W.: 1971, The Anatomy of Aging in Man and Animals.
Andrew, W.: 1944, Am. J. Anat. **74**, pp. 97-127.
Bertrand, H.A., Massaro, E.J., Yu, B.P.: 1978, Science **201**, pp. 1234-1235.
Bhaskar, S.N.: 1968, Geriatrics, **23**, pp. 137-149.
Bradley, A., Evans, M., Kaufman, M.H., and Robertson, E.: 1984, Nature **309**, pp. 255-256
Braun-Falco, O., and Thianprasit, M.: 1965, Arch. Klin. Exp. Dermatol. **221**. pp. 207-231.
Brizee, K.R., Ordo, J.M., Kaack, B.: 1974, J. Gerontol. **29**, pp.366-381.
Burnet, M.: 1974, Intrinsic mutagenesis: A genetic approach to aging. John Wiley, New York.
Calaby, J.H., and Taylor, J.M.: 1981, J. Mamm. **62**, pp. 329-341.
Chen, P.C., Lavin, M.F., Kidson, C., and Moss, D.: 1978, Nature, **176**, pp. 484-486.
Cichocki, T., and Ackermann, J.: 1967, Folia Histochem. Cytochem. (Krakow) **5**, pp. 145-150.
Clark, A.J. and Ganesan, A.: 1975, 'Lists of genes affecting DNA metabolism in Escherichia coli.' In Molecular Mechanisms for repair of DNA, Part B, by Hanawalt, P.C. and Setlow, R.B. Plenum Press, N.Y., pp. 431-437.
Cutler, R.G.: 1975, Proc. Natl. Acad. Sci. USA **72**, pp. 4664-4668.
Diamond, J.M.: 1982, Nature **298**, pp.115-116.
Donato, H. and Sohal, R.S.: 1978, Exp. Gerontol. **13**, pp. 171-179.

Eldred, G.E., Miller, G.V., Stark, W.S., and Feeney-Burns, L.: 1982, Science **216**, pp. 757-759.
Epstein, C. J., Martin, G.M., Schultz, A.L., and Motulsky, A.G.: 1966, Medicine **45**, pp. 177-221.
Geer, J.C., and Haust, M.D.: 1972, Monographs on Atherosclerosis, Vol 2.
Goldstein, J.L., and Brown, M.S.: 1983, 'Familial hypercholesterolemia.' In the Metabolic Basis of Inherited Disease ed. by Stanbury, J.B., Wyngaarden, J.B., Frederickson, D.S., Goldstein, J.L. and Brown, M.S., 5th ed., McGraw-Hill, N.Y., pp. 672-712.
Gordon, J.W.: 1983, Develop. Genet. **4**, pp. 1-20.
Harmon, D.: 1956, J. Gerontol. **11**, pp. 298-300.
Hodgson, C.: 1963, Arch. Dermatol.**87**, pp. 197-207.
Hookey, J.A.: 1931, Arch. Dermatol. Syphilol. **23**, pp. 946-959.
Horn, P.L., Turker, M.S., Ogburn, C.E., Disteche, C.M., and Martin, G.M.: 1984, J. Cell. Physiol. **121**, pp. 309-315.
Ives, P.J., Haensly, W.E., Maxwell, P.A., McArthur, H.: 1975, Mech. Ageing Devel. **4**, pp. 399-413.
Klass, M.R.: 1977, Mech. Ageing Devel. **6**, pp. 413-430.
Leopold, A.C.: 1975, Bioscience 25, pp. 659-662.
Leopold, A.C.:1978, 'The biological significance of death in plants.' In Biology of Aging, ed. by Behnke, J.A., Finch, C.E. and Moment, G.B. Plenum Press, pp. 101-114.
Martin, G.M., Sprague, C.A., Norwood, T. H., and Pendergrass, W.R.: 1974, Am. J. Pathol. **74**, pp. 137-154.
Martin, G.M.: 1977, Mech. Ageing Devel. **9**, pp. 385-391.
Martin, G.M.: 1978, 'Genetic syndromes in man with potential relevance to the pathobiology of aging.' In Genetic effects on aging, ed. by Bergsma, D. and Harrison, D.E., Alan R. Liss, N.Y., pp. 5-39.
Martin, G.M.: 1979, Mech. Ageing Devel. **9**, pp. 385-391.
Martin, G.M.: 1980, Adv. Pathobiol. **7**, pp. 5-20.
Martin, G.M.: 1982, Natl. Cancer Inst. Monogr. **60**: pp. 241-247.
Martin, G.M.: 1985, 'The Aging Process: Therapeutic Implications,' ed. by R.N. Butler and A.G. Bearn, Raven Press, New York, pp. 21-35.
Martin, G.M., Smith, A.C., Ketterer, D.J., Ogburn, C.E., and Disteche, C.M.: 1985, Isr. J. Med. Sci., In Press.
McKusick, V.A.: 1983, Mendelian Inheritance in Man, 6th Ed. Johns Hopkins University Press, Baltimore.
Medawar, P.B.: 1957, Uniqueness of the Individual.Methuen, London.
Miguel, J., Lundgren P.R., Johnson, J.E.: 1978, J. Gerontol. **33**, 5-19.
Munkres, K.D., Rana, R.S.: 1978a, Mech. Ageing Devel. **7**, pp. 399-406.
Munkres, K.D., and Rana, R.S.: 1978 b, Mech. Age. Develop. **7**, pp. 407-415.
Munnell, J.F., Getty, R.: 1968, J. Gerontol. **23**, pp.154-158.
Nandy, B.S., and Getty, R.: 1971, Exp. Gerontol. **6**, pp. 447-452.
Orgel, L.E.: 1963, Proc. Natl. Acad. Sci. USA **49**, pp. 517-521.
Orgel, L.E.: 1970, Proc. Natl. Acad. Sci. USA **67**, pp. 1476.
Parks, R.D., Scheerer, P.P., and Greene, R.R.: 1958, Surg. Gynecol. Obstet. **10**, pp. 413-420.
Robles, L.J.: 1978, Mech. Ageing Devel. **71**, pp. 53-64

Sacher, G.A.: 1975, 'Maturation and longevity in relation to cranial capacity in hominid evaluation.' In Primate Functional Morphology and Evolution. ed. by Tuttle, R., Mouton, The Hague, pp. 417-441.

Schneider, G.T., and Bechtel, M.: 1956, Obstet. Gynecol. **8**, pp. 713-719.

Shultz, L.D., Sweet, H.O., Davisson, M.T. and Coman, D.R.: 1982, Nature **297**, pp. 255-256.

Segre, D., and Segre, M.: 1976, J. Immunol. **116**, pp. 735-738.

Singh, R.M., and Mukherjee, B.: 1972, Acta Anat. (Basel) **83**, pp. 302-320

Shimasaski, H., Ueta, N., Privett, O.S.: 1980, Lipids **5**, 236-241.

Sokoloff, L.: 1972, 'The pathology of rheumatoid arthritis and allied disorders.' In: Hollander, J.L., D.J. McCarthy, Eds.Arthritis and Allied Conditions. A Textbook of Rheumatology. Lea and Febiger, 8th Ed., Philadelphia, pp. 309-332.

Strehler, B.L., Mark, D.D., Mildvan, A.S., Gee, M.V.: 1959, J. Gerontol. **14**, pp. 430-439.

Sundararaman, V., Cummings, D.J.L: 1976, Mech. Ageing Devel **5**, pp. 139-154.

Swift, M., Sholman, L., Perry, M., and Chase, C.:1976, Cancer Res. **36**, pp. 209-215.

Tanaka, Y., (Editor): 1984, Pathology of the Extremely Aged. Volume 1, Centenarians. Ishiyuku EuroAmerica, Inc., St. Louis.

Wilcox, H.H.: 1959, 'Structural changes in the nervous system related to the process of aging.' In: Birren J, Windle, I, Eds. The process of aging in the nervous system. Springfield, III, Thomas, pp. 16-39.

Zieg, J., Silverman, M., Hilman, M., and Simon, M.: 1977, Science **196**, p. 170-172.

REGULATORS OF GROWTH, DIFFERENTIATION AND THE REVERSAL OF MALIGNANCY

Leo Sachs
Department of Genetics, Weizmann Institute of Science,
Rehovot, Israel

ABSTRACT

Identification of normal growth and differentiation factors and how they interact in normal development, made it possible to identify the mechanisms that uncouple growth and differentiation so as to produce malignant cells. When cells become malignant, the malignant phenotype can again be suppressed. Results on the reversibility of malignancy in different types of tumors have shown, that in addition to genes for the expression of malignancy (oncogenes) there are other genes (soncogenes) that can suppress the action of oncogenes, that reversion does not have to restore all the normal controls, and that the stopping of cell multiplication by inducing differentiation to mature cells can by-pass the genetic abnormalities that give rise to malignancy.

INTRODUCTION

The multiplication and differentiation of normal cells is controlled by different regulatory molecules. These regulators have to interact to achieve the correct balance between cell multiplication and differentiation during embryogenesis and the normal functioning of the adult individual. The origin and further progression of malignancy requires genetic changes that uncouple the normal balance between multiplication and differentiation so that there are too many growing cells. This uncoupling can occur in various ways (Sachs, 1980,1982, 1985). What changes uncouple normal controls so as to produce malignant cells? When cells have become malignant, can malignancy again be suppressed so as to revert malignant back to non-malignant cells? Malignant cells can have different abnormalities in the controls for multiplication and differentiation. Do all the abnormalities have to be corrected, or can they be by-passed in order to reverse malignancy? These questions can now be answered.

B. Pullman et al. (eds.), Interrelationship Among Aging, Cancer and Differentiation, 35–42.

NORMAL GROWTH AND DIFFERENTIATION FACTORS

An understanding of the mechanism that controls multiplication (growth) and differentiation in normal cells would seem to be an essential requirement to elucidate the origin and reversibility of malignancy. The development of appropriate cell culture systems has made it possible to identify the normal regulators of growth (growth factors) for various types of cells, and also in some cell types the normal regulators of differentiation (differentiation factors). This approach has been particularly fruitful in identifying the normal growth factors for all the different types of blood cells, first for myeloid cells (See Sachs, 1974,1978,1980) and then for other cell types, including T lymphocytes (Mier and Gallo, 1980) and B lymphocytes (Möller, 1984), and in identifying the normal differentiation factors for myeloid cells (Sachs, 1974,1985) and B lymphocytes (Möller, 1984). The normal differentiation factor for myeloid cells is a DNA binding protein (Weisinger and Sachs, 1983). It will be interesting to determine how far this applies to normal differentiation factors for other cell types.

In cells of the myeloid series there are four different growth-inducing proteins. These are now called colony stimulating factors (CSF), or macrophage and granulocyte inducers - types 1 (MGI-1) (Sachs, 1980,1982; Nicola et al. 1983; Sachs and Lotem, 1984). Of the four growth factors, one (M) induces the development of clones with macrophages, another (G) clones with granulocytes, the third (GM) clones with both macrophage and granulocytes, and the fourth (also called Interleukin-3), clones with macrophages, granulocytes, eosinophils, mast cells, erythroid cells, or megakaryocytes. This multigene family represents a hierarchy of growth factors for different stages of blood cell development as the precursor cells become more restricted in their developmental program. Experiments with normal myeloid cell precursors have shown that in these cells the growth factors can induce both growth (cell viability and multiplication) and production of differentiation factors (Sachs, 1980; Lotem and Sachs, 1982,1983). This thus ensures the normal coupling of growth and differentiation, a coupling mechanism that may also apply to other cell types. Differences in the time of the switch-on of the differentiation factor would produce differences in the amount of cell multiplication before differentiation.

ONCOGENES AND THE UNCOUPLING OF NORMAL CONTROLS

Identification of these normal growth and differentiation factors and the cells that produce them, then made it possible to identify the different types of changes in the production or response to these normal regulators that occur in malignancy. In myeloid leukemic cells, different clones of malignant cells have been identified which have shown all the possible changes that can occur in the normal response to growth and differentiation factors (Sachs, 1978,1980,1982,1985). There are different leukemic clones that are 1. Independent of the normal growth factor for growth. 2. Constitutively produce their own growth factor. 3. Are blocked in the ability of the growth factor to induce production of the differentiation factor. 4. Are changed in their requirement for the normal growth factor, but can still respond normally to the normal differentiation factor, or 5. Are blocked in their ability to respond to the normal differentiation factor. These are thus different ways to uncouple the normal controls of growth and differentiation, and different ways for the cells to become malignant. Growth factors induce cell viability and cell multiplication. Independence of the normal growth factor, or constitutive production of their own growth factor, can also explain the survival and growth of metastasizing malignant cells in places in the body where the growth factor required for the survival of the normal cells is not present.

Studies on protein synthesis in these different types of myeloid leukemic cells have shown, that these different ways for the cells to become malignant were associated with changes from an induced to a constitutive expression of certain genes (Sachs, 1980; Liebermann et al. 1980). This indicated that changes from inducible to constitutive gene expression can produce asynchrony in the normal developmental program, uncouple normal controls, and result in malignancy (Sachs, 1980). The various types of changes that have been found in myeloid leukemic cells can serve as a model system to identify the different changes that can give rise to malignant cells.

The transformation of normal into malignant cells requires a number of genetic changes, and the genes involved in the expression of malignancy are now called oncogenes (Cooper, 1982; Land et al. 1983; Bishop, 1983). The changes of normal to oncogenes are in all cases associated with changes in the structure or regulation of the normal genes. As in the case of normal genes not all oncogenes have the same function. The sis oncogene is derived from the normal genes for platelet derived growth factor (Doolittle et al. 1983; Waterfield et al. 1983), the erb B oncogene from the gene for the receptor for epidermal growth factor (Downard et al. 1984), and the erb A oncogene from the gene for carbonic anhydrase that is involved in erythroid differentiation (Debuire et al. 1984). The origin and progression of malignancy can involve different genetic changes including changes in gene dosage (Sachs, 1974), gene mutations, deletions and gene re-arrangements (Land et al. 1983). The different genetic changes in the structure or regulation of the normal genes that control growth and differentiation, can thus produce in different ways the uncoupling of normal controls that is required for the origin and further evolution of malignancy.

SONCOGENES AND BY-PASSING OF GENETIC DEFECTS IN THE REVERSIBILITY OF MALIGNANCY

The change of normal into malignant cells involves a sequence of genetic changes. Evidence has, however, been obtained with various types of tumors including sarcomas (Sachs, 1974), myeloid leukemias

(Sachs, 1974,1978) and teratocarcinomas (Stewart and Mintz, 1981) that malignant cells have not lost the genes that control normal growth and differentiation. This was first shown in sarcomas by the finding that it was possible to reverse the malignant to a non-malignant phenotype with a high frequency in cloned sarcoma cells whose malignancy had been induced by chemical carcinogens, x-irradiation, or by a tumor-inducing virus (Rabinowitz and Sachs, 1968,1970a; Sachs, 1974). In sarcomas induced by chemical carcinogens or x-irradiation, this reversibility of malignancy included reversion to the limited life-span found with normal fibroblasts (Rabinowitz and Sachs, 1970b). A comparison of sarcomas with myeloid leukemias then showed that reversion of the malignant phenotype can be achieved by different mechanisms.

The chromosome studies on normal fibroblasts, sarcomas, revertants from sarcomas which had regained a non-malignant phenotype, and re-revertants, have indicated that the difference between these malignant and non-malignant cells is controlled by the balance between genes for expression (E) and suppression (S) of malignancy (Rabinowitz and Sachs, 1970a; Hitotsumachi et al. 1971; Yamamoto et al. 1973; Sachs, 1974; Bloch-Shtacher and Sachs, 1976,1977). When there is enough S to neutralize E malignancy is suppressed, and when the amount of S is not sufficient to neutralize E malignancy is expressed. These experiments indicated that in addition to genes for expression of malignancy (E) (oncogens), there are other genes, S genes (soncogenes), that can suppress the action of oncogenes. Suppression of the action of the Ki-ras oncogene in revertants (Noda et al. 1983), is presumably due to such suppressor genes. The balance between E and S genes, oncogenes and soncogenes, also seems to determine malignancy in other tumors including human retinoblastomas (Murphree and Benedict, 1984).

In the mechanism found with sarcomas reversion was obtained by chromosome segregation, resulting in a change in gene dosage due to a change in the balance of specific chromosomes. This reversion of malignancy by chromosome segregation, with a return to the gene balance required for expression of the non-malignant phenotype, occurred without hybridization between different types of cells. The non-malignant cells were thus derived from the malignant ones by genetic segregation. Reversion of the malignant phenotype associated with chromosome changes has also been found after hybridization between different types of cells (Ringertz and Savage, 1976; Evans et al. 1982; Kitchin et al. 1982; Stanbridge et al. 1982).

In addition to this reversion of malignancy by chromosome segregation, another mechanism of reversion was found in myeloid leukemia. In this second mechanism reversion to a non-malignant phenotype was also obtained in certain clones with a high frequency but, in contrast to the mechanism found with sarcomas, this reversion was not associated with chromosome segregation. Phenotypic reversion of malignancy in these leukemic cells was obtained by induction of the normal sequence of cell differentiation by the normal differentiation factor (Sachs, 1974,1978, 1980). In this reversion of the malignant phenotype, the stopping of cell multiplication in mature cells by inducing differentiation by-passes the genetic changes in the requirement for the normal growth factor that produced the malignant phenotype. Genetic changes which produce blocks in the ability to be induced to differentiate by the normal differentiation factor occur in the evolution of myeloid leukemia. But even these cells can be induced to differentiate by other compounds, either singly or in combination, that can induce the differentiation program by other pathways (Sachs, 1978,1980,1985). Also in these cases, the stopping of cell multiplication by inducing differentiation by these alternative pathways by-passes the genetic changes that block response to the normal differentiation factor. This by-passing of genetic defects is presumably also the mechanism for the reversion of malignancy by inducing differentiation in other types of tumors such as erythroleukemias (Friend, 1978; Marks and Rifkind, 1978) and neuroblastomas. There can, of course, also be cases of reversion in which all the oncogenes are lost, or the changes of normal genes into oncogenes are actually reversed.

It can, therefore, be concluded that the changes of normal genes into oncogenes that result in the expression of malignancy does not mean that this expression of malignancy can not again be suppressed. The results on the reversibility of malignancy have shown that there are different ways of reverting malignancy (Fig. 1), that reversion does not have to restore all the normal controls and that the stopping of cell multiplication by inducing differentiation to mature cells can by-pass genetic abnormalities that give rise to malignancy. The results have also shown that reversion of malignancy may be useful as an approach to cancer therapy (Sachs, 1974,1978; Lotem and Sachs, 1984).

ACKNOWLEDGMENTS

This research is now supported by a contract with the National Foundation for Cancer Research, Bethesda, and by grants from the Jerome A. and Estelle R. Newman Assistance Fund and the Julian Wallerstein Foundation.

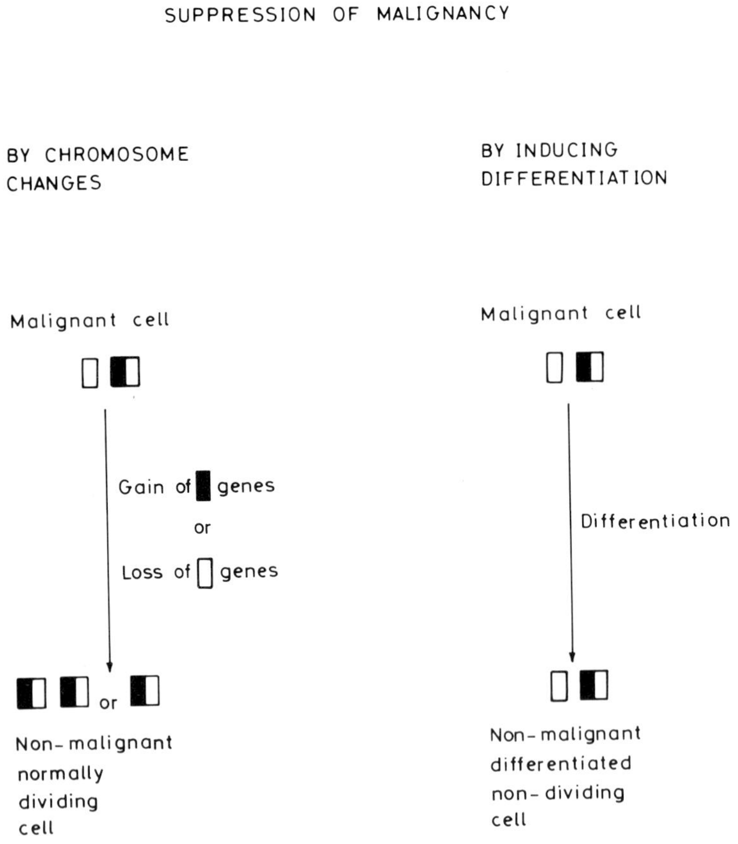

Fig. 1. Different ways of reverting malignancy. Malignancy can be suppressed A. By chromosome changes that change the balance between genes for the □ expression and ■ suppression of malignancy, or B. The stopping of cell multiplication in mature cells by inducing differentiation can by-pass the genetic changes that produce the malignant phenotype. Expressor genes are now called oncogenes and suppressor genes can be called soncogenes.

REFERENCES

Bishop, J.M.: 1983, Ann. Rev. Biochem. 52, pp. 301-354.
Bloch-Shtacher, N., and Sachs, L.: 1976, J. Cell. Physiol. 87, pp. 89-100.
Bloch-Shtacher, N., and Sachs, L.: 1977, J. Cell. Physiol. 93, pp. 205-212.
Cooper, G.M.: 1982, Science 218, pp. 801-806.
Debuire, B., Henry, C., Benaissa, M., Biserte, G., Claverie, J.M., Saule, S., Martin, P., and Stehelin, D.: 1984, Science 224, pp. 1456-1459.
Doolittle, R.F., Hunkapiller, M.W., Hood, L.E., Devare, S.G., Robbins, K.C., Aaronson, S.A., and Antoniades, H.N.: 1983, Science 221, pp. 275-277.
Downard, J., Yarden, Y., Mayers, E., Scrace, G., Totty, N., Stockwell, P., Ulrich, A., Schlessinger, J., and Waterfield, M.D.: 1984, Nature 307, pp. 521-527.
Evans, E.P., Burtenshaw, M.D., Brown,B.B., Hennion, R., and Harris, H.: 1982, J. Cell Sci. 56, pp. 113-130.
Friend, C.: 1978, Harvey Lectures 72, pp. 253-281, Academic Press, New York.
Hitotsumachi, S., Rabinowitz, Z., and Sachs, L.: 1971, Nature 231, pp. 511-514.
Kitchin, R.M., Gadi, I.K., Smith, B.L., and Sager, R.: 1982, Somat. Cell Genet. 8, pp. 677-689.
Land, H., Parada, L.F., and Weinberg, R.A.: 1983, Science 222, pp. 771-778.
Liebermann, D., Hoffman-Liebermann, B., and Sachs, L.: 1980, Develop. Biol. 79, pp. 46-63.
Lotem, J., and Sachs, L.: 1982, Proc. Nat. Acad. Sci. USA. 79, pp. 4347-4351.
Lotem, J., and Sachs, L.: 1983, Int. J. Cancer 32, pp. 127-134.
Lotem, J., and Sachs, L.: 1984, Int. J. Cancer 33, pp. 147-154.
Marks, P., and Rifkind, R.A.: 1978, Ann. Rev. Biochem. 47, pp. 419-448.
Mier, J.W., and Gallo, R.C.: 1980, Proc. Nat. Acad. Sci. USA. 77, pp. 6134-6138.
Möller, G. (Ed.): 1984, Immunological Reviews 78, pp. 7-224, Munksgaard, Copenhagen.
Murphree, A.L., and Benedict, W.F.: 1984, Science 223, pp. 1028-1033.
Nicola, N.A., Metcalf, D., Matsumoto, M. and Johnson, G.R.: 1983, J. Biol. Chem. 258, pp. 9017-9023.
Noda, M., Selinger, Z.,Scolnick, E.M., and Bassin, R.H.: 1983, Proc. Nat. Acad. Sci. USA. 80, pp. 5602-5606.
Rabinowitz, Z., and Sachs, L.: 1968, Nature 220, pp. 1203-1206.
Rabinowitz, Z., and Sachs, L.: 1970a, Nature 225, pp. 136-139.
Rabinowitz, Z., and Sachs, L.: 1970b, Int. J. Cancer 6, pp. 388-398.
Ringertz, N.R., and Savage, R.E.: 1976, Cell Hybrids, Academic Press, New York.
Sachs, L.: 1974, Harvey Lectures 68, pp. 1-35, Academic Press, New York.

Sachs, L.: 1978, Nature 274, pp. 535-539.
Sachs, L.: 1980, Proc. Nat. Acad. Sci. USA. 77, pp. 6152-6156.
Sachs, L.: 1982, Cancer Surveys, 1, pp. 321-342.
Sachs, L.: 1985, In "Molecular Biology of Tumor Cells", Nobel Conference, Stockholm, pp. 257-280, Raven Press, New York.
Sachs, L. and Lotem, J.: 1984, Nature 312, p. 407.
Stanbridge, E.J., Der, C.J., Doersen, C., Nishimi, R.Y., Peehl, D.M., Weissman, B.E., and Wilkinson, J.E.: 1982, Science 215, pp. 252-257.
Stewart, T.A., and Mintz, B.: 1981, Proc. Nat. Acad. Sci. USA. 78, pp. 6314-6318.
Waterfield, M.D., Scrace, G.T., Whittle, N., Stroobant, P., Johnsson, A., Wasteson, A., Westermark, B., Heldin, C.H., Huang, J.S., and Deuel, T.F.: 1983, Nature 304, pp. 35-39.
Weisinger, G., and Sachs, L.: 1983, EMBO J. 2, pp. 2103-2107.
Yamamoto, T., Rabinowitz, Z., and Sachs, L.: 1973, Nature New Biol. 243, pp. 247-250.

TRANSFORMING GENES OF HUMAN MALIGNANCIES

Alessandra Eva, Stuart A. Aaronson and Steven R. Tronick
Laboratory of Cellular and Molecular Biology
National Cancer Institute
Building 37, Room 1E24
Bethesda, Maryland 20205

Knowledge of the mechanisms by which normal cells become converted to the malignant state has increased within the past few years due to a series of remarkable discoveries. A wide variety of human tumors have been shown to possess discrete genetic sequences capable of inducing the transformation of appropriate assay cells. These genes were subsequently found to be the homologues of the cell-derived oncogene sequences responsible for the transforming activity of a group of well-characterized RNA tumor viruses. Investigations of the specific chromosomal translocations that occur in certain human tumors has also revealed that the movement of a specific cellular gene, again a homologue of a retroviral transforming gene, to a new regulatory environment alters its expression, thus contributing to tumor development.

In recent years, remarkable progress has also been made in deciphering the normal functions of cellular genes that give rise to oncogenes. In what follows, we summarize the properties of retroviral transforming genes, their products, and their relationships with cellular genes. In addition, we discuss some of the mechanisms by which cellular genes can become activated as transforming genes of human tumor cells.

VIRAL ONCOGENES AND PROTO-ONCOGENES

Proto-oncogenes comprise a small family of cellular genes that have been highly conserved throughout vertebrate evolution. These genes were initially discovered as the normal cellular progenitors of retroviral oncogenes by hybridization of viral oncogenes with normal cellular DNAs of a wide range of vertebrates (4, 19, 70). Their high degree of evolutionary conservation implies that they must serve very important functions in normal cellular growth processes.

If any of the thousands of genes that exist in our cells could acquire neoplastic properties when captured by one of these viruses, then we would never expect to observe the same sequence incorporated by more than one of the known two dozen or so acute transforming virus

B. Pullman et al. (eds.), Interrelationship Among Aging, Cancer and Differentiation, 43–57.

isolates. In fact, there is tremendous redundancy such that several virus isolates from the same or even different species have been shown to have captured identical onc sequences (8, 70). These findings imply that the number of cellular genes that can acquire transforming properties when incorporated within the retrovirus genome must be rather limited.

There is increasing knowledge of the biochemical functions of some oncogenes, as well as emerging evidence that a number of these genes code for proteins with related structural and functional properties. One class of oncogenes, which presently includes only v-sis, codes for a growth-factor-like molecule (Table 1). This protein is closely related to human platelet-derived growth factor (PDGF) (17, 68). We have recently shown that the normal human sequence corresponding to sis/PDGF can be activated as a transforming gene by causing it to be expressed in a cell type responsive to its growth stimulatory effects (27). These findings raise the possibility that genes for other normal growth factors or hormones might also act as oncogenes when inappropriately expressed in a suitable target cell.

Members of another class code for protein kinases which usually have specificity for tyrosine residues (Table 1) (4). These oncogene products are located in the cytoplasm or plasma membrane. Some members of this class code for proteins that have kinase activity specific for serine and threonine, but these genes still show significant homology to the tyrosine kinases.

Recent studies have shown strong homology between the protein sequence of the EGF receptor and the predicted product of erb B, implying that this oncogene is likely to be an altered form of the normal growth factor receptor (18). Receptors for a number of other growth factors and hormones appear to be tyrosine kinases, suggesting that additional relationships between known oncogenes and such receptors are likely to be found.

Another major class comprises the ras family, which encodes very closely related 21,000 dalton transforming proteins (Table 1) (1). These proteins possess GDP binding and GTPase activity and appear to possess protein kinase activity as well. Ras transforming proteins are also located in the plasma membrane. Recent evidence suggests that these proteins may act in an as yet undefined way to modulate the adenylate cyclase pathway.

A fourth class of viral oncogenes encode nuclear proteins whose functions may be linked with cell cycle regulation and DNA replication. For example, recent experiments have demonstrated a specific cell-cycle dependence for the expression of the proto-oncogenes c-myc and c-fos (32, 40). There is also evidence that these genes code for proteins which enhance expression of certain other cellular genes through direct or indirect effects on enhancer elements associated with such genes (31).

TABLE 1: Cellular and Viral Oncogenes

Oncogene	Species of Origin	Protein Product: Biochemical Function	Protein Product: Subcellular Localization
Class I			
sis	woolly monkey/ cat	related to PDGF	cytoplasm and plasma membrane
Class II			
src	chicken	tyrosine kinase	plasma membrane
yes	chicken	"	"
fps/fes	chicken/cat	"	cytoplasm
abl	mouse	"	plasma membrane
ros	chicken	"	cytoplasm
fgr	cat	"	?
erb B	chicken	(related to EGF receptor)	plasma membrane
fms	cat	serine/threonine kinase	cytoplasmic membrane
mos	mouse	"	cytoplasm
Class III			
H-ras/bas	rat/mouse	GTP binding/	plasma membrane
N-ras	human	GTPase activity/	" "
K-ras	rat	autokinase	" "
Class IV			
myc	chicken	DNA binding	nuclear matrix
n-myc	human	?	?
myb	chicken	?	nuclear matrix
ski	chicken	?	nucleus
fos	mouse	?	nucleus
Class V			
rel	turkey	?	?
erb A	chicken	?	cytoplasm
met	human	?	?
B-lym	chicken/human	?	?

The remaining oncogenes are grouped within an additional class (Class V) because, at present, little is known about their function or subcellular location.

ACTIVATION OF ONCOGENES IN HUMAN MALIGNANCIES BY CHROMOSOMAL TRANSLOCATION

Proto-oncogenes related to retroviral oncogenes are well conserved in human DNA. It is possible to map these genes to specific human chromosomes by testing for the presence of human DNA fragments related to a specific oncogene in somatic cell hybrids possessing varying numbers of human chromosomes as well as in segregants of such hybrids. The chromosomal assignments of proto-oncogenes indicate that such genes are distributed through the human genome. The results of chromosome mapping studies by our laboratory and others are summarized in Table 2.

TABLE 2. Chromosomal Assignments of the Human Proto-oncogenes

Chromosome Number	Proto-oncogene	Chromosomal Aberration	Tumor
1	N-*ras*	del 1p	Neuroblastoma
3	*raf-1*	t(3;14); t(3p); del 3p	Renal & lung CA, parotid tumors
4	*raf-2*	-	-
5	*fms*	del 5q	AML
6	K-*ras*-1; *myb*	$6q^-$; t(6:14)	ALL, ovarian CA
8	*mos*; *myc*	t(8;14); t(8;22); t(2;8)	Burkitt's lymphoma
		t(8;21)	AML
9	*abl*	t(9;22)	CML
11	H-*ras*-1	$11p^-$	Wilms' tumor
12	K-*ras*-2	12+	CLL
15	*fes*	t(15;17)	APL
20	*src*	-	-
22	*sis*	t(9,22)	CML

Nonrandom chromosomal aberrations occur in a number of malignant disorders. Most of these disorders affect hematopoietic tissues, and their consistent translocations are of importance because they are often specifically associated with morphologically well defined subtypes of leukemia and lymphomas. In particular, specific chromosomal translocations have been constantly associated with Burkitt's lymphomas and chronic myelogenous leukemia (CML) (54). The mapping of proto-oncogenes to chromosomes has indicated not only that chromosomes involved in such specific rearrangements harbor *onc*-related genes (Table 2) but also that some of these proto-oncogenes are located at the breakpoints of these translocations. It has been possible to show that, in the case of Burkitt's lymphoma with the t(8;14) translocation (46, 74), c-*myc* is translocated from chromosome 8 to chromosome 14 into the immunoglobulin μ chain switch region (13, 67, 21). In the other two Burkitt's lymphoma translocations, t(2;8) and t(8;22) (3), c-*myc* remains on chromosome 8, while the loci for the constant region of κ chain, located on chromosome 2, and of λ chain, located on chromosome 22, translocate to chromosome 8 into a region 3' to c-*myc* (12, 22).

It has been postulated that in Burkitt's lymphoma chromosome translocation could increase c-*myc* transcription and, as a consequence, induce transformation (11, 12, 22). It has, in fact, been shown that, whether it is structurally rearranged or not, the c-myc proto-oncogene involved in the translocation is transcriptionally active, while the c-myc proto-oncogene on the normal chromosome 8 is transcriptionally silent. Using somatic cell hybrids between Burkitt's lymphoma cells and either human lymphoblastoid cells or mouse plasmacytoma cells to analyze the expression of the translocated c-myc oncogene, it has also been shown that the translocated c-myc is transcribed in plasma cells but is suppressed in lymphoblastoid cells (11). It has been postulated that the translocated c-myc comes under the transcriptional control of enhancer elements associated with the immunoglobulin locus. Studies to date suggest that the activity of such an enhancer may depend on its interaction with trans-acting factors active in plasma cells and Burkitt's lymphoma cells but not present or active in lymphoblastoid cells (11).

Another specific chromosomal translocation has been associated with chronic myelogenous leukemia where chromosome 9 is translocated to chromosome 22, the Philadelphia chromosome (47, 53). The human c-abl gene was initially mapped on the long arm of chromosome 9 (38). It has subsequently been shown that in CML patients, the c-abl gene is translocated from chromosome 9 to chromosome 22 with the breakpoint occurring near the 5' end of c-abl gene (15, 34). More recent studies have shown an aberrant 8-kp c-abl transcript in leukemia cells of 5 of 6 patients who have CML and the Philadelphia chromosome (5, 26). Moreover, an altered human c-abl protein (p210) has been detected in K562 leukemia cells, established from a CML patient (41). This altered p210 c-abl protein was found to be phosphorylated on tyrosine in vivo and was also phosphorylated on tyrosine during in vitro kinase reactions. Thus, this translocation appears to alter the structure and function of the c-abl product.

The high degree of specificity of c-myc and c-abl rearrangements in Burkitt's lymphoma and CMLs, respectively, and the subsequent alteration of their biologic characteristics support the concept that these genes play a role in the etiology of such hematopoietic tumors. It seems likely that highly specific translocations associated with other hematopoietic malignancies may also activate as yet undiscovered oncogenes.

RAS PROTO-ONCOGENES OF HUMAN CELLS

Studies on the Kirsten and Harvey strains of murine sarcoma viruses (Kirsten-MSV, Harvey-MSV) led eventually to the identification of a small group of highly conserved cellular oncogenes, designated as the ras family. Early research on Harvey-MSV and Kirsten-MSV, both isolated from rats infected with either Moloney or Kirsten leukemia viruses, demonstrated that each was genetically related in non-helper virus-derived regions of their genomes (59). With the advent of molecular cloning techniques, it was possible to demonstrate that most of the non-helper derived sequences were contributed by endogenous rat retroviral information (20). However, an additional set of sequences in each viral genome, designated ras, was shown to be derived from the cellular genome and was not homologous to any known viral sequences. The ras sequences in Kirsten-MSV and Harvey-MSV were found to be only partially related and thus represent different members of a gene family (20).

Subsequent to the isolation of Kirsten-MSV and Harvey-MSV, two other murine sarcoma viruses, BALB-MSV and Rasheed-MSV, were shown to contain ras oncogenes derived from the BALB/c mouse and Fisher rat, respectively (1, 29). Another ras family member, designated N-ras, has been identified in mammalian cells but has not been identified to date as a transforming gene of any known retrovirus (61, 24, 36).

Three ras genes of human cells have been molecularly cloned and characterized in detail. The organization of their coding sequences is similar in that each gene contains four exons; however, the exons are distributed over a region anywhere from 4.8 kbp (H-ras) to 45 kbp (K-ras) in length (8, 56, 59). Pseudogenes of H-ras and K-ras have also been identified (designated H-ras-2 and K-ras-1) (8). The molecular cloning and nucleotide sequence analysis of ras genes of yeast has demonstrated a remarkable degree of evolutionary conservation of ras gene structure (14, 49).

RAS GENES ARE FREQUENTLY DETECTED AS HUMAN TRANSFORMING GENES

The involvement of the ras gene family in naturally occurring malignancies came to light in studies in which investigators asked whether DNAs of animal or human tumor cells possessed the capacity to directly confer the neoplastic phenotype to a susceptible assay cell. Some human tumor DNAs were shown to induce transformed foci in the continuous NIH/3T3 mouse cell line (39), which is highly susceptible to the uptake and stable incorporation of exogenous DNA (10, 69).

The first molecularly cloned human transforming gene, whose source was the T24 bladder carcinoma cell line, was demonstrated to be the activated homologue of the normal H-ras gene (28, 50, 56, 58). Subsequent analysis of oncogenes detected by transfection assays has established that the majority belong to the ras family. Thus, K-ras oncogenes have been observed at high frequency in lung and colon carcinomas (51). Carcinomas of the digestive tract, including pancreas, and gall bladder, as well as genitourinary tract tumors, and sarcomas have also been shown to contain ras oncogenes. N-ras appears to be the most frequently activated ras transforming gene in human hematopoietic neoplasms (25). These results are summarized in Table 3.

Not only can a variety of tumor types contain the same activated ras oncogene, but the same tumor type can contain different activated ras oncogenes. Thus, in hematopoietic tumors, we have observed

TABLE 3. Detection of ras Oncogenes in Human Tumors

Tumor Source	Percent Positive	ras Oncogene Activated		
		H-ras	K-ras	N-ras
Carcinoma	10-30	4/12	6/12	2/12
(lung, gastrointestinal, genitourinary)				
Sarcoma	~ 10	0/2	0/2	2/2
(fibrosarcoma, rhabdomyosarcoma)				
Hematopoietic	10-50	0/9	1/9	8/9
(AML, CML, ALL, CLL)				

different ras oncogenes (K-ras, N-ras) activated in lymphoid tumors at the same stage of hematopoietic cell differentiation, as well as N-ras genes activated in tumors as diverse in origin as acute and chronic myelogenous leukemia (25) (Table 1). These findings suggest that ras oncogenes detected in the NIH/3T3 transfection assay are not specific to a given stage of cell differentiation or tissue type.

Retroviruses that contain ras-related transforming genes are known to possess a wide spectrum of target cells for transformation in vivo and in vitro. In addition to inducing sarcomas and transforming fibroblasts (35), these viruses are capable of inducing tumors of immature lymphoid cells (48). They also can stimulate the proliferation of erythroblasts (37) and monocyte/macrophages (33), and can even induce alterations in the growth and differentiation of epithelial cells (71). Thus, the wide array of tissue types that can be induced to proliferate abnormally by these onc genes may help to explain the high frequency of detection of their activated human homologues in diverse human tumors.

MECHANISM OF ACTIVATION OF RAS ONCOGENES

The availability of molecular clones of the normal and activated alleles of human ras proto-oncogenes made it possible to determine the molecular mechanisms responsible for the malignant conversion of these genes. The genetic lesions responsible for activation of a number of ras oncogenes have been localized to single base changes in their p21 coding sequences. In the T24/EJ bladder carcinoma oncogene, a transversion of a G to a T causes a valine residue to be incorporated instead of a glycine into the 12th position of the predicted p21 primary structure (64, 52, 66, 6).

An activated H-ras oncogene of a human lung carcinoma derived cell line (72) was found to possess a single base alteration at codon 61. The change of an A to a T resulted in the replacement of glutamine by leucine in this codon. Thus, a single amino acid substitution was sufficient to confer transforming properties on the product of the Hs242 oncogene. These findings established that the site of activation in the Hs242 oncogene was totally different from that of the T24/EJ oncogene (70).

In subsequent studies, we and others have assessed the generality of point mutations as the basis for acquisition of malignant properties by ras proto-oncogenes by molecularly cloning and analyzing other activated ras oncogenes. Activation of an H-ras transforming gene of the Hs578T human breast carcinosarcoma line has been localized to a point mutation at position 12 changing glycine to aspartic acid in the amino acid sequence (42). Recently Wigler and co-workers (65) reported that the lesion leading to activation of the N-ras oncogene in a neuroblastoma line was due to the alteration of codon 61 from CAA to AAA causing the substitution in this case of lysine for glutamine. Another N-ras transforming gene, this one isolated from a human lung carcinoma cell line, SW1271, has been shown to result from a single point mutation of an A to a G at position 61 in the coding sequence resulting in the substitution of arginine for glutamine (73).

Investigators analyzing K-ras oncogenes (62, 7) have achieved strikingly similiar results. In two K-ras transforming genes so far analyzed, single point mutations in the 12th codon have been shown to be responsible for acquisition of malignant properties. Thus, mutations at positions 12 or 61 appear to be the genetic lesions most commonly responsible for activation of ras oncogenes under natural conditions in human tumor cells (Table 4).

The high frequency of detection of ras oncogenes in human tumors, coupled with the knowledge that ras proto-oncogenes can be activated as transforming genes by single point mutations, suggested that these genes might be preferential targets for somatic mutations. As an approach toward validating this assumption, we and others began to analyze animal tumor cells induced by specific carcinogens. Transforming DNAs from several independently derived mouse cell lines transformed by methylcholanthrene (MCA) had been shown to possess indentical patterns of susceptibility to restriction endonuclease inactivation (60), suggesting that the same transforming gene was inactivated in each case. Thus, we sought to analyze tumor cells induced in vivo for the presence of activated ras oncogenes (27). Two of four MCA-induced fibrosarcomas were shown to contain activated c-K-ras genes. Our experience and that of other workers in detecting ras oncogenes in chemically-induced tumors is summarized in Table 5 (23, 63, 2). It is

TABLE 4. Genetic Lesions that Activate ras Oncogenes of Human Tumors

ras Oncogene	Tumor	Base/Amino Acid Change	Codon No.	Reference
H-ras				
T24	Bladder carcinoma	Gly→Val (GGC→GTC)	12	Tabin et al. (64) Reddy et al. (52) Taparowsky et al. (65) Capon et al. (6)
Hs0578	Mammary carcinoma	Gly→Asp (GGC→GAC)	12	Kraus et al. (42)
Hs 242	Lung carcinoma	Gln→Leu (CAG→CTG)	61	Yuasa et al. (72)
K-ras				
Calu-1	Lung carcinoma	Gly→Cys (GGT→TGT)	12	Shimizu et al. (62) Capon et al. (7)
SW480	Colon carcinoma	Gly→Val (GGT→GTT)	12	Capon et al. (7)
N-ras				
SK-N-SH	Neuroblastoma	Gln→Lys (CAA→AAA)	61	Taparowsky et al. (65)
SW1271	Lung carcinoma	Gln→Arg (GGT←GTT)	61	Yuasa et al. (73)

TABLE 5. Ras Oncogene Activation in Carcinogen-induced Animal Tumors

Carcinogen	Species	Tumor	Oncogene
MCA	Mouse	Fibrosarcoma	K-ras (~50%)
DMBA	Mouse	Skin carcinoma	H-ras (~80%)
NMU	Rat	Mammary carcinoma	H-ras (~90%)

striking that these genes are very frequently activated as a result of exposure of animals to chemical carcinogens. In certain cases (63), it has been possible to show that the activation of ras oncogenes by chemical carcinogens is identical to that found to occur in human tumor cell lines. These studies provided model systems for the study of the multi-step process of chemical carcinogenesis and they also have indicated the importance of ras proto-oncogene activation in this process. It may now be possible to determine exactly when in the course of carcinogenesis that activation occurs.

OTHER ONCOGENES DETECTED BY DNA TRANSFECTION

It should be noted that not all oncogenes detected by NIH/3T3 transfection analysis are related to known retroviral oncogenes. For example, Cooper and coworkers have identified and molecularly cloned an oncogene, B-lym, from a B-cell lymphoma (30). B-lym appears to be activated in a large proportion of tumors at a specific stage of B-cell differentiation (16). Studies to date indicate that this oncogene is relatively small in size (<600 bp), and sequence analysis indicates that it possesses distant homology to transferrin (30). These investigators have also detected oncogenes which appear to be specifically activated in tumors at other stages of lymphoid differentiation (44) or in mammary carcinomas (43). None of these transforming genes appear to possess detectable homology with known retroviral onc sequences. Transforming genes unrelated to the ras family have also been detected in the NIH/3T3 transfection assay by several other laboratories (51, 9, 57).

In efforts to detect additional transforming genes, we have screened several human hematopoietic primary tumor DNAs for sequences capable of transforming NIH/3T3 by transfection. We found that the DNA of a non-Hodgkin's lymphoma induced morphologic transformation of NIH/3T3 cells (24). We observed that human repetitive sequences segregated with the transformed phenotype through several cycles of transfections (Fig. 1), that third-cycle transfectants were tumorigenic in nude mice, and that DNAs of these tumors contained human repetitive sequences and retained the ability to transform NIH/3T3 cells.

At least three types of morphologically distinguishable cells were observed in each of the various cycle transfectants and this different morphologic alteration was maintained following selection of clonal transformants in agar or microtiter plates (Fig. 2). These results suggest that this gene can induce a series of phenotypic changes in the same cell. By molecular hybridization with a variety of oncogenes, the transforming sequences of this non-Hodgkin's lymphoma were found not to be related to any of the known ras genes nor to a large number of other retroviral and cellular oncogenes, including B-lym. It remains to be established whether this transforming gene plays a critical role in the development of the tumor. Our identification of this human oncogene offers the hope of defining new, potentially relevant steps in the neoplastic conversion of specific human cell types.

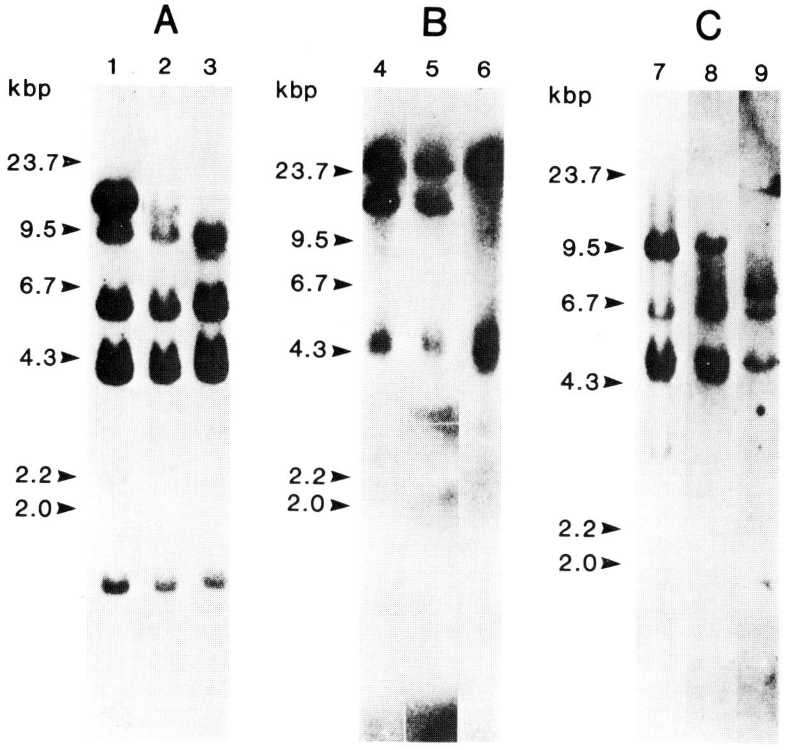

FIG. 1. Detection of human sequences in NIH/3T3 cells transformed by DNA of a 31-23-306 human primary non-Hodgkin's lymphoma. 30 μg of high molecular weight DNA was digested with EcoRI (A), Bam HI (B), and Hind III (C), electrophoresed, blotted and hybridized with a ^{32}P-labeled nick-translated 0.3 kbp cloned repetitive human DNA fragment purified from BLUR-8 (2 x 10^6 cpm/ml). The DNAs analyzed were from the following cell lanes: 57-5-2 a 31-23-306-derived second-cycle transfectant (lanes 1, 4 and 7); 66-3-9-3, a 31-23306-derived third-cycle trasfectant (lanes 2, 5, and 8) 66-1-1-2, a 31-23-306-derived third-cycle transfectant (lanes 3, 6, and 9).

IMPLICATIONS

The large fund of knowledge that has rapidly accumulated regarding the role of oncogenes in the neoplastic process has enabled investigators to conclude that by focusing on the limited set of cellular genes that we have described in this review, it may be possible to define the mechanisms of malignant transformation. This prospect is particularly encouraging when one considers that normal growth and differentiation

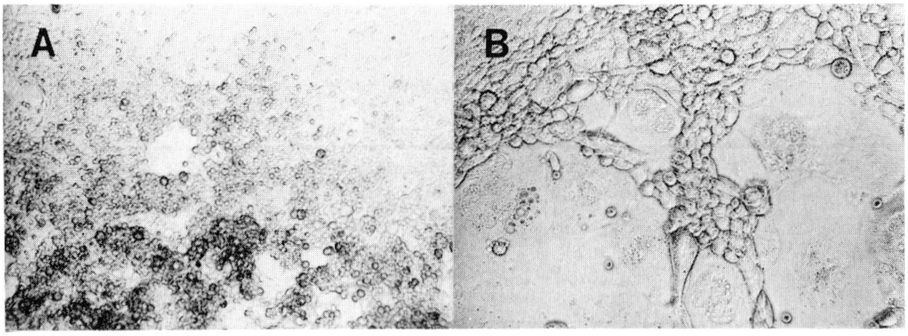

FIG. 2. Morphology of a transformed focus induced in NIH/3T3 cells by transfection of high molecular weight DNA from a human diffuse B-cell lymphoma. Panel A: 16X magnification. Panel B: 40 X magnification.

may involve interaction of many thousands of cellular genes and their products.

Although it has long been known from clinical observations that cancer arises as the result of a multi-step process, the number of these steps may be relatively limited. This is indicated by recent studies in which certain rodent embryo cells were transformed by the addition of two different oncogenes, whereas either one alone had no activity (45, 55). Furthermore, these studies have suggested classes of complementing oncogenes. It is now of critical importance to uncover additional cellular proto-oncogenes and define their normal functions and mechanims of activation.

The detailed information available regarding the structure and mechanisms of ras gene activation has made it possible to devise assays to rapidly distinguish activated from transforming ras alleles. Knowledge that some oncogene products act normally as either growth factors or their receptors may allow intervention by agents that interfere with receptor function. Thus, in addition to shedding light on basic mechanisms of normal and malignant growth, studies on cellular oncogenes are starting to provide a molecular armamentarium from which clinicians can select tools useful in the diagnosis, staging, and eventually even in the treatment of cancer.

REFERENCES

1. Andersen, P. R., Devare, S. G., Tronick, S. R., Ellis, R. W., Aaronson, S. A., and Scolnick, E. M.: 1981, Cell 26, pp. 129-134.
2. Balmain, A. and Pragnell, I. B.: 1983, Nature 303, pp. 72-74.
3. Bernheim, A., Berger, R., and Lenoir, G.: 1979, Cancer Genet. Cytogenet., 1, pp. 9-15.

4. Bishop, J. M.: 1983, In: Annual Review of Biochemistry, edited by E. E. Snell, P. D. Boyer, A. Meister, and C. C. Richardson, vol. 52, pp. 301-354, Academic Press, Palo Alto.
5. Canaani, E., Gale, R. P., Steiner-Saltz, D., Berrebi, A., Ashai, E., and Januszewicz, E.: 1984, Lancet 5, pp. 93-95.
6. Capon, D. J., Chen, E. Y., Levinson, A. D., Seeburg, P. H., and Goeddel, D. V.: 1983, Nature 302, pp. 33-37.
7. Capon, D. J., Seeburg, P. H., McGrath, J. P., Hayflick, J. S., Edman, U., Levinson, A. D., and Goeddel, D. V.: 1983, Nature 304, pp. 507-512.
8. Chang, E. H., Gonda, M. A., Ellis, R. W., Scolnick, E. M., Lowy, D. R.: 1982, Proc. Natl. Acad. Sci. USA 79, pp. 4848-4852.
9. Cooper, C. S., Park, M., Blair, D. G., Tainsky, M. A., Huebner, K., Croce, C. M., and Vande Woude, G. F.: 1984, Nature 311, pp. 29-33.
10. Cooper, G.M.: 1982, Science 217, pp. 801-806.
11. Croce, C. M., Erikson, J., Ar-Rushdi, A., Aden, D., and Nishikura, K.: 1985, Proc. Natl. Acad. Sci. USA (in press).
12. Croce, C. M., Thierfelder, W., Erikson, J., Nishikura, K., Finan, J., Lenoir, G., and Nowell, P. C.: 1983, Proc. Natl. Acad. Sci. USA 80, pp. 6922-6926.
13. Dalla-Favera, R., Bregni, M., Erikson, J., Patterson, D., Gallo, R. C., and Croce, C. M.: 1982, Proc. Natl. Acad. Sci. USA 79, pp. 4957-4961.
14. DeFeo-Jones, D., Scolnick, E. M., Koller, R., and Dhar, R.: 1983, Nature 306, pp. 707-709.
15. De Klein, A., Van Kessel, A. G., Grosveld, G., Bartram, C. R., Hagemeiger, A., Bootsma, D., Spur, N. R., Heisterkamp, N., Groffen, N., and Stephenson, J. R.: 1982, Nature 300, pp. 765-767.
16. Diamond, A., Cooper, G. M., Ritz, J., and Lane, M. A.: 1983, Nature 305, pp. 112-116.
17. Doolittle, R. F., Hunkapiller, M. W., Hood, L. E., Devare, S. G., Robbins, K. C., Aaronson, S. A., and Antoniades, H. N.: 1983, Science 221, pp. 275-277.
18. Downward, J., Yarden, Y., Mayes, E., Scrace, G., Totty, N., Stockwell, P., Ullrich, A., Schlessinger, G., and Waterfield, J.: 1984, Nature 307, pp. 521-527.
19. Duesberg, P. H.: 1983, Nature 304, pp. 219-226.
20. Ellis, R. W., DeFeo, D., Shih, T. Y., Gonda, M. A., Young, H. A., Tsuchida, N., Lowy, D. R., and Scolnick, E. M.:1981, Nature 292, pp. 506-511.
21. Erikson, J., Ar-Rushdi, A., Drwinga, L., Nowell, P. C., and Croce, C. M.: 1983, Proc. Natl. Acad. Sci. USA 80, pp. 820-824.
22. Erikson, J., Nishikura, K., Ar-Rushdi, A., Finan, J., Emanuel, B., Lenoir, G., Nowell, P. C., and Croce, C. M.: 1983, Proc. Natl. Acad. Sci. USA 80, pp. 7581-7585.
23. Eva, A. and Aaronson, S.A.: 1983, Science 220, pp. 955-956.
24. Eva, A. Aaronson, S. A.: 1984, In: Genes and Cancer, edited by J. M. Bishop, J. D. Rowley, and M. Greaves, pp. 373-382. Alan R. Liss, Inc., New York.

25. Eva, A., Tronick, S. R., Gol, R. A., Pierce, J. H., Aaronson, S. A.: 1983, Proc. Natl. Acad. Sci. USA 80, pp. 383-387.
26. Gale, R. P. and Canaani, E.: 1984, Proc. Natl. Acad. Sci. USA 81, pp. 5648-5652.
27. Gazit, A., Igarashi, H., Chiu, I., Srinivasan, A., Yaniv, A., Tronick, S., Robbins, K., and Aaronson, S.: 1984, Cell 39, pp. 89-97.
28. Goldfarb, M., Shimizu, K., Perucho, M., and Wigler, M.: 1982, Nature, 296:404-409.
29. Gonda, M. A., Young, H. A., Elser, J. E., Rasheed, S., Talmadge, C. B, Nagashima, K., Li, C. C., and Gilden, R. V.: 1982, J. Virol. 44, pp. 520-524.
30. Goubin, G., Goldman, D. S., Luce, J., Neiman, E., and Cooper, G. M.: 1983, Nature 302, pp. 114-119.
31. Green, M. R., Treisman, R., and Maniatis, T.: 1983, Cell 35, pp. 137-148.
32. Greenberg, M., and Ziff, E.: 1984, Nature 311, pp. 433-437.
33 Greenberger, J. S.: 1979, J. Nat. Cancer Inst. 62, pp. 337-344.
34. Groffen, J., Stephenson, J. R., Heisterkamp, N., DeKlein, A., Bartram, C. R., and Grosfeld, G.: 1984, Cell 36, pp. 93-99.
35. Gross, L.: 1970, Oncogenic Viruses, 2nd edit., Pergamon Press., Oxford.
36 Hall, A., Marshall, C. J., Spurr, N. K., and Weiss, R. A.: 1983, Nature 303, pp. 396-400.
37. Hankins, W. D., and Scolnick, E. M.: 1981, Cell 26, pp. 91-97.
38. Heisterkamp, N., Groffen, J., Stephenson, J. R., Spurr, N. K., Goodfellow, P. N., Solomon, B., Garrit, B., and Bodmer, W. F.: 1982, Nature 299, pp. 747-749.
39. Jainchill, J. L., Aaronson, S. A., and Todaro, G. J.: 1969, J. Virol. 4, pp. 549-553.
40. Kelly, K., Cochran, B. H., Stiles, C. D., and Leder, P.: 1983, Cell 35, pp. 603-610.
41. Konopka, J. B., Watanabe, S. M., Witte, O. N.: 1984, Cell 37: pp. 1035-1042.
42. Kraus, M., Yuasa, Y., and Aaronson, S. A.: 1984, Proc. Natl. Acad. Sci. USA 81, pp. 5384-5388.
43. Lane, M. A., Sainten, A., and Cooper, G. M.: 1981, Proc. Natl. Acad. Sci. USA 78, pp. 5185-5189.
44. Lane, M. A., Sainten, A., Cooper, G. M.: 1982, Cell 28, pp. 873-880.
45. Land, H., Parada, L. F., and Weinberg, R. A.: 1983, Nature 304, pp. 596-602.
46. Manolov, G. and Manolova, Y.: 1972, Nature 237, pp. 33.
47. Nowell, P. C. and Hungerford, D. A.: 1960, Science 132, pp. 1497.
48. Pierce, J. H. and Aaronson, S. A.: 1982, J. Exp. Med. 156, pp. 873-887.
49. Powers, S., Kataoka, T., Fasano, O., Goldfarb, M., Strathern, J., Broach, J., and Wigler, M.: 1984, Cell 36, pp. 607-612.
50. Pulciani, S., Santos, E., Lauver, A. V., Long, L. K., Robbins, K. C., and Barbacid, M.: 1982, Proc. Natl. Acad. Sci. USA 79, pp. 2845-2849.

51. Pulciani, S., Santos, E., Lauver, A. V., Long, L. K., Aaronson, S. A., and Barbacid, M.: 1982, Nature 300, pp. 539-542.
52. Reddy, E. P., Reynolds, R. K., Santos, E., and Barbacid, M.: 1982, Nature 300, pp. 149-152.
53. Rowley, J. D.: 1973, Nature 243, pp. 290-291.
54. Rowley, J. D.: 1983, Cancer Invest. 1, pp. 267-276.
55. Ruley, H. E. and Fried, M.: 1983, Nature 304, pp. 604-606.
56. Santos, E., Tronick, S. R., Aaronson, S. A., Pulciani, S., and Barbacid, M.: 1982, Nature 298, pp. 343-347.
57. Schechter, A. L., Stern, D. F., Vaidyanathan, L., Decker, S. J., Drebin, J. A., Green, M. I., and Weinberg, R. A.: 1984, Nature 312, pp. 513-516.
58. Shih, C., Weinberg, R. A.: 1982, Cell 29, pp. 161-169.
59. Shih, T. Y., Williams, D. R., Weeks, M. O., Maryak, J. M., Vass, W. C., and Scolnick, E. M.: 1978, J. Virol. 27, pp. 45-55.
60. Shilo, B. Z. and Weinberg, R. A.: 1981, Nature 289, pp. 607-609.
61. Shimizu, K., Birnbaum, D., Ruley, M. A., Fasano, O., Suard, Y., Edlund, L., Taparowsky, E., Goldfarb, M., Wigler, M.: 1983, Nature 304, pp. 497-500.
62. Shimizu, K., Goldfarb, M., Perucho, M., Wigler, M.: 1983, Proc. Natl. Acad. Sci. USA 80, pp. 383-387.
63. Sukumar, S., Notario, V., Martin-Zanca, D., and Barbacid, M.: 1983, Nature 306, pp. 658-661.
64. Tabin, C., Bradley, S. M., Bargmann, C. I., Weinberg, R. A., Papageorge, A. G., Scolnick, E. M., Dhar, R., Lowy, D. R. and Chang, E. H.: 1982, Nature 300, pp. 143-149.
65. Taparowsky, E., Shimizu, K., Goldfarb, M., Wigler, M.: 1983, Cell 34, pp. 581-586.
66. Taparowsky, E., Suard, Y., Fasano, O., Shimizu, K., Goldfarb, M., Wigler, M.: 1983, Nature 300, pp. 762-765.
67. Taub, R., Kirsch, I., Morton, C., Lenoir, G., Swan, D., Tronick, S., Aaronson, S., and Leder, P.: 1982, Proc. Natl. Acad. Sci. USA 79, pp. 7837-7841.
68. Waterfield, M. D., Scrace, G. T., Whittle, N., Stroobant, P., Johnsson, A., Wasteson, A., Westermark, B., Heldin, C.-H., Huany, J. D., and Deuel, T. F.: 1983, Nature 304, pp. 35-39.
69. Weinberg, R.A.: 1982, Cell 30, pp. 3-4.
70. Weiss, R. A., Teich, N., Varmus, H., and Coffin, J., editors: 1982, Molecular Biology of Tumor Viruses (2nd edit.): RNA Tumor Viruses. Cold Spring Harbor Laboratory, New York.
71. Weissman, B. E., and Aaronson, S. A.: 1983, Cell 32, pp. 599-606.
72. Yuasa, Y., Srivastava, S. K., Dunn, C. Y., Rhim, J. S., Reddy, E. P., and Aaronson, S. A.: 1983, Nature 303, pp. 775-779.
73. Yuasa, Y., Gol, R. A., Chang, A., Chiu, I.-M., Reddy, E. P., Tronick, S. R., and Aaronson, S. A.: 1984, Proc. Natl. Acad. Sci. USA 81, pp. 3670-3674.
74. Zech, L., Haglund, V., Nilsson, N., and Klein, G.: 1979, Int. J. Cancer 17, pp. 47-52.

CANCER CELLS AS PROBES OF EMBRYONIC DEVELOPMENT

G. Barry Pierce, M.D. and Robert S. Wells, M.D.
Department of Pathology, University of Colorado
Health Sciences Center, 4200 E. Ninth Avenue,
Denver, CO 80262 USA

ABSTRACT: Five embryonic fields have been shown to regulate appropriate carcinomas. Specificity for the reaction has been established in two of these fields, suggesting that a close developmental correspondence between the field and the cancer cell is required. Inhibitors of mitosis are present in two of these embryonic fields, and in one case it is likely that the inhibitor plus cell contact is required for regulation to occur.

Five embryonic fields regulate carcinomas. This statement may startle those raised with the classical dogma that cancer cells are uncontrolled and lawless, and that cancer cells only beget cancer cells.

Studies of the pathogenesis of teratocarcinomas of animals (1) and plants (2) proved for the first time that cancer cells could differentiate, that the differentiated progeny were benign (3), and that the process could be experimentally manipulated (4). These observations led to the idea that enhancing or directing differentiation of malignant cells might serve as an alternative to cytotoxic therapy for patients with malignant tumors with metastasis (5).

Additional facts are necessary before considering how the embryo regulates embryonal carcinoma. In this regard, Stevens (6) demonstrated the origin of embryonal carcinoma cells (the stem cells of teratocarcinoma) from primordial germ cells. The primordial germ cells may be conceptualized as the stem cell of the species, and this normal stem cell gave rise to an embryonal carcinoma cell, the stem cell of the teratocarcinoma. These malignant cells proliferate and differentiate and caricaturize the process

B. Pullman et al. (eds.), Interrelationship Among Aging, Cancer and Differentiation, 59–65.

of early embryonic development by making embryoid bodies (3). This was the clue that eventually led to the experiments first performed by Brinster (7) and confirmed by others (8,9) that chimeric mice could be produced by injecting embryonal carcinoma cells into blastocysts.

What happened to the primordial germ cells of the genital ridges that Stevens transplanted into the testes of adult animals (6). The cells of the transplant that were to form embryonic testes did so in their new environment, but by seven days after transplantation recognizable embryonal carcinoma cells were present in the fetal testes that contained primordial germ cells.

Other studies indicated that normal cells contain the genetic information necessary for the expression of the malignant phenotype (10). It turns out that even invasion and metastasis, considered by most to be the ultimate manifestations of the malignant phenotype, are not the exclusive prerogative of malignant cells. Neural crest cells invade from the dorsum of the neural tube and eventually settle as melanocytes in the skin, even the skin of the appendages. The cells of the reticulo-endothelial system colonize various sites. Rapid cell division is characteristic of cells of the marrow, testis, and the epithelial lining of the gut. When these traits are expressed simultaneously, the malignant phenotype results.

One forgets that Bovari discarded the idea that malignant tissue developed, as do all other tissues, by differentiation, because the process of differentiation occurrs too rapidly to explain carcinogenesis, a slowly evolving process (11). It is now known that initiation occurs rapidly in carcinogenesis, but the expression of the malignant phenotype may take a prolonged time. The latter is under epigenetic control, however, and requires an appropriate environment plus time. I wonder what Professor Bovari would have thought of Stevens' experiment? Would expression of the malignant phenotype in less than seven days change his mind about the mechanism of cancer?

In 1979, c-onc was discovered (12, 13). In other words, the genetic information responsible for the expression of the malignant phenotype of RNA tumor viruses was shown to be a normal cellular gene. Derepression of these c-onc loci is believed to play an important role in development, and presumably the loci are repressed when rapid growth is no longer needed. Oncologists then considered how this genetic information could be activated in carcinogenesis. Of course it could be activated by mutation; the majority of oncologists _know_ that cancer is caused by mutation. On the other hand, a minority of oncologists are epigeneticists and believe cancer to be caused by altered regulation of the genome. During carcinogenesis c-onc loci would be derepressed, and Schimke (14) believes they might then undergo amplification leading to the malignant state. These malignant cells when placed in the appropriate embryonic environment would have their c-onc repressed once again leading to the development of normal cells from the malignant ones. On the other hand, there is no reason to believe that mutated genes could not be repressed leading to the malignant phenotype.

How good is the evidence that embryonic fields can regulate cancer cells?

As mentioned previously, Brinster (7) put embryonal carcinoma cells from a black strain of mouse into blastocysts of a genetically white strain and then put the injected blastocyst into the uteri of animals made pseudopregnant. One of these animals threw a chimeric pup, as evidenced by its black and white coat. This experiment was confirmed in two laboratories that used isoenzyme analysis to prove that the offspring of blastocysts injected with embryonal carcinoma cells were in fact chimeric (8,9). They were composed of cells derived from the embryo and from the cancer cell. This was an important extension of the demonstration that cancer cells could differentiate. The differentiated progeny of cancer cells might be the equivalent of benign tumors cells that would be unresponsive to usual homeostatic regulation. Clearly the cells derived from the embryonal carcinoma cells were responsive to normal homeostatic mechanisms of the blastocyst to the point that they and their progeny took part in normal development.

Embryonal carcinoma cells are the normal counterpart of inner cell mass cells. The best evidence for this statement is the observation that when inner cell masses are cultured _in vitro_ some spontaneously transform and become embryonal carcinoma cells (15,16). Somehow the blastocyst was able to regulate the embryonal carcinoma cells so that they became inner cell mass cells. Then they took part in normal development. This implies that there might be specificity to the reaction and that the blastocyst would not be able to regulate other kinds of malignant tumors.

Assays were developed to determine the mechanism whereby the blastocyst regulates embryonal carcinoma cells (17,18). These assays took advantage of two of the phenotypic markers of cancer cells: their ability to grow as tumors or as neoplastic colonies *in vitro*. With these assays, it was shown that tumor and colony formation were regulated and that the reaction was specific in that leukemia cells, melanoma cells, and sarcoma cells were not regulated by the blastocyst (19). Neither blastocele fluid, inner cell mass cells, nor trophectoderm alone could regulate embryonal carcinoma cells, but the combination of blastocele fluid and contact of the cancer cells with trophectoderm proved regulatory of colony formation (20). Logistical problems have prevented analysis of the effects of blastocele fluid and inner cell mass on the regulation. Indirect evidence suggests the presence of molecules in blastocele fluid that are inhibitory of cell division of the cancer cells and of inner cell mass cells. It is our postulate that contact with the inhibited cell then results in regulation.

Because of the paucity of cells and of blastocele fluid for analysis, other embryonic fields were tested to see if they could regulate their closely related cancers. Neuroblastoma, melanoma, and adenocarcinoma of the breast were chosen for these experiments.

C1300 neuroblastoma cells were injected into tissue fragments representing various regions of the neural crest migration pathway (between 8 1/2 - 14 days gestation)(21). These fragments of tissues with the cancer cells were then injected into the testis of appropriate adult mice to see if the neuroblastoma cells could in fact form tumors. Tumors were consistently produced from the region of the crest formation as well as from fragments of brain and heart containing neuroblastoma cells, but abrogation of tumor formation by somites was observed. The best abrogation of tumor formation was found in the region of the adrenal primordium (unpublished).

We were not surprised that the neural crest migratory route was regulatory to a degree, because others have shown it to cause differentiation of migrating neural crest cells (22,23,24). Neither was it a surprise to observe that the best regulation of tumor formation occurred at the endpoint of the migratory route. Tissue culture studies have thus far shown no evidence for the necessity of a diffusible factor in the regulation of tumor formation (unpublished).

Because neural crest cells were regulated in the adrenal primordium at the time that neural crest migratory cells normally arrive in that primordium, it was decided to test the regulatory ability of mouse embryonic skin on melanoma cells. Melanocyte precursors arrive in the skin of the back of the mouse embryo on the 10th day of fetal life and in the limb bud on the 14th day. Dr. K. Graves developed a technique to inject mouse embryos _in utero_ with 300 melanoma cells (unpublished). Three hundred melanoma cells injected subcutaneously in newborn mice caused tumors in 80% of cases. This number injected in the skin of the back on day 10 or in the limb bud on day 14 caused tumors in only 30% of animals that survived the procedure. Melanomas occurred in 84% of cases when 300 melanoma cells were injected in the skin of the back on the 14th day. The data suggested the presence of inhibition of tumor growth in the skin of the back at 10 days, with lessening of this effect by the 14th day, and absence of the effect in the newborn. It is of interest that tumors grew at the 84% level in the skin of the back on the 14th day and only at the 30% level in the skin of the limb bud. This suggested that there might be localized effects involved.

These data were corroborated in studies in which tumor cells were injected into fragments of skin taken from the back or limb bud at the above mentioned times. The fragments were then injected into appropriate animals to see if the neuroblastoma cells could produce tumors.

The next series of experiments involved determination of whether or not a diffusible factor might be involved in the regulation. To this end, skin of the back from 10 and 14 day old mouse embryos and skin from the limb bud at 14 days was cultured _in vitro_ and conditioned medium was obtained after 48 hours. This medium was tested at 1, 5, 10, and 20% concentrations in normal tissue culture medium, and a dose response inhibition of growth of the melanoma cells was observed. The factor is being purified and at this point little is known about it.

Similar tissue culture experiments have been performed with 17 day gestation rat breast primordia and adenocarcinoma of the breast. A diffusible factor inhibitory of growth of adenocarcinoma cells of the breast has been implicated.

Dr. Sachs and his associates have demonstrated regulation of leukemia cells injected into the 10 day mouse placenta (25,26). Most of the animals injected with leukemia cells died, as would be expected, but two of the animals survived and had circulating leukocytes carrying the isoenzyme markers of the leukemia. This indicated that in these animals all of the leukemia cells had been induced to differentiate.

In summary, five embryonic fields have been shown to be capable of regulating carcinomas. In two of these fields specificity for the reaction has been demonstrated, and it can be postulated that for an embryonic field to regulate a carcinoma there must be a close correspondence between the field and the cancer cell. For example, embryonal carcinoma cells, the malignant counterpart of inner cell mass cells, are regulated in the blastocyst. The presence of mitotic inhibitors has been demonstrated in two of these fields and the necessity for a mitotic inhibitor and cell contact has been demonstrated in one of them. It seems probable that an embryonic field must exist that is capable of regulating any type of carcinoma. Of equal importance is the observation that it is possible to use the phenotypic markers of malignant cells as probes of normal embryonic development. Using embryonal carcinoma cells as probes it has been possible to show that regulation of inner cell mass is dependent upon an inhibitor in the blastocele fluid and contact with trophectoderm. Possibly the mechanism of differentiation will yield to these kinds of analyses.

ACKNOWLEDGMENTS

This work was supported in part by a gift from R. J. Reynolds Industries, Inc., NIH grant CA-15823, NIH grant CA-35367, and NIH grant CA-36069.

REFERENCES

1. Pierce, G.B., and Dixon, F.J.: 1959, Cancer 12, pp. 573-583.
2. Braun, A.C.: 1969, "The Cancer Problem; a Critical Analysis and Modern Synthesis", New York, Columbia University Press.
3. Pierce, G.B., Dixon, F.J., Jr., and Verney, E.L.: 1960, Lab. Invest. 9, pp.583-602.
4. Pierce, G.B., and Verney, E.L.:1961, Cancer 14, pp. 1017-1029.
5. Pierce, G.B.: 1961, "Teratocarcinoma. A Problem in Developmental Biology", Fourth Canadian Cancer Conference (R. Begg, ed.), Vol. 4, Academic Press,

New York, pp. 119-137.
6. Stevens, L.C.: 1967, J. Nat. Cancer Inst. 38, pp. 549.
7. Brinster, R.L.: 1974, J. Exp. Med. 140, pp. 1049-1056.
8. Papaioannou, V.E., McBurney, M.W., Gardner, R.L., and Evans, R.L.: 1975, Nature (London) 258, pp. 70-73.
9. Mintz, B., and Illmensee, K.: 1975, Proc. Natl. Acad. Sci. USA 72, pp. 3585-3589.
10. Pierce, G.B. and Johnson, L.D.: 1971, In Vitro 7, pp. 140-145.
11. Boveri, T.: 1929, "The Origin of Malignant Tumors", (M. Boveri, ed.), Baltimore, Williams and Wilkins Co.
12. Stehelin, D., Guntaka, R.V., Varmis, H.E., and Bishop, J.M.: 1976, J. Mol. Biol. 101, pp. 349-365.
13. Purchio, A.F., Erikson, E., Brugge, J.S., Erikson, R.L.: 1978, Proc. Natl. Acad. Sci. USA 75, pp. 1567-1571.
14. Shimke, R.T.: 1984, Cancer Res. 44, pp. 1735-1742.
15. Evans, M.J. and Kaufman, M.H.: 1981, Nature 292, pp. 154-156.
16. Martin, G.R.: 1981, Proc. Natl. Acad. Sci. USA 78, pp. 7634-7638.
17. Pierce, G.B., Lewis, S.H., Miller, G., Moritz, E., and Miller, P.: 1979, Proc. Natl. Acad. Sci. USA 76, pp. 6649-6651.
18. Wells, R.S.: 1982, Cancer Res. 42, pp. 2736-2741.
19. Pierce, G.B., Pantazis, C.G., Caldwell, J.E., and Wells, R.S.: 1982, Cancer Res. 42, pp. 1082.
20. Pierce, G.B., Aguilar, D., Hood, G., and Wells, R.S.: 1984, Cancer Res. 44, pp. 3987-3996.
21. Podesta, A.H., Mullins, J., Pierce, G.B., and Wells, R.S.: 1984, Proc. Natl. Acad. Sci. USA 81, pp. 7608-7611.
22. LeDouarin, H.: 1980, Current Topics Dev. Biol. 16, pp. 31-85.
23. Weston, J.A.: 1981, In: "Advances in Neurology" (Ricardi, V.M. and Mulverhill, J.J., eds.), Raven Press, New York, Vol. 29, pp. 77-94.
24. Thiery, J.P., Dubond, J.L. and DeLouvie, A.: 1983, Dev. Biol. 93, pp. 324-343.
25. Gootwine, E., Webb, C.G. and Sachs, L.: 1984, Dev. Biol. 101, pp. 221-224.

ALTERATIONS IN EPIDERMAL DIFFERENTIATION IN SKIN CARCINOGENESIS

Stuart H. Yuspa, M.D.
Laboratory of Cellular Carcinogenesis and Tumor Promotion
National Cancer Institute
Bethesda, Maryland

The induction of carcinomas on mouse skin by initiation and promotion protocols involves three operationally distinct stages. Initiation appears to be a genetic change in epidermal basal cells which provides the capability for altered cells to proliferate under conditions where normal cells are obligated to differentiate. Tumor promotion stimulates the clonal expansion of initiated cells through differential growth responses of normal and initiated cells with each promoter exposure. The result of these two stages is the formation of benign tumors, epidermal papillomas. The third stage, malignant conversion, occurs as a result of further genetic changes in papilloma cells either spontaneously at a low frequency or by the action of genotoxic carcinogens at a higher frequency. The specific genes involved in any of the stages of skin carcinogenesis have not been elucidated, but activation of the $\underline{ras}^{Ha}$ gene may be associated with initiation or malignant conversion.

INTRODUCTION

The initiation-promotion model of epidermal carcinogenesis in mice has provided the basis for many conceptual advances in the understanding of the biology of carcinogenesis. Initiation by a single exposure with a low dose of carcinogen (Stage I) results in the permanent alteration of some epidermal cells. In the absence of further treatment, these initiated cells do not develop into tumors, but the cellular changes are heritable. As a result of subsequent repeated exposure of initiated skin to promoting agents (Stage II), multiple benign papillomas develop. The promotion process is reversible; papillomas do not develop after insufficient promoter treatment or if the interval between treatments is prolonged. Many papillomas regress when the promoting stimulus is removed. Other papillomas progress irreversibly to carcinomas either spontaneously or as a result of further exposure to a tumor initiator; this process has been termed malignant conversion (Stage III).

The irreversibility of both initiation and malignant conversion suggests that these processes involve genetic alterations. The rever-

B. Pullman et al. (eds.), Interrelationship Among Aging, Cancer and Differentiation, 67–81.

sibility of promotion of papilloma formation indicates an epigenetic mechanism. Promoting agents produce different responses in initiated and normal cells with regard to effects on growth and terminal differentiation. This may form the basis for selective growth of initiated cells. In this report, I will summarize studies performed during the last several years which have clarified the cellular and molecular basis for stages in epidermal carcinogenesis.

PHENOTYPE OF INITIATION

Selection of Initiated Cells In Vitro

The principal lesions which develop as a result of initiation-promotion protocols are papillomas, benign tumors with varying potential for malignant conversion. The initiation of papilloma formation has an apparent one-hit dose-response pattern (2), and the tumors which form are monoclonal (3), suggesting that the papilloma represents a clonal expansion of an initiated cell. As such, the phenotype of initiation should be completely represented by the papilloma. Papillomas resemble normal epidermis in that both are composed of multiple layers of cells undergoing an orderly progression of differentiation; however in papillomas, each stratum is generally represented in greater abundance than in normal skin. Autoradiography of papillomas from animals which had received [^{3}H]-thymidine shortly before sacrifice revealed that the labeling index was 10-fold higher than normal skin. The most striking alteration, however, was the presence of thymidine labeled cells in strata far above the basement membrane zone, a finding not observed in normal epidermis or epidermis in a non-neoplastic hyperplastic state where DNA synthesis is confined to a single layer of basal cells (4). This result implied that in epidermis, initiated cells have an altered response to differentiation signals and are not obligated to cease proliferation after migration away from the basement membrane. In cultured keratinocytes, epidermal differentiation is modulated by extracellular Ca^{2+} and the observed changes in differentiating keratinocytes cultured in high Ca^{2+} medium (> 0.1 mM Ca^{2+}) are similar to those found _in vivo_ when normal basal cells migrate from the basement membrane. That is, in low Ca^{2+} medium (< 0.1 mM Ca^{2+}) cells proliferate as basal cells and in high Ca^{2+} medium, proliferation ceases and cells terminally differentiate (5). This _in vitro_ model seemed ideal to test the hypothetical link of initiation to an altered response to differentiation signals.

Basal epidermal cells were cultured in low Ca^{2+} medium, exposed to carcinogens and then changed stepwise at weekly intervals into higher Ca^{2+} medium over 4 weeks. Carcinogen exposure _in vitro_ resulted in the focal persistence of keratinocytes which proliferated in high Ca^{2+} medium and could be easily recognized because they stained dark red with rhodamine (6). A number of characteristics of this assay suggest it selects for initiated cells: 1) cultures of basal cells isolated from initiated mouse skin and selected in high Ca^{2+} medium yield identical foci (7,8); 2) the number of foci increases with higher doses of carcinogen in vivo or in vitro (6,8,9); 3) stronger initiators yield more foci than weaker initiators for both in vivo and in vitro exposures (8,9); 4) for benzo(a)pyrene, the number of foci directly correlates with the extent of DNA binding after in vitro exposure (10); 5) the focus forming potential of initiated skin persists for at least 10 weeks after initiation (8); 6) spontaneous foci develop in cultures from SENCAR mice, which are sensitive to papilloma induction by promotion alone (7,8); 7) foci are not tumorigenic when first formed, but cell lines derived from foci may progress to produce carcinomas upon in vivo testing (6,9,11).

It should be emphasized that the foci which are selected in this assay are not differentiation resistant. These cells change morphology in response to Ca^{2+}, stratify and cells in the upper strata terminally differentiate (Fig. 1). The distinguishing feature for which we select is the continued proliferation of substrate attached cells in high Ca^{2+} which results in expanding foci. Autoradiographs of cross sections of [^{3}H]-thymidine labeled foci reveal the growth fraction of substrate attached cells is nearly 100 percent. Thus, foci which develop from carcinogen exposure demonstrate an alteration in the regulatory pathway which functions to inhibit proliferation in normal cells when they respond to a differentiation signal.

Characterization of Foci and Derived Cell Lines

Twenty-six primary foci have been tested for tumorigenicity by inserting intact foci on a plastic carrier into the flank of nude mice (9). No tumors developed in more than one year. Immunofluorescence analysis using anti-mouse keratin antibodies indicated that all primary foci studied (12/12) contained a keratin cytoskeleton. Thirty cell lines have been established by growing single foci to confluence and then subculturing or by cloning from subconfluent foci. Cell lines were established from foci derived from SENCAR and Balb/c mouse cells, from adult and newborn cultures, from foci resulting after in vivo initiation or after in vitro exposure to carcinogens, and from spontaneous foci. There were no distinguishing characteristics which could predict the origin or biological potential (e.g., tumorigenicity) of cell lines. Most had a unimodal DNA content suggesting a clonal origin, and 2N or 4N DNA content was observed in most cases. Almost all retained their keratin cytoskeleton. The majority were not tumorigenic although several became tumorigenic at later passages and produced carcinomas (11). All lines were resistant to growth inhibitory and differentiation inducing effects of high Ca^{2+} medium and thus had characteristics consistent with the initiated phenotype. No lines which retained their keratin cytoskeleton were capable of anchorage independent growth. Several lines have been evaluated for their relative levels of RNA transcripts for five known oncogenes by slot blot hybridization. No altered levels of transcripts for ras^{Ha}, ras^{Ki}, myc, fos, or abl could be detected relative to normal cultured keratinocytes (12).

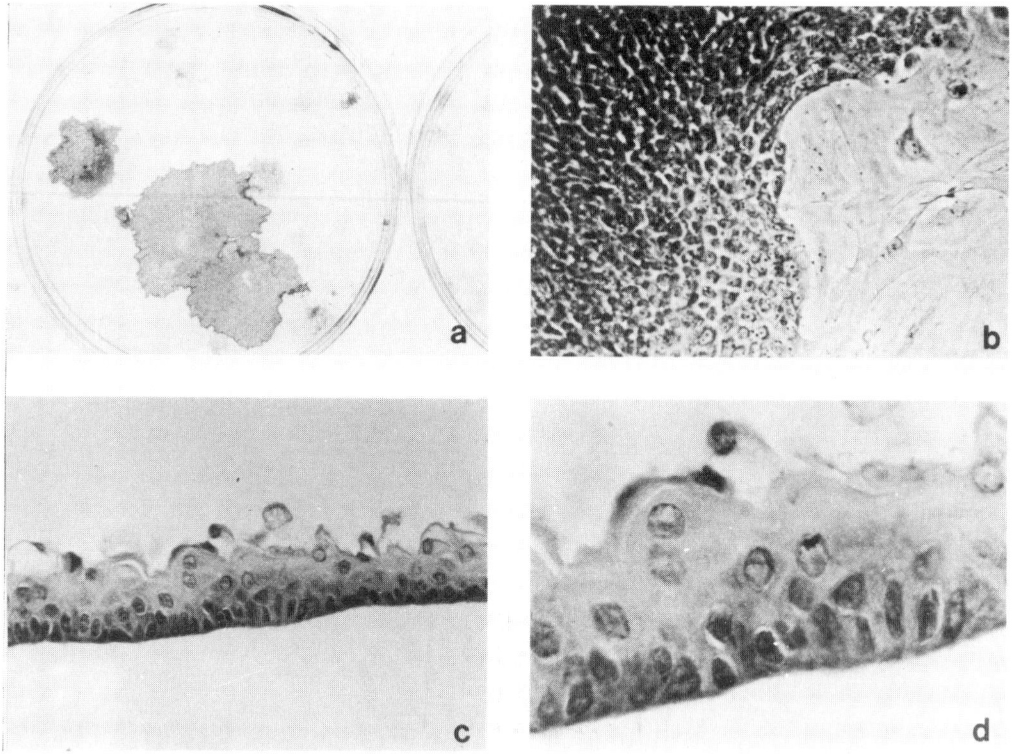

FIGURE 1

Characteristics of foci produced by carcinogen treatment in primary mouse keratinocytes. Foci were produced by MNNG treatment on day 3 in culture as described in (6). a) appearance of a focus in a rhodamine stained dish after 4 weeks in high Ca^{2+} medium; b) higher power (100X) magnification of the focus margin under phase contrast microscopy. Note the dense growth of polygonal cells in the focus (left) and the background of large, quiescent (< 1 percent labeling index) senescent keratinocytes which form a lawn on the culture dish; c) focus released intact by dispase treatment and sectioned and stained with hematoxylin and eosin; d) higher magnification of a cross section through a focus showing stratification and maturation.

During the last year we have established conditions to isolate viable cells from mouse skin papillomas induced by initiation and promotion of SENCAR mice. Viable cells were plated onto dishes coated with collagen-fibronectin-albumin in dermal fibroblast conditioned medium with 0.05 mM Ca^{2+} and 10 ng/ml EGF. Six cell lines were established from separate primary isolates and designated PA, PB, PC, PD, PE, and PF. We have compared the properties of these papilloma cells to those of normal keratinocytes and to our putative initiated cell lines described above. These comparisons are shown on Table 1. All three cell types grow rapidly in low Ca^{2+} medium. Normal cells will not grow at clonal density in high Ca^{2+} medium and terminally differentiate in this medium yielding a large number of cornified cells. This is associated with high activity of the enzyme epidermal transglutaminase which catalyzes the formation of the cornified envelope (13). At clonal density, papilloma lines grew better in low Ca^{2+} medium than in high Ca^{2+} medium. When switched to high Ca^{2+} medium, 4 of the 6 lines produced large numbers of cornified envelopes. In these respects, papilloma derived lines differed from putative initiated cell lines which always grew better in high Ca^{2+} medium and did not show an increase in transglutaminase or cornified cells in high Ca^{2+} conditions.

The phorbol ester tumor promoter, 12-O-tetradecanoyl phorbol-13-acetate (TPA), induces terminal differentiation in normal keratinocytes (see below) and will not stimulate clonal growth of normal cells in vitro. Exposure to TPA stimulated clonal growth of all papilloma lines and some "initiated" cell lines and did not induce differentiation in any of these two sets of lines. In this regard "initiated" cell lines and papilloma cell lines were similar. In addition (not shown), papilloma lines did not show enhanced mRNA transcripts for $\underline{ras}^{Ha}$, $\underline{ras}^{Ki}$, _myc_, _fes_, _raf_, and _abl_, above the level of messages found in normal keratinocytes or "initiated" cell lines. In tumorigenicity testing, 2/6 papilloma lines produced squamous cell carcinomas after 4-6 million cells were injected into nude mice. About 20 percent of "initiated" cell lines yielded carcinomas when tested under similar conditions.

Significance

The elucidation of the cellular and molecular changes which define the initiated state is a high priority in cancer research. The knowledge obtained is essential for comprehending carcinogenesis in general and practical for its potential to identify early markers of neoplasia in order to recognize lesions which are still manageable. An understanding of the initiated phenotype is required for logical analysis of the process of tumor promotion, and the availability of initiated cells provides a critical target to evaluate promoter responses. We have developed a model which has many similarities to initiation _in vivo_ and provides a method to isolate putative initiated cells. Inherent in the model is a positive selection procedure for the cells of interest. These are the elements required for a model system useful for a genetic approach to understand initiation. Our findings have supported the hypothesis that alterations in response to differentiation signals are early events in epithelial carcinogenesis. Viral oncology has provided insights into a number of cancer-related genes which were recognized because of their ability to stimulate unregulated proliferation in their target cells. However, those test systems have not required an interruption in a differentiation program to achieve a state of high proliferation. Thus, many of the current oncogenes may reflect only a subclass of genes involved in cancer related proliferative events. Our current model, like epithelial cancer _in vivo_, requires a different

CHARACTERISTIC	NORMAL	PAPILLOMA	INITIATED
Growth <0.1 mM Ca^{2+}	+ +	+ +	+ +
>0.1 mM Ca^{2+}			
Clonal Growth	−	+ or −	+ +
Terminal Differentiation	+	−	−
Cornification - TGAse	↑	↑	↔
TPA			
Clonal Growth	↔	↑	↑ or ↔
Terminal Differentiation	+	−	−
Cornification - TGAse	↑	↔	↔
Tumorigenicity in Nude Mice (lines positive / lines tested)	−	2/6	5/26

Table 1. Comparative characteristics of normal keratinocytes, papilloma cell lines, and "initiated" (altered-differentiation) cell lines.

class of regulatory alterations. We anticipate that this model will reveal other classes of genes whose functions or control must be altered to achieve the neoplastic state.

CELLULAR BASIS FOR TUMOR PROMOTION

Considerable experimental data support an epigenetic basis for phorbol ester-mediated tumor promotion in mouse skin. Initiation-promotion protocols yield primarily benign tumors while repeated application of genotoxic initiating agents yield many malignant tumors. Promoters do not enhance malignant conversion of papillomas while initiators are effective converting agents. Phorbol esters are not mutagenic, and their clastogenic action has been demonstrated in only a few specialized cell types. Most, if not all, of the effects of phorbol esters are mediated via activation of cell membrane receptors (protein kinase C) suggesting a second messenger complex is involved in the responses. A number of aspects of tumor promotion indicate that cell selection is involved in the process. Papillomas are monoclonal, and are likely to represent expanded clones of initiated cells. Repeated promoter application is required for promotion to be effective in yielding tumors, which is to be expected if a selective pressure must be maintained. Epidermal response patterns change with repeated phorbol ester exposure, supporting a selective expansion of a keratinocyte subpopulation with a particular response pattern. Having been strongly influenced by this established biology for skin tumor promotion in vivo, we felt that the keratinocyte culture model might reveal a cellular basis for phorbol ester action. The achievement of this understanding would naturally lead to appropriate experiments to understand the molecular events involved.

Heterogeneity in Epidermal Responses to Phorbol Esters

When cultured basal cells (low Ca^{2+} medium) are exposed to 12-O-tetradecanoylphorbol-13-acetate (TPA) or other phorbol esters, they respond in a heterogeneous manner (14). A portion of the population is induced to differentiate by the promoter. This is detected by morphological changes, measurement of the formation of cornified squames and increase in the activity of the differentiation-associated enzyme epidermal transglutaminase. As terminally differentiated cells slough from the culture dish, transglutaminase activity decreases. The basal cells remaining after a brief TPA exposure are transiently resistant to terminal differentiation when cultured in high Ca^{2+} medium or when exposed to TPA a second time. However, these cells are stimulated to proliferate by the second TPA exposure. Thus, two types of responses occur in epidermal basal cells exposed to phorbol esters. In one population, terminal differentiation is induced. In a second population, TPA stimulates proliferation and produces a transient block in response to a differentiation stimulus. This latter subpopulation cannot be induced to differentiate by a second TPA exposure if the interval between exposures is short. This is similar to the response of mouse skin in vivo after several TPA exposures where proliferative stimulation is the principal response. If the culture interval between TPA exposures is prolonged (10 days), then heterogeneity in responsiveness is restored. This is similar to findings in vivo when treatment intervals are prolonged. The net effect of this type of heterogeneity is to select for and to expand the proliferating population with each promoter exposure, thus remodeling the target tissue. Clonal selection of carcinogen-altered (differentiation-altered) cells would result if normal basal cells were induced to differentiate while altered (initiated) cells were among those which are stimulated to proliferate. Our studies of the effects of TPA on "initiated" and papilloma cell lines suggest this is the case.

The biological basis for the heterogeneous responses of keratinocytes to phorbol esters appears to be the maturation state of the basal cell at the time of exposure since 1) proliferative responders give rise to differentiative responders; 2) advancing the maturation of basal cells by culture in high Ca^{2+} medium leads to loss of proliferative responses and enhancement of differentiative responses (15); 3) altered keratinocytes which fail to mature in response to high Ca^{2+} medium are resistant to the differentiation inducing effects of phorbol esters.

Pharmacological differences in phorbol ester receptors have been measured in keratinocytes at different maturation states (16). Examination of phorbol ester binding in basal cells showed two binding sites with K_d of 5.5 and 100 nM for phorbol dibutyrate. When maturation is induced by switching basal cells to high Ca^{2+} medium, conditions which enhance the differentiation response to TPA, there is a substantial increase in the number of low affinity binding sites. Thus, the heterogeneity of epidermal responses could be related to two classes of receptors, one leading to proliferative responses and the other to differentiative responses.

Molecular Consequences of Phorbol Ester Exposure

The orderly expression of keratin genes in mouse skin is markedly disturbed following TPA exposure. Using cDNA probes for differentiation-specific keratins (59 and 67 kd) and proliferation-specific keratins (50, 55b and 60 kd) (17), specific transcript levels were measured in total RNA extracted from TPA treated and control epidermis at multiple times over a seven-day interval after treatment *in vivo*. There is a rapid loss of the differentiation-specific keratin mRNA's within 12 hours of TPA exposure, and this is coupled with a relative increase in the proliferation-related keratin mRNA's. These data are most consistent with a rapid loss of cells actively synthesizing the messages for the differentiation-related keratins. *In vivo*, the number of cell layers in the fully mature stratum corneum increases substantially within 24 hours of TPA exposure (18). Taken together with the keratin expression data, it seems likely that *in vivo* the maturing cells are accelerated through their differentiation program to the terminal stage by exposure to TPA. These conclusions from *in vivo* studies are consistent with the effects of these agents on the terminal differentiation of cultured keratinocytes.

We have also examined epidermis *in vivo* and epidermal cells in culture for modified expression of oncogene transcripts after TPA exposure. We have been unable to find a significant change in expression of *ras*^{Ha}, *ras*^{Ki}, *fos*, *myc*, *abl*, or *raf* at multiple times after a single exposure to TPA or other phorbol esters and related compounds (12). Chronic TPA application leads to a significant reduction in *abl* expression.

Significance

These studies support an epigenetic mechanism for tumor promotion in mouse skin based on selective clonal expansion of initiated cells. The data are specific for phorbol esters and mouse epidermis and thus describe only one avenue (heterogenous response patterns of target cells) by which clonal expansion could be achieved. However, the basic biology involved in promotion is similar in many target tissues (e.g., liver, breast, bladder). Thus these studies provide a model for exploration of cellular responses which would lead to selective clonal expansion following exposure to other promoters specific for a particular organ site. Since promotion has a benign lesion as an endpoint and is reversible, the elucidation of the mechanisms involved is critical for controlling potentially manageable lesions in carcinogenesis. This is particularly important for identifying antipromoting agents and exploring their mechanism of action. These studies have provided important new insights for understanding the regulation of epidermal differentiation. The phorbol ester receptor, protein kinase C, appears to be the major regulator of epidermal differentiation. Preliminary studies suggest that extracellular Ca^{2+} acts through a similar mechanism. These revelations offer a point of focus toward understanding the alterations in the regulation of epidermal differentiation associated with the initiation phase of carcinogenesis.

THE ROLE OF RAS ONCOGENES IN SKIN CARCINOGENESIS

The isolation from human and experimental tumors of activated cellular homologues of retroviral oncogenes and the demonstration of their transforming activity in 3T3 cells has stimulated substantial interest in the possibility that activation of oncogenes plays a causal role in carcinogenesis. For our purposes, it was particularly noteworthy that these activated oncogenes were commonly found in human tumors originating in lining epithelia. When a sufficient data base formed from such studies, the vast majority of isolated oncogenes were members of the *ras* gene family. Balmain and Pragnell (19) reported that an activated *ras*^{Ha} gene could be isolated from a transplantable mouse skin carcinoma and Balmain et al. (20) subsequently found an activated form of *ras*^{Ha} in about 60 percent of primary papillomas and carcinomas. Having defined a cellular phenotype associated with epidermal transformation, we were interested to know if an activated form of *ras* could by itself produce this characteristic phenotype. The substantial advantage of using sarcoma viruses for introducing the activated *ras* oncogene was utilized since high frequency of transduction was likely and transfection methodology for primary epidermal cells was not established. In addition, viruses carrying other transforming genes could be studied.

Characteristics of Keratinocytes Infected with Harvey or Kirsten Sarcoma Viruses

The effects of an activated *ras* oncogene on primary epidermal cells were studied by infecting basal cells in low Ca^{2+} medium with Harvey or Kirsten sarcoma virus and monitoring the effects by parameters we have previously used to characterize cells exposed to chemical initiators or tumor promoters (Table 2). Infection with either Kirsten or Harvey sarcoma virus causes a marked stimulation of basal cell proliferation with values for both [^{3}H]-thymidine incorporation into DNA and labeling indices reaching 3- to 5-fold above control within a few days. Basal cell number increased proportionally. Challenge of infected basal cell cultures with high (0.5 mM) Ca^{2+} medium resulted in a marked morphological change resembling a more advanced stage of differentiation, but these cultures persisted for long periods without stratifying and sloughing from the culture dish. Specific immunoprecipitation of labeled cell extracts with monoclonal antibodies indicated that infected cultures express much higher levels of the ras gene product (the p21 protein) than controls, and studies with temperature sensitive mutants showed that p21 function is required for the persistence of virus infected cells in high Ca^{2+} medium. Although virus infected cells do not terminally differentiate in response to high Ca^{2+} medium, their proliferation rate is markedly reduced and to the same extent as control cells. Expression of p21 by infected keratinocytes remains as elevated in high Ca^{2+} as in low Ca^{2+}. Thus the proliferative block associated with the Ca^{2+} signal cannot be attributed to Ca^{2+} induced changes in expression of p21. Virus infected cells as well as controls exhibit a rise in the differentiation associated enzymatic activity of epidermal transglutaminase when switched from low to high Ca^{2+}, yet cornified envelope production remains low. These results (21) indicate that the expression of an altered ras gene is closely linked to epidermal basal cell proliferation and possibly a partial disruption of an intermediate stage of epidermal differentiation prior to events leading to cell death.

The proliferative capacity of virus infected cells and chemically initiated cells differ under culture conditions that induce terminal differentiation in normal cells. Specifically, while virus infected cells are growth arrested in high Ca^{2+} medium, chemically initiated cells continue to proliferate. This seems inconsistent with the inference of the work of Balmain et al. that ras gene activation is sufficient to initiate skin carcinogenesis.

Identification of the Maturation State Achieved by Virus Infected Cells in High Ca^{2+} Medium

We have now conducted additional analyses of the properties of keratinocytes expressing an activated ras gene transduced by either Harvey or Kirsten sarcoma virus infection (22). To identify the differentiation state achieved by virus infected cells, two epidermal marker proteins were assayed. Previous studies have shown that the pemphigoid antigen, a 220 kd protein found in the basement membrane, is synthesized exclusively by basal cells in low Ca^{2+} and the pemphigus antigen, a 210 kd glycoprotein found on the cell surface is exclusively synthesized by differentiating cells maintained in high Ca^{2+} medium (23). However, virus infected cells did not stop synthesizing pemphigoid antigen nor begin synthesizing pemphigus antigen when switched to high Ca^{2+} medium, suggesting that these cells are blocked at a late stage in basal cell maturation when signaled to differentiate by Ca^{2+}. Support for this conclusion was drawn from the finding that when cells maintained in high Ca^{2+} medium were returned to low Ca^{2+}, normal cells were incapable of resuming basal cell growth, while virus infected cells resumed their basal cell morphology. Of considerable interest was the response of virus infected cells to phorbol esters (Table 3). Normal basal cells

1. Infected basal cells (< 0.1 mM Ca^{2+}) proliferate at a rate 3-5X greater than uninfected cells with a corresponding increase in cells/dish.

2. When challenged by > 0.1 mM Ca^{2+}, infected cells cease proliferating, assume a squamous morphology, but do not progress through the entire differentiation program and thus remain attached to the culture dish.

3. Transglutaminase activity increases substantially but few cornified cells are produced.

4. Studies with TS mutants indicate a functioning _ras_ gene product is required for viral-induced changes.

5. Expression of the p21 protein is very high in virus-infected cells and is not modulated by extracellular Ca^{2+}.

6. Viral effects require a high multiplicity of infection and are not focal.

Table 2. Cellular phenotype of keratinocytes infected with Kirsten or Harvey sarcoma virus.

are highly sensitive to the induction of ornithine decarboxylase (ODC) by TPA but rapidly lose sensitivity once they are exposed to > 0.1 mM Ca^{2+} (24). In contrast, virus infected cells remain responsive to ODC induction by TPA even after several days in high Ca^{2+} medium. Phorbol ester exposure may even reverse the maturation state of virus infected cells in high Ca^{2+} as suggested by the finding that elevated transglutaminase activity, associated with maturation of both normal and virus infected keratinocytes in high Ca^{2+} medium, is reduced in the virus infected (but not normal) cells by exposure to TPA. Furthermore, the proliferative block imposed by high Ca^{2+} growth conditions on virus infected cells can be partially reversed by TPA exposure.

Significance

These studies provide new information concerning the role of an activated _ras_ gene in epidermal tumorigenesis. Although these studies are confined to the effects of v-_ras_, both v-_ras_^{Ha} and v-_ras_^{Ki} produce similar effects. Thus, it is likely that the results are relevant for other activated forms of the _ras_ gene. The data suggest that an activated _ras_ gene could provide the proliferative signal required to form an epidermal tumor, but the altered cells would growth arrest in response to a differentiation signal (such as high Ca^{2+} _in vitro_ or migration from the basement membrane _in vivo_). Thus _ras_ activation alone would be insufficient for tumor formation since papillomas are characterized by

RESPONSE	KERATINOCYTES IN 0.5 mM Ca^{2+} MEDIUM FOR 72 HOURS			
	Control		Virus Infected	
	-TPA	+TPA	-TPA	+TPA
Ornithine Decarboxylase (nmoles CO_2/μg Protein/hr)	0.14	0.18	0.22	2.55
Transglutaminase (CPM/μg Protein/ 30 min x 10^{-2})	9.2	10.4	7.5	3.2
DNA Synthesis (CPM ^{3}H-TdR/μg DNA)	511	299	489	953

Table 3. Responses to TPA of Kirsten or Harvey sarcoma infected or control keratinocytes in 0.5 mM Ca^{++} medium.

proliferating cells in the maturing strata. However, activation of a ras gene in primary keratinocytes does appear to block the maturation program in a non-terminal stage of differentiation, probably a late basal cell stage in which proliferation normally stops but can still be stimulated. Most importantly such cells respond to phorbol ester tumor promoters by undergoing a phenotypic reversion to a less mature cell type (high ODC, low transglutaminase) and can re-enter the proliferative pool. Thus activation of a ras gene could result in a conditionally initiated cell in which the full expression of its biological potential depends on either a complimentary genetic alteration or exposure to tumor promoters.

In many mouse strains at least two classes of skin papillomas develop in protocols involving initiation and promotion. Promoter dependent tumors require repeated promoter exposures for both expression and maintenance and regress upon withdrawal of the promoting stimulus. Autonomous papillomas are independent of promoter exposure for maintenance once established and are more likely to progress to carcinomas. The biological changes we have defined for keratinocytes expressing an activated ras gene are most consistent with ras activation being involved in the formation of promoter-dependent tumors. In contrast the characteristics of chemically altered cells, selected by virtue of their differentiation resistance, are most consistent with changes in autonomous papillomas. Whether the latter cell type requires more than one genetic change (of which ras activation could be a component and thus explain the isolation of an activated ras from 60 percent of papillomas) or an alteration in a different genetic element remains to be determined.

ACKNOWLEDGEMENT

I am greatly indebted to a number of colleagues in these studies including Molly Kulesz-Martin, Henry Hennings, Ulrike Lickti, Hideki Kawamura, James Strickland, Dennis Roop, and Rune Toftgard. The excellent work of Sandy White in preparing this manuscript is also appreciated.

REFERENCES

1. Hennings,H.,Shores,R.,Wenk,M.L.,Spangler,E.F.,Tarone,R.,and Yuspa,S.H.: 1983,Malignant conversion of mouse skin tumors is increased by tumour initiators and unaffected by tumour promoters, Nature,304,pp.67-69.

2. Burns,F.,Albert,R.,Altshuler,B.,Morris,E.: 1983,Approach to risk assessment for genotoxic carcinogens based on data from the mouse skin initiation-promotion model,Environ.Health Perspectives,50, pp.309-320.

3. Reddy,A.L.,and Fialkow,P.J.: 1983,Papillomas induced by initiation-promotion differ from those induced by carcinogen alone,Nature (Lond.),304,pp.69-71.

4. Yuspa,S.H.: 1984,Molecular and cellular basis for tumor promotion in mouse skin. In: Fujiki,H.,Hecker,E.,Moore,T.E.,Sugimura,T.,and Weinstein,I.B.(eds.): "Cellular Interactions of Environmental Tumor Promoters," Tokyo,Japan Scientific Societies Press,pp.315-326.

5. Hennings,H.,Michael,D.,Cheng,C.,Steinert,P.,Holbrook,K.,and Yuspa, S.H.: 1980,Calcium regulation of growth and differentiation of mouse epidermal cells in culture,Cell,19,pp.245-254.

6. Kulesz-Martin,M.,Koehler,B.,Hennings,H.,and Yuspa,S.H.: 1980, Quantitative assay for carcinogen altered differentiation in mouse epidermal cells,Carcinogenesis,1,pp.995-1006.

7. Yuspa,S.H.,and Morgan,D.L.: 1981,Mouse skin cells resistant to terminal differentiation associated with initiation of carcinogenesis,Nature,293,pp.72-74.

8. Kawamura,H.,Strickland,J.E.,and Yuspa,S.H.: in press,Association of resistance to terminal differentiation with initiation of carcinogenesis in adult mouse epidermal cells,Cancer Res.

9. Kilkenny,A.E.,Morgan,D.,Spangler,E.F.,and Yuspa,S.H.: 1985, Correlation of initiating potency of skin carcinogens with potency to induce resistance to terminal differentiation in cultured mouse keratinocytes,Cancer Res.,45,pp.2219-2225.

10. Nakayama,J.,Yuspa,S.H.,and Poirier,M.C.: 1984,Benzo(a)pyrene-DNA adduct formation and removal in mouse epidermis in vivo and in vitro: relationship of DNA binding to initiation of skin carcinogenesis,Cancer Res.,44,pp.4087-4095.

11. Kulesz-Martin,M.,Kilkenny,A.E.,Holbrook,K.A.,Digernes,V.,and Yuspa, S.H.: 1983,Properties of carcinogen altered mouse epidermal cells resistant to calcium induced terminal differentiation,Carcinogenesis,4,pp.1367-1377.

12. Toftgard,R.,Roop,D.R.,and Yuspa,S.H.: 1985,Protooncogene expression during two-stage carcinogenesis in mouse skin, Carcinogenesis,6, pp.655-657.

13. Hennings,H.,Steinert,P.,and Buxman,M.M.: 1981,Calcium induction of transglutaminase and the formation of ε(γ-glutamyl)lysine cross-links in cultured mouse epidermal cells,Biochem.Biophys. Res.Commun.,102,pp.739-745.

14. Yuspa,S.H.,Ben,T.,Hennings,H.,and Lichti,U.: 1982,Divergent responses in epidermal basal cells exposed to the tumor promoter 12-O-tetradecanoyl phorbol-13-acetate,Cancer Res.,42,pp.2344-2349.

15. Yuspa,S.H.,Ben,T.,and Hennings,H.: 1983,The induction of epidermal transglutaminase and terminal differentiation by tumor promoters in cultured epidermal cells,Carcinogenesis,4,pp.1413-1418.

16. Dunn,J.A.,Yuspa,S.H.,and Blumberg,P.M.: 1983,Specific binding of [20-^{3}H]phorbol 12,13-dibutyrate to proliferating and differentiated mouse primary keratinocytes,Proc.Am.Assoc.,Cancer Res.,24,p.111.

17. Roop,D.R.,Hawley-Nelson,P.,Cheng,C.K.,and Yuspa,S.H.: 1983,Keratin gene expression in mouse epidermis and cultured epidermal cells, Proc.Natl.Acad.Sci. USA,80,pp.716-720.

18. Astrup,E.G.,and Iversen,O.H.: 1983,Cell population kinetics in hairless mouse epidermis following a single topical application of 12-O-tetradecanoyl-phorbol-13-acetate II,Virchows Arch. (Cell Pathol.),42,pp.1-18.

19. Balmain,A.,and Pragnell,I.B.: 1983,Mouse skin carcinomas induced in vivo by chemical carcinogens have a transforming Harvey-ras oncogene,Nature,303,pp.72-74.

20. Balmain,A.,Ramsden,M.,Bowden,G.T.,and Smith,J.: 1984,Activation of the mouse cellular Harvey-ras gene in chemically induced benign skin papillomas,Nature,307,pp.658-660.

21. Yuspa,S.H.,Vass,W.,and Scolnick,E.: 1983,Altered growth and differentiation of cultured mouse epidermal cells infected with oncogenic retrovirus: contrasting effects of viruses and chemicals, Cancer Res.,43,pp.6021-6030.

22. Yuspa,S.H.,Kilkenny,A.E.,Stanley,J.,and Lichti,U.: 1985,Harvey and Kirsten sarcoma viruses block keratinocytes in an early, phorbol ester responsive, stage of terminal differentiation,Nature,314, pp.459-462.

23. Stanley,J.R.,and Yuspa,S.H.: 1983,Specific epidermal protein markers are modulated during calcium-induced terminal differentiation, J.Cell Biol.,96,pp.1809-1814.

24. Lichti,U.,Patterson,E.,Hennings,H.,and Yuspa,S.H.: 1981,The tumor promoter 12-O-tetradecanoylphorbol-13-acetate induces ornithine decarboxylase in proliferating basal cells but not in differentiating cells from mouse epidermis,J.Cell Physiol.,107,pp.261-270.

VIRUS TRANSFORMATION AS A FUNCTION OF AGE, DIFFERENTIATION AND HEREDITARY FACTORS

W. H. Kirsten
Department of Pathology
University of Chicago
Chicago, IL, USA

Neoplastic transformation of cell cultures has become a powerful tool to analyze the multiple steps involved in carcinogensis. Such in-vitro transformation of mammalian cells has been shown to occur after exposure to DNA- or RNA-oncogenic viruses, ionizing irradiation and chemical carcinogens (1-4). In most instances, dispersed monolayer cultures of freshly explanted cell strains or established cell lines have been used for such assays. Multiple parameters have been defined to assess in-vitro transformations (5). Much less precise information is available for cultures of differentiated cells or embryonal, fetal or neonatal organ cultures and their responses to carcinogenic stimuli. In particular, the interrelationship between the differentiated state of a target organ and susceptibility to carcinogenesis remains largely unexplored. Even the in-vitro responses of dispersed cells derived at various ages from the host deserve further attention if a correlation between potential target cells and host age is to be explained at the mechanistic level.

The purpose of this paper is to examine the relationship between donor age, stage of differentiation and transformability by an acutely transforming mammalian retrovirus, Kirsten murine sarcoma virus (K-MSV) (6-8). Like other acutely transforming retroviruses, the genome of K-MSV has captured, during replication in rat cells, discrete segments of host DNA and such transduced cellular (onc) sequences account for the ability of the virus to induce neoplastic cell transformation in vivo and in vitro (9-10). The virus induces a wide spectrum of mesenchymal tumors, such as fibro-, osteo- and hemagiosarcomas, whereas tumors of epithelial origin are not part of the response pattern to viral infection in mice or rats (6-8). Lymphomas or leukemias may occasionally arise but these are presumably accounted for by the presence of a "helper" mouse leukemia virus which invariably accompanies laboratory stocks of K-MSV.

Earlier in-vivo assays of K-MSV in inbred strains of mice and rats have revealed (a) that the tumor response, expressed as multiple histogenically diverse malignant mesenchymal neoplasms, is a function of

B. Pullman et al. (eds.), Interrelationship Among Aging, Cancer and Differentiation. 83–86.

virus dose and host age at inoculation; (b) that infection of adult hosts results in a single tumor most of which regress without metastates and (c) that a late response (after 6-8 months) following sub-optimal virus doses at birth is an occasional renal sarcoma (6). These in-vivo observations led us to study the phenomenon of decreasing susceptibility with increasing host age in cell and organ cultures.

The results of a series of experiments with secondary or tertiary fibroblast-like cell cultures from whole embryos or subcutaneous tissues of weanling or adult rats of the inbred Wistar/Furth strain are summarized in Table 1. As expected from the in-vivo experience, the susceptibility to transformation by K-MSV decreased significantly with increasing host-cell age, irrespective of tissue origin. For practical purposes, complete resistance to transformation was observed for cells derived from 28-day old rats, even with virus doses exceeding those used for the standard assays for embryonal or neonatal cells. The exception to this finding were cell cultures from renal mesenchyme which had shown a late tumor response with low virus doses given to neonatal rats. Renal mesenchymal cultures were transformed in vitro well beyond the weanling age, perhaps because of the comparatively long developmental process of the renal mesenchyme in vivo.

These studies were extended to organ cultures derived from either the metanephric rudiment or odontogenic primodium of 15-19 day old rat embryos. Both of these differentiating organ rudiments can be maintained in vitro for brief time periods under suitable culture conditions (11-13). The responses of the nephrogenic and odontogenic rudiments to infection with K-MSV were similar. The mesenchymal components are readily transformed, provided tissue organization is maintained (14-15). The latter observation is particularly intriguing because of the need for continued inductive interaction between the differentiating ureteric bud and mesenchymal cap in the case of the kidney as well as odontoblasts for the incisor rudiment. In both rudimentary systems, "micro-sarcomas" developed in virus-infected organ cultures. The malignant potential of the micro-sarcomas could be confirmed by transplantation assays in isologous hosts. It is well established that the tissue components of the kidney and incisor rudiments are required to interact during normal embryogenesis (11-13). Experimentally, they can be separated by trypsinization and grown in the presence of impermeable filter membranes. Under such conditions, virus infection fails to induce sarcomatous transformations. Thus, tissue integrity is a prerequisite for neoplastic transformation as it is for physiologic differentiation.

The murine sarcoma viruses are capable of inducing in-vitro transformation of human cells (16-18). Interestingly enough, fibroblast-like cells from certain cancer-prone patients transform more effectively than diploid fibroblasts from apparently normal donors (17-18). Repeated passages of K-MSV in human fibroblast cell lines alter the virus which is no longer capable of transforming mouse or rat cells even after readaptation to these cell lines (19). Results from a survey of diploid human cells can be summarized as follows. K-MSV induces morhpologic transformation in monolayer cultures of normal diploid human cells and in selected cases from patients considered cancer-prome by clinical -epidemiologic evidence. Significantly greater transformation freqencies were obtained for fibroblast-like cells from patients with Down's syndrome and from samples of the colonic submucosa of patients with polyposis coli. In contrast to experiences with rodent cells, however, none of the human cell strains tested became fully transformed even after single-cell cloning and prolonged in-vitro propagation of the initial transformants. Assays for colony growth in agar, transplantation or other parameters suggest that the transformation of human cells is abortive.

The experiments summarized here demonstrate limitations imposed by in-vitro assays for carcinogenicity, at least for a retrovirus. Although viral transformation systems have contributed significantly to an understanding of the mechanism(s) of cellular transformation, they fail to overcome resistance factors related to age and cessation of differentiation. As model systems, they are limited to the embryonic and neonatal periods of organogenesis. Even variations in virus dose or the use of methods to enhance virus absorption do not overcome the resistance imposed by age. The cellular or organismal factors that determine such resistance remain to be explored.

References

1. Berwald, Y. and Sachs, L.: 1963, Nature (London) 200, pp. 1182-1184.
2. Borek, C. and Sachs, L.: 1966, Nature (London) 210, pp. 276-278.
3. Tooze, J. (ed.): 1980, Cold Spring Harbor, NY.
4. Bishop, J.M.: 1978, Annu. Rev. Biochem. 47, pp. 35-88.
5. McPherson, I.: 1970, Adv. Cancer Res. 13, pp 169-216.
6. Kirsten, W.H. and Mayer, L.: 1967, J. Nat. Cancer Inst. 39, pp 311-335.
7. Somers, K. and Kirsten, W.H.: 1969, Int. J. Cancer 4, pp. 679-704.
8. Kirsten, W.H., Schauf, V. and McCoy, J.: 1970, Bibl. Haematol. 36, pp. 246-250.
9. Bishop, J.M. and Varmus, H.: 1982, In RNA Tumor Viruses. Cold Spring Harbor Lab., pp. 999-1108.
10. Ellis, R.W., DeFeo, D., Shih, T.Y., Gonda, M.L., Young, H.A., Tsuchida, N., Lowry, D.R., and Scolnick, E.M.: 1981, Nature (London) 292, pp. 506-511.
11. Grobstein, C.: 1953, Science 119, pp. 52-55.
12. Main, J.H.P. and Dawe, C.J.: 1967, J. Nat. Cancer Inst. 39, pp. 153 -157.
13. Weis, T.P. and Kirsten, W.H.: 1962, Arch. Path. 74, pp. 380-386 .
14. Schwartz, S.A. and Kirsten, W.H.: 1973, J. Dent. Res. 53, pp. 509-515.
15. Schwartz, S.A. and Snead, M.L.: 1982, Arch. Oral Biol. 27, pp. 9-12.
16. Aaronson, S.A. and Todaro, G.J.: 1970, Nature (London) 225, pp. 458-459.
17. Aaronson, S.A. and Weaver, C.A.: 1971, J. Gen. Virol. 13, pp.

Table 1. In-Vitro Transformation of W/Fu Rat Cells by K-MSV as a Function of Donor Age

Source	Age of Donor (Days postpartum)	No/Foci/Dish (in triplicate cultures)
Whole Embryo	-	89-99
Subcutaneous	1	62-71
Subcutaneous	7	19-27
Subcutaneous	28	-
Renal mesenchyme	7	58-71
Renal mesenchyme	28	19-26
Renal mesenchyme	56	4-11

245-252.
18. Klement, V., Freedman, M.H., McAllister, R.M., Nelson-Rees, W.A., and Huebner, R.J.: 1971, J. Nat. Cancer Inst. 47, pp. 65-73.
19. Aaronson, S.A.: 1971, Nature (London) 230, pp. 445-447.

NEGATIVE CONTROL OF VIRAL AND CELLULAR ENHANCER ACTIVITY BY THE PRODUCTS OF THE IMMORTALIZING E1A GENE OF HUMAN ADENOVIRUS-2

Emiliana BORRELLI, René HEN and Pierre CHAMBON

Laboratoire de Génétique Moléculaire des Eucaryotes du CNRS,
U184 de Biologie Moléculaire et de Génie Génétique INSERM,
Faculté de Médecine - 11 rue Humann
67085 Strasbourg-Cédex - France

ABSTRACT The products of the Adenovirus-2 E1A gene, 12S and 13S mRNAs, have been shown to stimulate transcription from all the early Adenovirus genes as well as from cellular genes. In a transient assay, we show that both E1A products repress transcription activation mediated by viral enhancers such as polyoma, SV40 and E1A, when cotransfected in HeLa cells and by the cellular enhancer of the immunoglobulin heavy chain, when cotransfected in the appropriate cell type. The repression occurs at the transcriptional level as showed by nuclear run-on experiments and it is possible to relieve it by the simple addition, in cotransfection experiments, of plasmid containing enhancer elements. Thus, this repression involves an interaction between the enhancer elements and a trans-acting factor(s), possibly the E1A products themselves. The repression of both cellular and viral enhancers with different specificity by the E1A products may be related to common enhancer sequence or factors.

INTRODUCTION

Enhancers are promoter elements, unique to eukaryotes, which potentiate transcription from other homologous or heterologous "natural" or "substitute" promoter elements (such as the TATA box), either in the presence or absence of upstream (distal) promoter elements (refs 1-5 and references therein). Viral and cellular enhancers exhibits species and/or cell specificity (1,2,4,6-8). For example a viral enhancer like the simian virus 40 (SV40) enhancer is active in most cell types, while the polyoma virus enhancer functions better in mouse cells than in primate cells. These differences are due to variations in enhancer activity rather than to a complete blocking of enhancer functions. On the other hand a cellular enhancer, such as the immunoglobulin heavy chain gene enhancer, is active only in lymphoid cells, exhibiting a strong tissue specificity. These observation suggest that the enhancer activity may be positively regulated by factors acting in trans. Recent studies from our laboratory have in fact shown that stimulation of transcription <u>in vitro</u> by the SV40 enhancer involves the

B. Pullman et al. (eds.), Interrelationship Among Aging, Cancer and Differentiation, 87–99.

formation of a stable complex between a trans-acting factor and the enhancer element (9-11). No example of negative regulation of an enhancer element has yet been described. However, we previously have found (12) that when HeLa cells were co-transfected with a recombinant containing the rabbit β-globin gene linked to the polyoma virus (Py) enhancer and a recombinant expressing the adenovirus-2 (Ad-2) early region 1A (E1A) gene products, there was a marked decrease in the amount of RNA initiated from the β-globin gene promoter. This suggested that the E1A products might block the activity of the Py enhancer. Such a possibility is intriguing, as it is known that the E1A products encoded in 13S and 12S messenger RNAs are involved in the activation of transcription from all other early viral transcription units (13-15), as well as in cell immortalization and cell transformation (16-19). Moreover, transcription from cellular promoters can be stimulated by the E1A products when recombinants containing these promoters are either co-transfected with the E1A transcription unit into HeLa cells or transfected into 293 cells that express constitutively an integrated E1A gene (refs 20-26 and references therein).

We demonstrate here that the E1A products repress the activity of the SV40, Py, Ad-2 E1A and immunoglobulin heavy chain enhancers, and that this inhibition probably involves an interaction between the enhancer elements and a trans-acting factor(s), possibly the E1A products themselves.

MATERIALS AND METHODS

Cell Culture. Thymidine kinase deficient mouse plasmocytoma MOPC11 (39) lymphoid cells and HeLa cells were maintained in Dulbecco's medium supplemented with 10% fetal calf serum.

Plasmids. (see Fig. 1 and 3) pE1ASV [4.6 kilobases (kb), see ref. 12] contains the 1,570 left-most nucleotides of Ad-2. The E1A transcription unit from positions -500 to + 1,070 with respect to the E1A cap site is linked to the 135 bp SV40 HpaI (2,604)/BamHI (2,469) fragment containing the SV40 polyadenylation signal (hatched box). pE1A$^-$ (3.4kb, see ref. 12) and pCR (4kb) were derived from pE1ASV and contain, respectively, the E1A promoter (positions -500 to +130) or coding region (positions +130 to + 1,070). pE1AG (5.2kb, a gift of C. Goding and C. Kēdinger) contains the E1A promoter from positions -500 to -9 linked by a HindIII linker to the rabbit β-globin gene (from positions -9 to +1,650 with respect to the β-globin cap site). pβ(244+)β (13.1kb, see ref. 37) contains the polyoma viral enhancer [fragment PvuII (5,262)-BclI (5,021)] and two copies of the rabbit 4.5kb chromosomal fragment including the rabbit β-globin gene (38). pSVBA34 (7.7kb, see ref. 5) contains the SV40 enhancer (coordinates 113-270) linked to a fragment of the Ad-2 major late promoter (Ad-2MLP) (-34 to +33 with respect to the major late cap site) and to the SV40 large-T antigen coding region (coordinates 2,533-5,227). SE1 (3kb, see ref. 11) contains the SV40 enhancer (coordinates 113-270) cloned between the PvuII and BamHI sites of pBR322. E1AE (4.5kb) contains the E1A enhancer sequences from

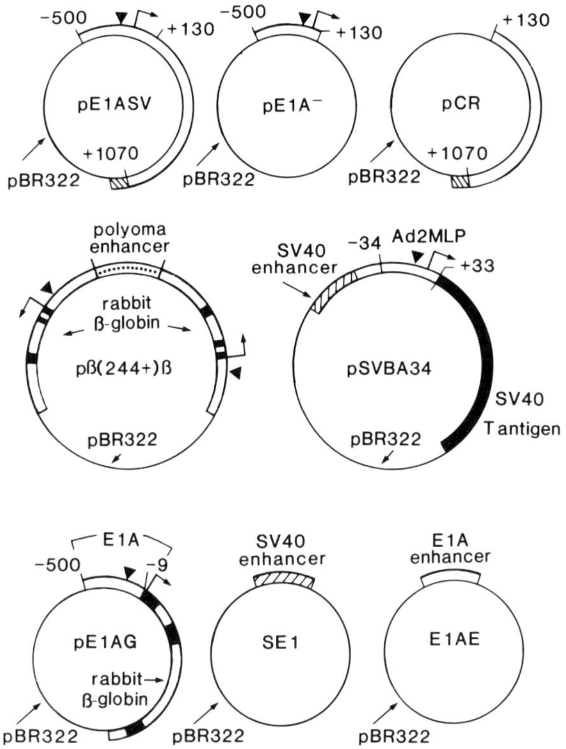

Fig. 1

Structure of the recombinants (see Materials and Methods). None of the recombinants is drawn to scale. Solid boxes in pE1AG and pβ(244$^+$)β indicate β-globin exons.

position -500 to -145 (with respect to the E1A cap site) cloned between the EcoRI and the BamHI sites of pBR322 (see ref. 11). The recombinant pTCT contains the conalbumin promoter region from position -102 to +62 linked to the SV40 large-T antigen coding region (coordinates 2,533-5,255) (3). In pTCTMI (a generous gift of B. and C. Wasylyk) the Ig heavy chain enhancer fragment EcoRI-PstI is inserted at position -102 of recombinant pTCT.

Transfection. HeLa cells and MOPC11 lymphoid cells (at 60% confluency) were transfected using the calcium phosphate procedure (5) with the following modification : the cell monolayer was washed after 24 hours and the cells collected 16 hours later.

mRNA analysis. Cytoplasmic RNA was purified from cell lysed in 10 mM Tris-HCl pH7.9, 10 mM NaCl, 2 mM $MgCl_2$, 0.3% Nonidet P40. For nuclease S1 mapping, 10 μg of cytoplasmic RNA were dissolved in 10 μl of 10 mM Pipes, pH6.5/0.4 M NaCl containing single stranded probes (see legend to Fig. 2).

Nuclear run-on. pβ(244+)β was transfected into HeLa cells with increasing amounts of ME1ASV; nuclei were isolated, incubated in vitro (15), in the presence of heparin and ammonium sulphate with [^{32}P]-CTP and cold GTP, UTP and ATP and, after purification, RNA was hybridized to the DNA fragments, as shown in Fig. 4.

RESULTS

We used the recombinant plasmid pE1ASV (Fig. 1, ref. 12), containing the entire E1A transcription unit including the enhancer sequences (23-25), to study the effect of the E1A products on the transcription

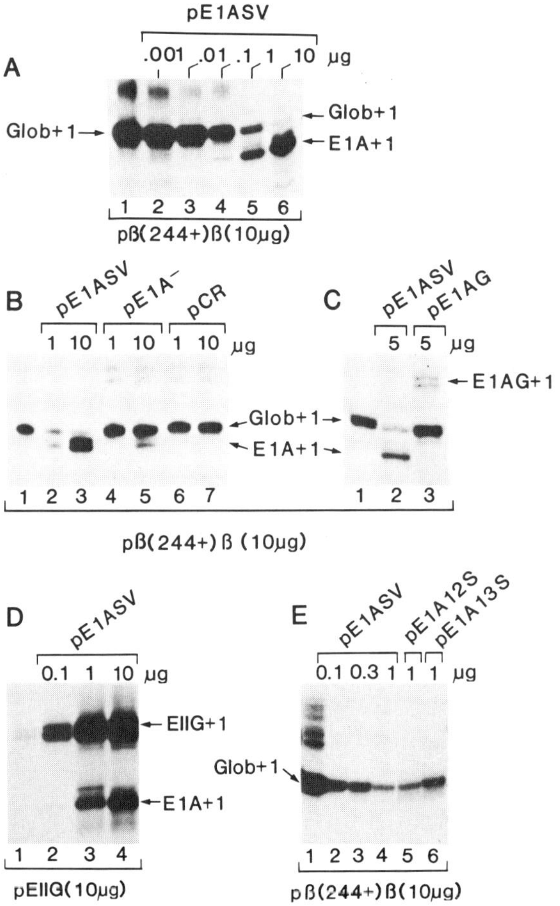

Fig. 2 : Quantitative S1 nuclease analysis of cytoplasmic RNA extracted from HeLa cells co-transfected with 10 µg recombinant pβ(244+)β and pE1ASV, pE1A^{-}, pCR, pE1AG, pE1AG, pE1A12S or pE1A13S (A-C, E); D, 10 µg recombinant pEIIG (see text and ref. 15) was co-transfected with increasing amounts of pE1ASV. The recombinants co-transfected and the amount of DNA transfected are indicated above the gel lanes. The total amount of transfected DNA was kept constant at 20 µg by the addition of pBR322. Cytoplasmic RNA was prepared (5) and 10 µg RNA hybridized with excess 5'-end labelled E1A and globin single-stranded probes (5, 12), except in C, lane 3, and E, where the globin probe alone was used. The E1A probe is an EcoRI (-500)/Sau3A(+130) fragment and the β-globin probe a BstN1 fragment as described previously (38). After S1 nuclease treatment, the E1A-protected fragment (E1A+1), corresponding to RNA transcribed from pE1ASV or pE1A, was 130 nucleotides in length. The globin-protected fragment (Glob+1), corresponding to RNA transcribed from pβ(244+)β, was 134 nucleotides long, whereas RNA transcribed from pE1AG(E1AG+1) or pEIIG(EIIG+1) protected a 140-nucleotide long fragment. The decrease in RNA transcribed from the β-globin genes of pβ(244+)β was estimated by densitometric scanning of autoradiograms corresponding to several experiments similar to A. Average decreases of 6-, 40- and 200-fold were obtained with 100 ng, 1 µg and 10 µg of co-transfected pE1ASV, respectively.

of two enhancer-containing recombinants. The first of these, pβ(244+)β, contains the Py enhancer linked to two rabbit β-globin genes; the second, pSVBA34 (ref. 5), contains the SV40 enhancer (72bp repeat region) linked to a fragment of the Ad-2 major late promoter (Ad-2MLP) and the SV40 large-T antigen coding region (Fig. 1). When increasing amounts of pE1ASV were co-transfected into HeLa cells with a constant amount (10 μg) of the pβ(244+)β recombinant, a progressive decrease in RNA transcribed from the β-globin genes was observed (Fig. 2A, Glob+1; B, lanes 1-3). A sixfold decrease was obtained with 100 ng pE1ASV (Fig. 2A, lane 4), corresponding to a ratio of 1 molecule of pE1ASV plasmid to 30 molecules of the polyoma β-globin recombinant. In parallel experiments, the stimulation of transcription from the EIIa transcription unit by the E1A products was assayed by co-transfecting pE1ASV and EIIG (a chimeric recombinant containing the EIIa promoter and the rabbit β-globin coding sequence; see ref. 15). The results show clearly that the decrease in RNA transcribed from pβ(244^+)β and the activation of the EIIa promoter occur in the same range of pE1ASV concentration (Fig. 2D).

The low molar ratio at which inhibition occurs suggests that it is not due to competition between the promoters of the two co-transfected recombinants. This hypothesis is supported by the results of co-transfection experiments with recombinant $pE1A^-$, containing the E1A promoter region, but not the E1A coding sequences (Figs. 1 and 2B, lanes 4, 5), or recombinant pE1AG, in which the E1A coding region is replaced by the β-globin coding region (Figs 1 and 2C, lane 3). In neither case was there a decrease in the amount of RNA transcribed from the β-globin mRNA start site of pβ(244+)β. In addition, in vitro nuclear run-on transcription experiments have shown previously that $pE1A^-$ is more efficiently transcribed than pE1ASV (see ref. 15). Thus, competition between the promoters of the co-transfected recombinants cannot account for the decrease in transcription from the β-globin promoters of pβ(244+)β. To exclude the possibility that the E1A coding region might contain a sequence trapping factors required for Py enhancer activity, we constructed recombinant pCR, containing only the E1A coding sequences (Fig. 1). No decrease in β-globin RNA is observed when pCR is co-transfected with pβ(244+)β (Fig. 2B, lanes 6, 7).

We next co-transfected pE1ASV with pSVBA34, a recombinant in which transcription from the +33 to -34 Ad-2MLP element is stimulated by the SV40 enhancer (Fig. 1). We found a decrease in RNA initiated from the Ad-2MLP, similar to those obtained previously with pβ(244+)β (Fig. 5A and B, lanes 1 and 2 and data not shown). To examine the individual roles of the 12S and 13S mRNA products of the E1A transcription unit, we co-transfected Py or SV40 enhancer-containing recombinants with cDNA-derived E1A recombinants producing either the 13S or the 12S mRNA (see ref. 15). Both E1A 12S and 13S mRNA products can inhibit Py and SV40 enhancer activity (Fig. 2E, lanes 5 and 6 and data not shown).

To investigate whether repression by the E1A products could also affect cellular enhancers, we performed co-transfection experiments using a chimeric recombinant containing the immunoglobulin heavy chain enhancer (6) activating transcription from the conalbumin promoter, in lymphoid cells. Conditions used were similar to those used to repress

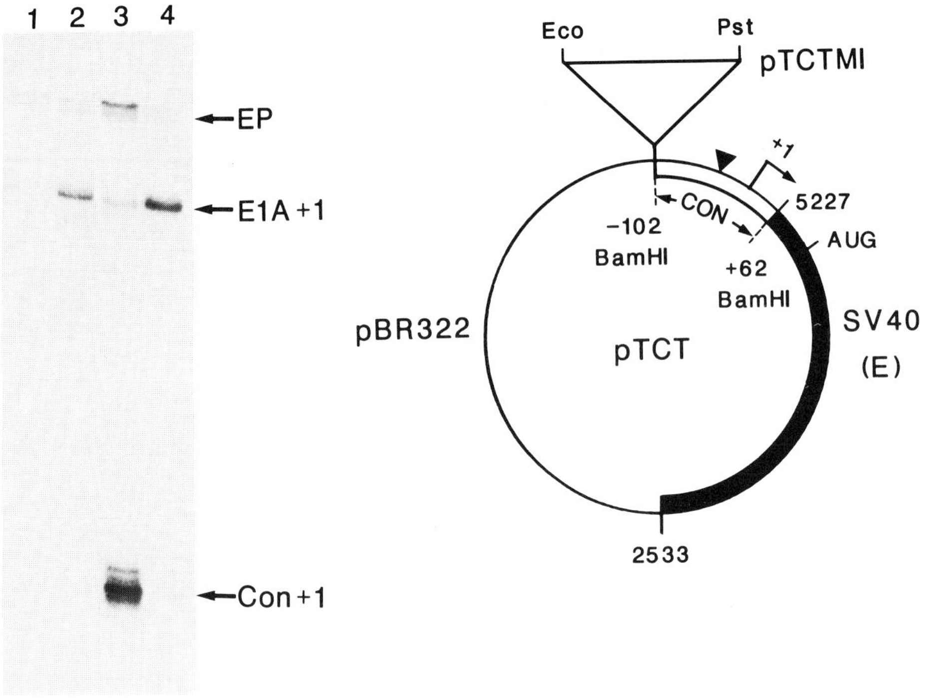

Fig. 3 : Quantitative S1 nuclease analysis of cytoplasmic RNA extracted from lymphoid cells (MOPC11) transfected with 10 μg of the recombinants pTCT (lanes 1, 2) and pTCTMI (lanes 3, 4). In lanes 1 and 3, pTCT and pTCTMI were co-transfected with the recombinant pE1A⁻ (5 μg) and in lanes 2 and 4 with the recombinant pE1ASV (5 μg). The E1A probe was as in legend to Fig. 2 and the conalbumin probe was the BamH1 fragment (-102/+62) yielding a 62 bp protected fragment after S1 nuclease digestion (3).

the viral enhancers. As shown in Fig. 3, no transcription was detectable starting from the conalbumin promoter (Con+1) in pTCT (the recombinant that does not contain the enhancer), either in absence (lane 1) or in presence (lane 2) of the E1A products. On the contrary a high level of transcription was detectable in the presence of the Ig enhancer (pTCTMI, lane 3), that is almost completely abolished in presence of the E1A products (lane 4). These data suggest that the Ig enhancer is the target for the E1A product repression. We can exclude competition between promoters because the same amount of recombinant pE1A⁻ was transfected in lanes 1 and 3 (see above).

To rule out the possibility that the E1A products would not inhibit transcription activation by the enhancers, but rather interfere with the stability or the transport of the RNA produced, we performed run-on transcription experiments on isolated nuclei (Fig. 4). Co-transfection of 10 μg pβ(244+)β with as little as 0.1 μg of ME1ASV (a

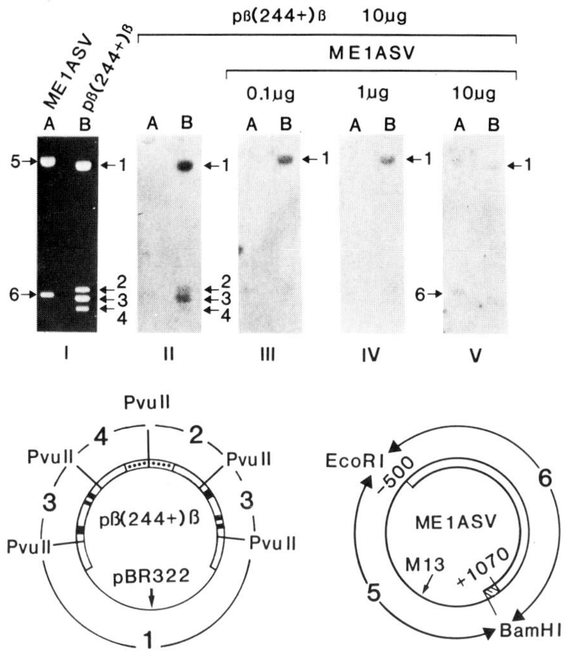

Fig. 4 : Nuclear run-on transcription from nuclei of cells transfected with recombinants pβ(244+)β and ME1ASV. Lower part: ME1ASV was derived from pE1ASV by replacing the pBR322 sequences with Mp11 sequences using the polylinker single EcoRI and BamHI sites. Upper part: restriction digests (2 μg DNA) of pβ(244+)β (lanes B) and ME1ASV (lanes A) were electrophoresed on a 1.5% agarose gel and blotted onto nitrocellulose filters as indicated. I, Ethidium bromide-stained agarose gel. Fragments 1-6 (see lower part) are 6.5, 1.8, 1.65, 1.5, 7.2, 1.7 kb in length, respectively. Fragment 3 contains the two β-globin genes of pβ(244+)β. II-V, Autoradiograms of the nitrocellulose blots after hybridization with labelled nuclear run-on RNAs prepared from HeLa cells transfected with pβ(244+)β alone (II) or with ME1ASV (as indicated).

recombinant in which the pBR322 sequence of pE1ASV was replaced by M13 sequences) resulted in inhibition of transcription of the β-globin genes (Fig. 4II-V, fragment 3). As, in these nuclear run-on transcription experiments, the RNA synthesized corresponds only to elongation of RNA chains previously initiated *in vivo* (see ref. 15), we conclude that the E1A products repress the enhancer-mediated activation of β-globin gene transcription. In fact, a decrease in transcription from the other sequences of the recombinant pβ(244+)β is also observed (Fig. 4, fragments 1, 2, 4). The residual level of transcription obtained with pβ(244+)β when repression is maximal (10 μg ME1ASV, Fig. 4V), corresponds to the level obtained with a recombinant lacking the enhancer sequence (data not shown). This suggests strongly that the E1A products repress all enhancer-activated transcription in recombinant pβ(244+)β, including transcription initiated not only from the β-globin promoters, but also from pBR322 "substitute" promoter elements (3).

To investigate whether repression of the enhancer-mediated stimulation of transcription by the E1A products is a trans-acting process involving an interaction with the enhancer, we attempted to relieve the repression by co-transfecting increasing amounts of a third recombinant containing only a purified enhancer element cloned in pBR322. Increasing amounts of SE1 (a recombinant in which the SV40 enhancer is cloned

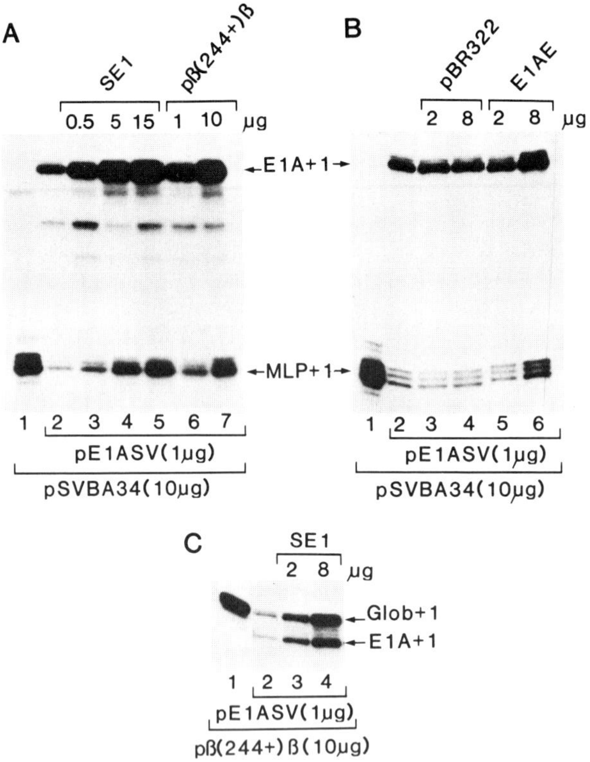

Fig. 5 : Co-transfection with recombinants containing the Py virus or SV40 enhancers relieves repression of enhancer-mediated transcription by the E1A products. The experimental procedure is described in Fig. 2 legend. A, B, lanes 1, 10 μg pSVBA34 was transfected alone; A, B, remaining lanes, pSVBA34 and pE1ASV were transfected in the amounts indicated. In A, lanes 3-7, B, lanes 3-6, a third recombinant was co-transfected. The nature of this recombinant and the amount of DNA transfected are indicated above the gel lanes. C, 10 μg pβ(244+)β was transfected either alone (lane 1) or with 1 μg of pE1ASV (lanes 2-4). A third recombinant, SE1, was co-transfected in lanes 3 and 4 as indicated. RNA transcribed from pE1ASV (E1A+1) and pβ(244+)β was analysed by quantitative S1 nuclease mapping as indicated in Fig. 2 legend. The probe used for analysing RNA initiated from pSVBA34 at the Ad-2MLP cap site was the 5' end-labelled single-stranded HindIII-XbaI fragment (see Ref. 5) giving a protected fragment of 93 nucleotides (MLP+1).

in pBR322, see Fig. 1) in co-transfections where transcription of pSVBA34 (Fig. 5A, lanes 2-5) or pβ(244+)β (Fig. 5C, lanes 2-4) is repressed by pE1ASV, derepresses transcription initiated either from the Ad-2MLP (MLP + 1 in Fig. 5A, compare lanes 1-5) or from the β-globin promoter (Glob+1 in Fig. 5C, compare lanes 1-4). The molar ratios of SE1:pSVBA34 are 0.13, 1.3 and 3.9 in lanes 3-5 of Fig. 5A, whereas those of SE1 to pβ(244+)β are 0.9 and 3.6 in lanes 3 and 4 of Fig. 5C. Similarly, the addition of the Py enhancer in the form of the recombinant pβ(244+)β to a transfection also containing pSVBA34 or pE1ASV relieves the repression of transcription from the Ad-2MLP (Fig. 5A, compare lanes 1, 2, 6 and 7; the molar ratios of pβ(244+)β to pSVBA34 are 0.17 and 1.7 in lanes 6 and 7, respectively). In contrast, no "depression" is observed when pBR322 is transfected with pSVBA34 and pE1ASV (Fig. 5B, lanes 3, 4). We conclude from these results that repression of enhancer activity by the E1A products may involve an interaction between the enhancer and either the E1A products or some cellular factor(s) induced by the E1A products, or a combination of both.

In the derepression experiments (Fig. 5A and C), the amount of RNA initiated at the E1A mRNA start site (E1A+1) increases in parallel with that of RNA initiated from either the Ad-2MLP (Fig. 5A, lanes 3-7) or the β-globin promoter (Fig. 5C,lanes 3, 4). These results suggest that E1A products can also repress transcription stimulated by the E1A enhancer(s) (23-25). This hypothesis is supported by the increase of RNA initiated from the Ad-2MLP (MLP+1) and E1A promoter (E1A+1) when a recombinant containing the E1A enhancer region inserted in pBR322 (E1AE, see Fig. 1) is transfected with pSVBA34 and pE1ASV (Fig. 5B, compare lanes 1, 2, 5 and 6).

DISCUSSION

The results described here demonstrate that the stimulation of transcription mediated by enhancers can be negatively regulated. Because it is possible to compete the negative effect of the E1A products by co-transfecting recombinants containing only an enhancer element, and also because repression of the enhancer activity does not interfere with the ability of the E1A products to stimulate transcription from the β-globin promoter (data not shown), the enhancer itself is probably the target for the negative regulation. Future studies will determine whether the E1A products themselves, or a cellular product(s) induced by them, or some combination of the two, interact with the enhancer sequence and/or with other enhancer factors. Using an extensive set of SV40 enhancer mutants (M. Zenke, T. Grundström, H. Matthes, M. Wintzerith and P. Chambon, manuscript in preparation), we will determine which enhancer sequences are involved in E1A repression. In this respect, the only sequence feature which is apparently shared by the SV40, Py virus and Ad-2 enhancers is the core sequence (23, 27). *In vivo* studies have clearly shown that the SV40 enhancer can generate an altered chromatin structure over its own sequence (28). Recent *in vitro* studies demonstrate that a transcriptional factor(s) interacts specifically with the SV40 enhancer (9-11). We are now investigating whether the E1A products interfere with either of these enhancer functions.

At present it is not known whether the E1A products inhibit the activity of the E1A enhancer during lytic infection or whether the inhibition could be observed only with an isolated E1A transcription unit. However, inhibition of the E1A enhancer activity by the E1A products provides a means by which the expression of the E1A transcription unit can be autoregulated during lytic infection. Indeed, it has been reported that E1A transcriptional activity is highest during the early phase of infection (13, 14). On the other hand, there is evidence that the E1A products stimulate transcription of the E1A transcription unit early in adenovirus infection (29,30). It appears, therefore, that the E1A products can act as both negative and positive regulators of their own transcription units during lytic infection, and as positive regulators of all other viral transcription units (see refs 13-15). Note that the SV40 early gene product, the large-T antigen, also regulates it own synthesis, but by a more "classical" repression mechanism that does not involve an interaction with the SV40 enhancer (31). In addition, like the E1A products, the large-T antigen is a pleiotropic protein also activating transcription, as it has been reported recently that it stimulates the activity of the SV40 late promoter (32). It is not known whether inhibition of cellular enhancers by the E1A products occurs during productive infection or during cell transformation by adenoviruses. Our results with the Ig enhancer suggest that not only "constitutive" viral enhancers can be repressed by the E1A products, but also the immunoglobulin tissue-specific enhancer. This enhancer is recognized by lymphoid specific factor(s). In addition in vitro and in vivo competition experiments suggest that it shares some common factors with the SV40 enhancer (11, 40). Consequently the target of E1A repression might be the sequence recognized by these common factors or the factors themselves. We are now investigating whether the immunoglobulin gene in its natural chromosome location could be repressed by the Ad-2 E1A products. As yet, the expression of only one defined cellular gene has been shown clearly to be repressed by E1A products. The expression of a class I major histocompatibility complex (MHC) gene is repressed in rat cells transformed by the E1A transcription unit of the highly oncogenic adenovirus-12 (Ad-12), but not by that of the weakly oncogenic adenovirus-5 (ref. 19). Since the E1A products of Ad-12 also inhibit activation of transcription by SV40 and Py enhancers (our unpublished observations), we are now investigating the possibility that this class I MHC gene contains an enhancer whose activity can be repressed by the Ad-12 E1A products.

Gene expression in eukaryotic cells may be controlled by regulatory factors repressing enhancer activity. This has not yet been reported. However, the Py enhancer is not active in undifferentiated embryonal carcinoma cells, although mutations in the region containing the Py enhancer restore its activity in these cells (33-35). It has been suggested that embryonal carcinoma cells express a protein with properties similar to those of the E1A products (36). This protein may repress the activity of the Py enhancer. Thus, in addition to demonstrating that the adenovirus E1A products have both negative and positive regulatory functions, our study raises the possibility that cellular enhancers are regulated not only positively as suggested previously (1,2,4,6-8), but also negatively, which would significantly increase the combinatorial possibilities for the control of gene expression at the transcriptional level in eukaryotes.

We thank W. Schaffner, P. Sassone-Corsi, C. Goding, C. Kedinger and C. & B. Wasylyk for gifts of recombinants; B. Boulay, C. Kutschis, B. Chambon and C. Werlé for illustration and preparation of the manuscript and B. Heller and M. Acker for cell culture. E.B. receives a fellowship from Université Louis Pasteur Strasbourg. This work was supported by CNRS (ATP 3582), INSERM (PRC124026), Ministère de l'Industrie et de la Recherche (82V1283), Fondation pour la Recherche Médicale and Association pour le Développement de la Recherche sur le Cancer.

REFERENCES

1. Yaniv, M. 'Enhancing elements for activation of eukaryotic promoters'. Nature **297**,17-18 (1982).
2. Khoury, G. and Gruss, P. 'Enhancer elements' Cell **33**, 313-314 (1983).
3. Wasylyk, B., Wasylyk, C., Augereau, P. and Chambon, P. 'The SV40 72bp repeat preferentially potentiates transcription starting from proximal natural or substitute promoter elements'. Cell **32**, 503-514 (1983).
4. Chambon, P., Dierich, A., Gaub, M-P., Jakowlev, S., Jongstra, J., Krust, A., LePennec, J-P., Oudet, P. and Reudelhuber, T. 'Promoter elements of genes coding for proteins and modulation of transcription by estrogens and progesterone' Recent Prog. Horm. Res. **40**, 1-42 (1984).
5. Hen, R., Sassone-Corsi, P., Corden, J., Gaub, M-P. and Chambon, P. 'Sequences upstream from the TATA box are required in vivo and in vitro for efficient transcription from the adenovirus-2 major late promoter'. Proc. Natl. Acad. Sci. USA **79**, 7132-7136 (1982).
6. Banerji, J., Olson, L. and Schaffner, W. 'A lymphocyte specific cellular enhancer is located downstream of the joining region in immunoglobulin heavy chain genes' Cell **33**, 729-740 (1983)
7. Gillies, S.D., Morrison, S.L., Oi, V.T. and Tonegawa, S., 'A tissue-specific transcription enhancer element is located in the major intron of a rearranged immunoglobulin heavy chain gene'. Cell **33**, 717-728.
8. Chandler, V.L., Maler, B.A. and Yamamoto, K.R. 'DNA sequences bound specifically by glucocorticoid receptor in vitro render a heterologous promoter responsive in vivo'. Cell, **33**, 489-499 (1983).
9. Sassone-Corsi, P., Dougherty, J., Wasylyk, B. and Chambon, P. 'Stimulation of in vitro transcription from heterologous promoters by the SV40 enhancer'. Proc. Natl. Acad. Sci. USA **81**, 308-312 (1984).
10. Wildeman, A.G., Sassone-Corsi, P., Grundström, T., Zenke, M. and Chambon, P. 'Stimulation of in vitro transcription from the SV40 early promoter by the enhancer involves a specific trans-acting factor. EMBO J. **3**, 3129-3133.
11. Sassone-Corsi, P., Wildeman, A. and Chambon, P. 'A trans-acting factor is responsible for the simian virus 40 enhancer activity in vitro'. Nature **313**, 458-463 (1985).
12. Sassone-Corsi, P., Hen, R., Borrelli, E., Leff, T. and Chambon, P. 'Far upstream sequences are required for efficient transcription from the Adenovirus-E1A transcription unit'. Nucleic Acids Res. **11**, 8735-8745 (1983).
13. Flint, S.J. 'Expression of adenoviral genetic information in productively infected cells'. Biochim. Biophys. Acta **651**, 175-208 (1982).
14. Cross, F.R. and Darnell, J.E. 'Cycloheximide stimulates early adenovirus transcription if early gene expression is allowed before treatment'. J. Virol. **45**, 683-692 (1983).

15. Leff, T., Elkaim, C.R., Goding, C., Jalinot, P., Sassone-Corsi, P. Perricaudet, M., Kédinger, C. and Chambon, P. 'The individual products of the adenovirus 12S and 13S EIa mRNAs stimulate viral EIIa and EIII expression at the transcriptional level'. Proc. Natl. Acad. Sci. USA **81**, 4381-4385 (1984).
16. Pettersson, U. and Akusjärvi, G. in Advances in Viral Oncology **3** (ed. Klein, G.) 83-131 (Raven, New York,1983).
17. Ruley, H.E. 'Adenovirus early region 1A enables viral and cellular transforming genes to transform primary cells in culture'. Nature **304**, 602-606 (1983).
18. Montell, C., Coutois, G., Eng, C. and Berk, A. 'Complete transformation by Adenovirus-2 requires both E1A proteins'. Cell **36** 951-961 (1984).
19. Schrier, .I., Bernards, R., Vaessen, R.T.M.J., Houweling, A. and Van der Eb, A.J. 'Expression of class I major histocompatibility antigens switched off by highly oncogenic adenovirus 12 in transformed crat cells'. Nature **305**, 771-775 (1983).
20. Green, M.R., Triesman, R. and Maniatis, T. 'Transcriptional activation of cloned human β-globin genes by viral immediate-early gene products'. Cell **35**, 137-148 (1983).
21. Treisman, R., Green, M.R. and Maniatis, T. 'cis and trans activation of globin gene transcription in transient assays'. Proc. Natl. Acad. Sci. USA **80**, 7428-7432 (1983).
22. Svensson, C. and Akusjärvi, G. 'Adenovirus-2 early region 1A stimulates expression of both viral and cellular genes'. EMBO J. 3, 789-793 (1984).
23. Hen, R., Borrelli, E., Sassone-Corsi, P. and Chambon, P. 'An enhancer element is located 340 base pairs upstream from the Adenovirus-2 E1A capsite'. Nucleic Acids Res. **11**, 8748-8760.
24. Hearing, P. and Shenk, T. 'The Adenovirus type 5 E1A transcription control region contains a duplication enhancer element'. Cell 33, 695-703 (1983).
25. Imperiale, M.J., Feldman, L.T. and Nevins, J.R. 'Activation of gene expression by adenovirus and herpes virus regulatory genes acting in trans and by a cis-acting adenovirus enhancer element'. Cell **35**, 127-136 (1983).
26. Gaynor, R.B., Hillman, D. and Berk, A.J. 'Adenovirus early region 1A protein activates transcription of a nonviral gene introduced into mammalian cells by infection or transfection'. Proc. Natl. Acad. Sci. USA **81**, 1193-1197 (1984).
27. Weiher, H., König, M. and Gruss, P. 'Multiple point mutations affecting the simian virus 40 enhancer'. Science **219** 626-631 (1983).
28. Jongstra, J., Reudelhuber, T.L., Oudet, P., Benoist, C., Chae, C.B., Jeltsch, J.M., Mathis, D.J. and Chambon, P. 'Induction of altered chromatin structures by the SV40 enhancer and promoter elements'. Nature **307**, 708-714 (1984).
29. Berk, A.J., Lee, F., Harrison, T., Williams, J. and Sharp, P.A. 'Pre-early adenovirus-5 gene product regulates synthesis of early viral messenger RNAs'. Cell **17**, 935-944 (1979).

30. Nevins, J.R. 'Mechanism of activation of early viral transcription by the adenovirus E1A gene product'. Cell **26**, 213-220 (1981).
31. Myers, R.M., Rio, D.C., Robbins, A.K. and Tjian, R. 'SV40 gene expression is modulated by the cooperative binding of T antigen to DNA'. Cell **36**, 381-389 (1984).
32. Keller, J.M. and Alwine, J.C. 'Activation of the SV40 late promoter direct effects of T antigen in the absence of viral DNA replication'. Cell **36** 381-389, (1984).
33. Vasseur, M., Katinka, M., Herbomel, P., Yaniv, M. and Blangy, B. 'Physical and biological features of polyoma virus mutants able to infect embryonal carcinoma cell lines'. Journal of Virology **43**, 800-808 (1982).
34. Linney, E. and Donerly, S. 'DNA fragments from F9 PyEC mutants increase expression of heterologous genes in transfected F9 cells'. Cell **35**, 693-699 (1983).
35. Herbomel, P., Bourachot, B. and Yaniv, M. 'Two distinct enhancers with different cell specificities coexist in the regulatory region of polyoma'. Cell **39**, 653-662.
36. Imperiale, M., J. Kao, H.T., Feldman, L.T., Nevins, J.R. and Strickland, S. 'Common control of the heat shock gene and early adenovirus genes : evidence for a cellular E1A-like activity'. Molec. Cell. Biol. **4**, 867-874 (1984).
37. DeVilliers, J. and Schaffner, W. 'A small segment of polyoma virus DNA enhances the expression of a cloned rabbit β-globin gene over a distance of at least 1400 base pairs'. Nucleic Acids Res. **9**, 6251 6264 (1981).
38. Van Ooyen, A., Van den Berg, J., Mantei, N. and Weissman, C. 'Comparison of total sequence of a cloned rabbit β-globin gene and its flanking regions with a homologous mouse sequence'. Science **206**, 337-344 (1979).
39. Stanton, L.W., Yang, J.R., Eckhardt, L.A., Harris, L., Birshtein, B.R. and Marcu, K.B. 'Products of a reciprocal chromosome translocation involving the c-myc gene in a murine plasmacytoma'. Proc. Natl. Acad. Sci. USA **81**, 829-833 (1984).
40. Mercola, M., Goverman, J., Mirell, C. and Calame, K. 'Immunoglobulin heavy chain enhancer requires one or more tissue-specific factors'. Science **227**, 266-270 (1985).
41. Ephrussi, A., Church, G.M., Tonegawa, S. and Gilbert, W. 'B lineage specific interactions of an immunoglobulin enhancer with cellular factors in vivo'. Science **227**, 134-140 (1985).

EXPRESSION OF TWO MURINE GENE FAMILIES IN TRANSFORMED CELLS AND EMBRYOGENESIS

Jeffrey A. Moshier, Richard A. Morgan and Ru Chih C. Huang
The Johns Hopkins University
Department of Biology
Baltimore, Maryland 21218
U.S.A.

ABSTRACT. Mouse clones IAP81 and mR2 have been shown to contain different repetitive DNA elements. These clones were used to investigate the expression of their respective repetitive DNA families in myelomas, embryonal carcinoma cells (EC) and, in one case, pre-implantation embryos. IAP genes which code for intracisternal A-type viral particles are highly transcribed in myeloma and, to a lesser extent, in EC cells. While IAP transcripts of three sizes (7.2 kb, 5.3 kb and 3.8 kb) were found in myelomas, only two of the IAP RNA species (7.2 kb and 5.3 kb) were detected in EC cells. Very little IAP RNA could be detected in EC cells following differentiation. Abundant IAP RNA was visualized in 1, 2 and 4-8 cell stage embryos by <u>in situ</u> hybridization. The amount of IAP RNA decreases drastically in embryos at the morula stage. In similar studies using the mR2 clone, a 1.4 kb transcript was detected in myeloma and EC cells. This transcript is not detected in differentiated EC cells or liver and brain of adult mice.

1. INTRODUCTION

The similarity between malignant and embryonic cells has long been recognized (1,2). Proliferation of stem cells occurs not only in embryos but also in many adult tissues, such as bone marrow and skin epithelium, that undergo constant renewal during the life span of the organism. It is also known that stem cells are multi-potent in their developmental capacities i.e., they maintain a critical balance between differentiation and extended multiplication. When normal stem cell differentiation is interrupted the cell becomes malignant. Embryonal carcinomas (EC) and plasmacytomas (myelomas) are two types of stem cell neoplasms. EC cells have retained their developmental multi-potency and can be induced to differentiate by injection into a developing blastocyst <u>in vivo</u> (3,4), by treatment with certain chemicals such as retinoic acid <u>in vitro</u> (5) or by simply growing EC cells in aggregates (6). A series of biochemical, immunological and physiological changes accompany this developmental switch.

In the past several years, we have cloned several genes which are

B. Pullman et al. (eds.), Interrelationship Among Aging, Cancer and Differentiation, 101–116.

expressed both in transformed cells and early embryos. These cloned sequences were used to probe for the expression of several gene families during development and in response to cellular differentiation. The intracisternal A-particle (IAP) genes are one of the gene families we studied extensively. IAPs are retrovirus-like structures present in mouse plasmacytomas (7), embryonal carcinomas (8) and other neoplasms (9). They are also detected in early mouse pre-implantation embryos (10) but not in other mouse tissues (11). These particles are encoded by an unknown number of the approximately 1000 IAP DNA elements scattered throughout the mouse genome (12). The major RNA species packaged in IAP virions in myelomas is 3.8 kb in length; however, two additional polyadenylated IAP RNA species (7.2 kb and 5.3 kb) have been detected in the nuclei of MOPC-315 (11). Recently an abundant 73,000 dalton protein (p73) was isolated from IAP particles. Anti-sera directed against p73 was used to probe for the expression of IAP antigens during early embryogenesis (13). IAP antigens first appear at the zygote stage of development, exhibit peak expression on 2 to 8 cell embryos and was not detected on embryos at the morula stage, blastocyst stage or on cells of the inner cell mass. Our investigation of the transcription of IAP genes in early embryos was undertaken to complement and extend these protein studies.

During the cloning of small nuclear U2 RNA genes from mouse, we isolated another repetitive DNA sequence (in clone mR2) which is located approximately 700 base pairs upstream from the U2 gene in clone p61-41. Preliminary hybridization studies indicate that the repetitive sequence in mR2 shares no significant sequence homology with IAP genes or other reported interspersed gene families in mouse. As in the case with IAP genes, the expression of the repetitive DNA family represented in the mR2 clone appears to be regulated during growth and development.

In this communication, we report the experimental results which demonstrate that the gene families homologous to IAP81 and mR2 clones are transcribed in myelomas and in undifferentiated embryonal carcinomas (EC). We show that the predominant IAP RNA species in EC cells is 5.3 kb in size and that mR2 DNA hybridizes to a prominant 1.4 kb RNA transcript in EC cells and myelomas. The amounts of IAP RNA and the 1.4 kb RNA homologous to mR2 are greatly reduced in EC cells following differentiation.

2. MATERIALS AND METHODS

Eukaryotic Cells. Myeloma MOPC-315 cells were maintained in stationary suspension culture in Delbecco's modified Eagle's (DME) medium and 10% fetal calf serum (14). Myeloma TEPC-15 cells were grown in DME plus 10% horse serum. Embryonal carcimona cell lines PCC3 and PCC4 were generously provided by Dr. A. Rizzino, University of Nebraska, and Dr. S. Rosenberg, University of Maryland, respectively. The F9 embryoid cells were a gift of Dr. M. Edidin, The Johns Hopkins University. EC cell lines PCC3 and PCC4 were grown in DME medium, 15% fetal calf serum, 0.4% glucose, 4 mM L-glutamate and 5% CO_2 and subcultured before reaching 80% confluency. PCC3 cells were induced to differentiate by growing in DME medium supplemented with 1×10^5 M retinoic acid (6).

F9 embryoid cells were grown in Eagle's minimal essential medium (MEM), 15% fetal calf serum and 0.24% glucose.

Nucleic Acid Techniques. The cloning and restriction site mapping of IAP81, p61-41 and mR2 were described previously (15,16). DNA manipulations were performed according to standard protocols described by Maniates et al. (17). Total cellular RNA was isolated from myeloma and embryonal carcinoma tissue culture cells in the presence of guanidinium isothiocyanate (18). Polyadenylated RNA was recovered by oligo-dT cellulose chromatography (17). The protocols described by Marzluff and Huang (19) were followed for the *in vitro* synthesis of RNA by isolated myeloma nuclei and chromatin in the presence of [γ-S] ribonucleotides. Transcripts initiated *in vitro* were fractionated from endogenous RNAs and transcripts merely elongated *in vitro* by mercury-agarose or mercury-cellulose affinity chromatography (19).

In Situ Hybridization to Mouse Embryos. Pre-implantation stage embryos were obtained from naturally mated Balb/c mice. The morning the copulation plug was found was considered day 1 of gestation. One cell, two cell and four to eight cell stage embryos were flushed from the oviduct on days 1, 2 and 3, respectively. The morula and blastocyst stages were obtained from the uterus on day 4.

Embryos were fixed in ethanol/acetic acid (3:1; Carnoy's fixative) for one hour, washed with PBS and then pre-embedded in a drop of 1% agar. The agar block containing the embryos was dehydrated through a graded series of alcohol, a solution of 50% absolute alcohol (saturated with eosin Y) and 50% toluene (1 hour), 100% toluene (1 hour) and embedded in paraffin. Each paraffin block contained embryos at a specific stage of development. Blocks were then sectioned at 5 microns and the sections containing embryos were mounted on subbed slides (0.01% chromium potassium sulfate, 0.1% gelatin). All of the slides were mounted with a row of sectioned embryos at each of the pre-implantation stages of development. The slides were deparaffinized, hydrated, treated with 0.2 N HCl (20 min at 20°C), rinsed and dried. ^{3}H-labeled IAP81 DNA (6 X 10^6 cpm/µg) in 3 X SSC containing 40% formamide was applied to each slide (5 µl), covered with a cover glass and the edges sealed with paraffin. After incubation at 44°C for 18 hours, the slides were uncovered, rinsed in 2 X SSC at room temperature and washed with 2 X SSC at 55°C for 1 hour. Slides were dehydrated through a graded series of alcohol and air dried according to Harrison et al. (20).

The slides were dipped in Kodak NT B2 emulsion (diluted 1:1 with distilled H_2O), dried, packaged and exposed for 2-8 weeks at 4°C. Autoradiographs were developed for 4 min at 19°C in Kodak 10-19, fixed, washed for 30 min in running H_2O and air dried. The tissue was viewed unstained under a phase contrast microscope.

3. RESULTS

3.1. Genomic organization of two mouse repetitive DNA families.

In several of our previous reports (15,21) we have described the isolation and characterization of several bacteriophage Charon 4A recombinant DNA clones containing endogenous intracisternal A-particle genes. Extensive studies have been made of one of these clones designated IAP81. The 17.7 kb mouse DNA insert in IAP81 contains one complete 7.2 kb type I IAP gene and one type I variant gene spaced approximately 5.0 kb apart. (IAP gene types described below.) IAP81 has many structural features in common with other retroviruses, such as: long terminal repeats (LTR); viral polymerase (POL); and virion structural proteins (GAG), see Fig. 1. The POL gene region was identified by nucleic acid homology and the GAG gene has been proposed to lie 5' to the POL gene on the basis of <u>in vitro</u> translation experiments (22,23). However, sequences analogous to retroviral ENV genes, which code for envelop proteins, have not been identified for any IAP proviruses.

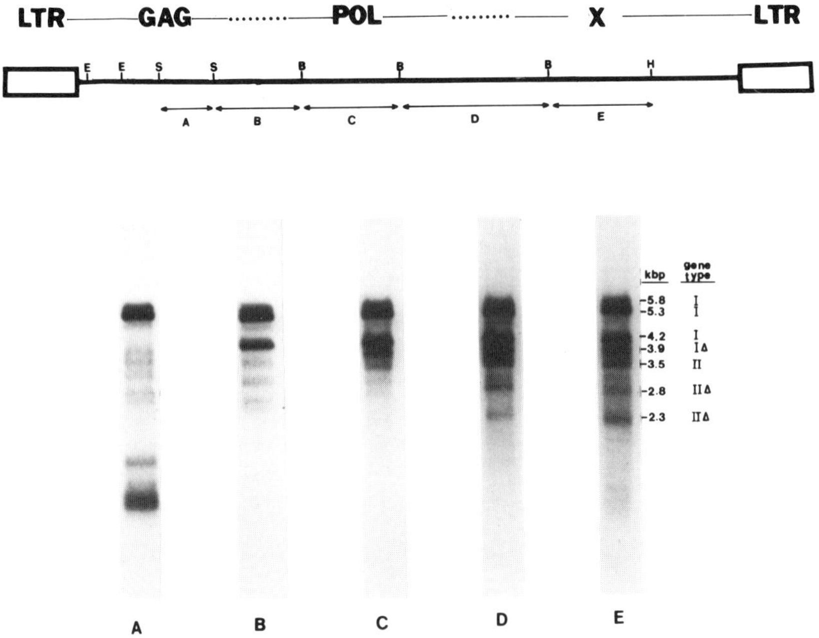

Figure 1. Restriction map and genomic organization of IAP81. (Top) Shown is the restriction endonuclease map of IAP81. Restriction enzyme sites are defined as follows: E, EcoRI; S, SstI; B, BamHI; and H, HindIII. (Bottom) Organization of IAP proviruses in Balb/c genome. Southern blots of EcoRI plus HindIII mouse genomic DNA were hybridized to [^{32}P] nick-translated probes from IAP81. A-E represent identical Southern blots using specific DNA fragments (arrows A-E below restriction map) derived from IAP81. The size and gene type of each EcoRI plus HindIII fragment are indicated (right).

We have used IAP81 as a hybridization probe to examine the organization of IAP genes in mus musculus. For this study, genomic DNA was restricted with EcoRI plus HindIII and probed with DNA fragments derived from different regions of IAP81 (Fig. 1. A-E). This analysis resolves IAP genes into the following distinct proviral classes (24); three type I genes (5.8 kb, 5.3 kb and 4.2 kb), one type I variant (3.9 kb), one type II gene (3.5 kb) and two type II variants (2.8 kb and 2.3 kb).

In addition to IAP genes, we have recently cloned from the mouse genome a member of another highly repeated DNA family. This particular repeated DNA element is located approximately 700 nucleotides upstream from a small nuclear U2 RNA gene in clone p61-41 (Fig. 2). The RsaI fragment, designated Rsa2a, containing the repetitive element was inserted into a M13 vector to construct subclone mR2. Figure 3A (lane 1) is an autoradiograph of ^{32}P-labeled mR2 DNA hybridized to a Southern blot of HindIII restricted mouse liver DNA. Although two bands (5.0 kb and 3.2 kb) stand out above the background it appears that this repetitive family is dispersed throughout the genome. In a preliminary experiment, the mR2 probe did not hybridize to any of the restriction fragments characteristic of MIF-1 (25), Bam5 (26), 1.3 kb EcoRI (25) or R (27) mouse repetitive DNA families (data not shown).

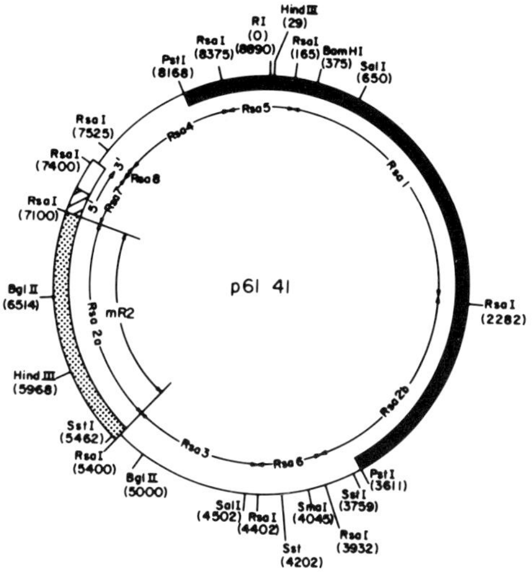

Figure 2. Restriction map of p61-41 containing a member of a repetitive DNA family upstream from a small U2 RNA gene. Open bar, U2 RNA gene; hatched bar, 5' flanking region proximal to the U2 gene; stippled bar, Rsa2a fragment harboring a sequence repeated throughout the mouse genome; blackened bar, pBR322 vector. The Rsa2a fragment was inserted into M13 vector to construct the subclone mR2.

The boundaries of the repetitive sequence in mR2 were further refined by probing Southern blots of restricted mR2 DNA with [^{32}P] nick-translated myeloma genomic DNA (Fig. 3). The autoradiograph in

Fig. 3 demonstrates that the fragments designated I and II in the schematic contain sequences belonging to the repetitive DNA family. Fragment III which is closest to the U2 gene in clone p61-41 is present in the genome at a low copy number. The 1.8 kb band which hybridizes to mR2 DNA is derived from sequences between the BglII site in the mouse insert and an EcoRI site in the vector. This preliminary mapping indicates that the maximum length of the repeat element (RE) in mR2 is 1100 bp and that it is located at least 700 bp upstream from the U2 gene in p61-41.

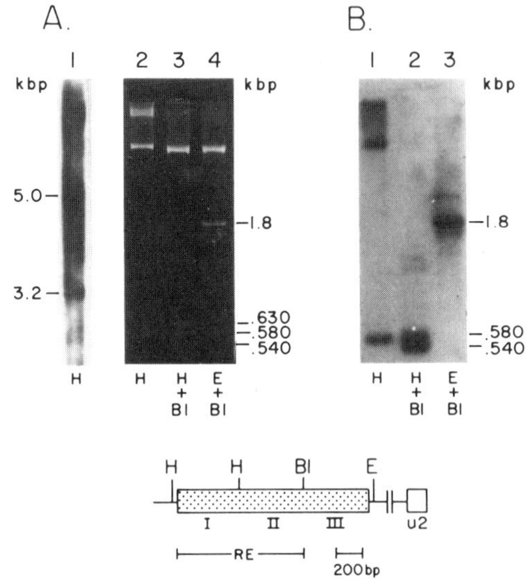

Figure 3. Reciprocal hybridization between mR2 and myeloma genomic DNA. (A) Lane 1. Autoradiograph of [^{32}P] nick-translated mR2 DNA to a Southern blot of HindIII (H) digested myeloma genomic DNA. Lanes 2-4. Ethidium bromide stained agarose gel of restricted mR2 DNA. H, HindIII; Bl, BglII; E, EcoRI. (B) Lanes 1-3. Autoradiograph of [^{32}P] nick-translated myeloma genomic DNA hybridized to a Southern blot of (A). Lanes 2-4. RE, repetitive element; U2, U2 RNA gene; kbp, kilobase pairs; bp, base pairs.

3.2. Transcription of IAP genes and sequences homologous to mR2 in mouse myeloma and embryonal carcinoma (EC) cells.

Murine EC cells are a model system to study the events involved in embryogenesis and differentiation. Since IAP antigens are found on the cell surface of pre-implantation embryos but are not detected on embryos after implantation (13), we decided to test whether the transcription of IAP genes is also developmentally regulated. Northern blot analysis was performed with poly A^+ RNA isolated from mouse EC cells, PCC3 and PCC4, and an F9 EC cell line which has spontaneously differentiated into embryoid bodies. The results of this experiment are shown in Fig. 4A. Three major IAP RNA species (7.2 kb, 5.3 kb and 3.8 kb) were found in myeloma. Of the three major IAP RNA species, the predominant IAP RNA in EC cells is the 5.3 kb class, as also described (8). The amount of IAP RNA is much greater in myeloma than in EC cells. IAP RNA levels decreased drastically when PCC3 cells were differentiated by treatment with retinoic acid in vitro, thus mimicing the shut off of IAP production in developing embryos.

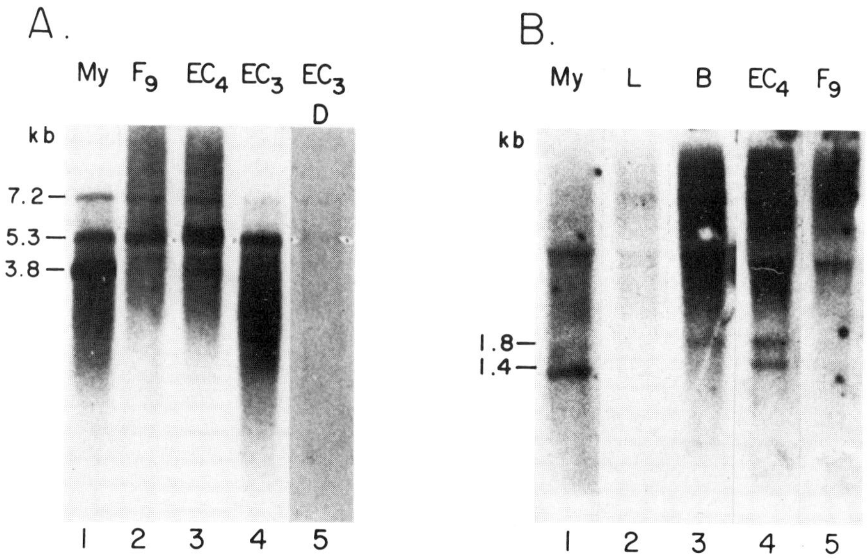

Figure 4. Northern blot analysis of transcripts homologous to IAP81 and mR2. (A) Poly A^{+} RNA from the following cell types: 1, 0.2 µg MOPC-315; 2, 2.0 µg F9 embryoid bodies; 3, 2.0 µg PCC4; 4, 2.0 µg PCC3; and 5, 5.0 µg differentiated PCC3 were subject to electrophoresis in a formaldehyde-agarose gel, transferred to nitrocellulose and hybridized with IAP-specific DNA probes. (B) Total RNA (15 µg) from the following cell types: 1, MOPC-315; 2, liver; 3, brain; 4, PCC4; and 5, F9 embryoid bodies were treated as in (A) and hybridized to mR2 DNA.

A similar Northern blot experiment was conducted using mR2 DNA as probe. Figure 4B shows that large, heterogeneous-size transcripts homologous to mR2 are detected to differing extents in every cell type examined. These RNAs are most abundant in brain and greatly diminished in liver. The 1.4 kb RNA which is so prominent in myeloma is not detected in liver or brain. Furthermore, this RNA is present in EC cell line PCC4 but not in F9 embryoid bodies. We have found this 1.4 kb transcript in PCC3 undifferentiated EC cells but not in PCC3 cells induced to differentiate (16).

3.3. Detection of IAP RNA in early embryos.

Mouse embryos at early stages of development were assayed for IAP expression by in situ hybridization. By counting the silver grains accumulated per embryonic cell, we found the highest amount of IAP RNA at the 2 cell stage (Fig. 5 and Table 1). This correlates with the appearance of cell surface IAP antigens during mouse embryogenesis (13). The number of silver grains per cell is significantly reduced at the morula-blastocyst stage (Table 1). The number of silver grains over morula-blastocyst embryos varied between different preparations of Balb/c embryos. Thus we were not able to establish the precise time at which IAP RNA synthesis is shut down.

Table 1

Quantitative Analysis of IAP RNA During Mouse Early Embryogenesis by In Situ Hybridization

Developmental Stage	Average Number of Grains per Cell	Number of Cells Counted
1 cell stage	19.0	3
2 cell stage	24.0	4
4-8 cell stage	14.9	10
Morula-blastula	8.3	6

3.4. Transcription of IAP genes and sequences.

We have previously established that myeloma nuclei and chromatin can be used as active transcriptional units for initiation and processing of a variety of RNA species (19). The DNA families represented by IAP81 and mR2 are also active in these in vitro systems. RNA synthesis was conducted with [α-^{32}P] CTP in the presence of [γ-S] ATP or [γ-S] GTP to specifically labeled initiated transcripts. RNA synthesized in vitro was fractionated on a mercury-column to separate the initiated RNA (Hg bound) from elongated transcript (Hg unbound). The use of α-amanitin in the in vitro transcription reaction enabled us to analyze which RNA polymerase may be responsible for the transcription of IAP genes. Figure 6 shows that myeloma nuclei synthesize large amounts of IAP RNA in the absence of α-amanitin. However, the presence of α-amanitin (2 μg/ml) inhibited IAP RNA synthesis by approximately 90%. Thus RNA polymerase II appears to be responsible for the transcription of IAP genes in vitro.

We have mapped the initiation site of IAP RNAs synthesized in vitro to the long terminal repeat (LTR) of the IAP gene (11). Several LTR sequences from different IAP gene clones have been reported recently (28). A diagram indicating the regulatory sequences essential for the transcription of IAP genes is displayed in Fig. 7.

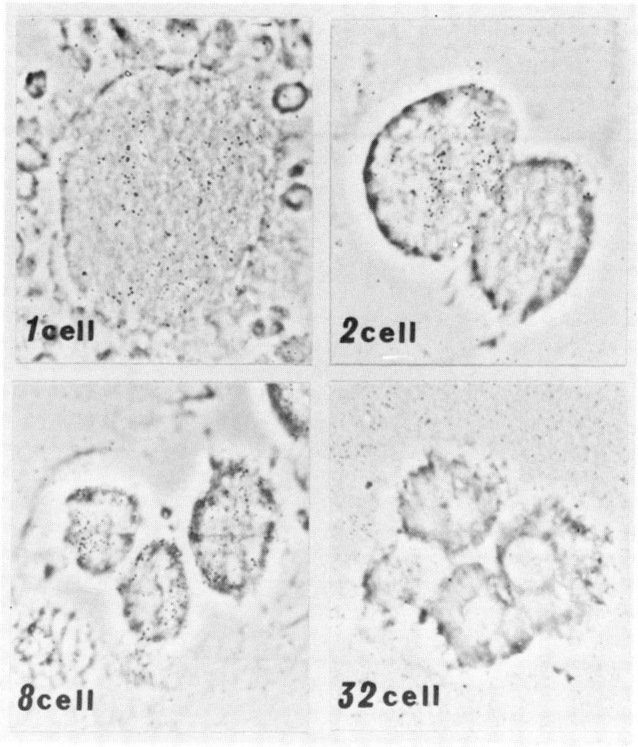

Figure 5. In situ hybridization to mouse embryos. Pre-implantation embryos were isolated and prepared for autoradiography as described in Materials and Methods. Shown is the resultant autoradiograph with cell stages as indicated.

Since transcripts homologous to mR2 DNA were also found to be abundant in myeloma cells, we have used isolated myeloma nuclei and chromatin to examine whether these repetitive gene transcripts are initiated in vitro. RNAs initiated in vitro with [γ-S] GTP or [γ-S] ATP in myeloma TEPC-15 nuclei or chromatin were hybridized to Southern blots containing Rsa2b and BglII restricted Rsa2a from p61-41 (Fig. 8). The 1650 bp DNA fragment contains Rsa2b which is derived from pBR322 and undigested Rsa2a. The autoradiographs in Fig. 8B indicate that the repetitive gene initiates with both [γ-S] GTP and [γ-S] ATP in nuclei, with neither one significantly preferred over the other. The fact that this repeated DNA represents a gene family with at least two different promoters (typified by transcripts enhanced in either brain or embryo) may explain this observation. In the chromatin transcription system (Fig. 8C) [γ-S] GTP is the initiation nucleotide of choice under the conditions employed. It may be that certain transcription factors have been extracted or altered in chromatin relative to isolated nuclei. How this observation correlates to transcripts of the repetitive family *in vivo* is unknown at present.

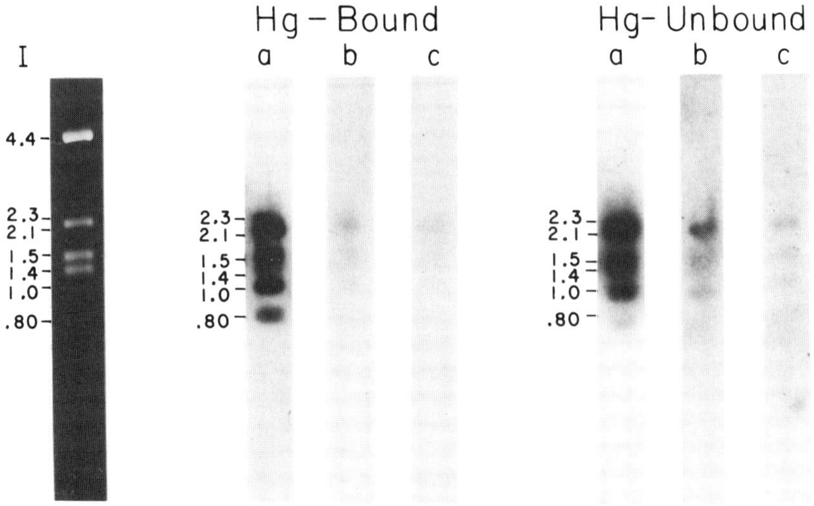

Figure 6. Inhibition of IAP RNA transcription in myeloma nuclei by α-amanitin. RNA was synthesized in MOPC-315 nuclei *in vitro* with α-amanitin at the concentrations 0 µg/ml (lanes a), 2 µg/ml (lanes b), and 100 µg/ml (lanes c). [γ-S] ATP was used as one of the substrates and as an affinity probe for the initiated RNA as described in the text. The mercury-bound fractions (*in vitro*-initiated RNA) and mercury-unbound fractions (*in vitro*-elongated RNA) were hybridized to IAP81 DNA digested with BamHI, EcoRI amd HindIII (Ref. 11).

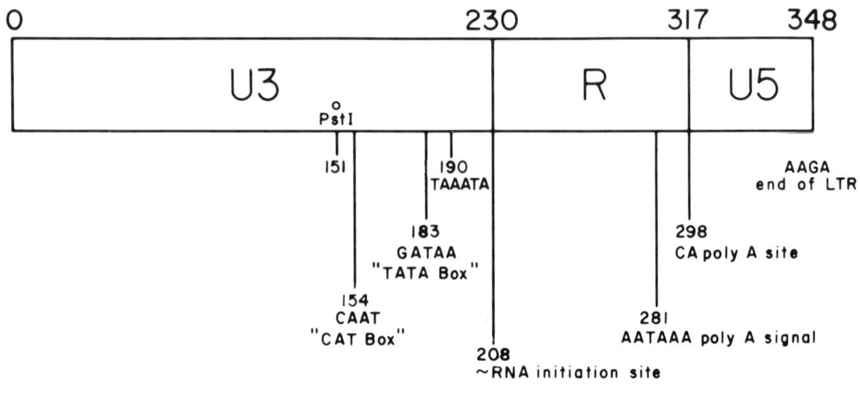

Figure 7. Diagram of the long terminal repeat of IAP81. Shown are the regulatory sequences as identified by DNA sequence analysis.

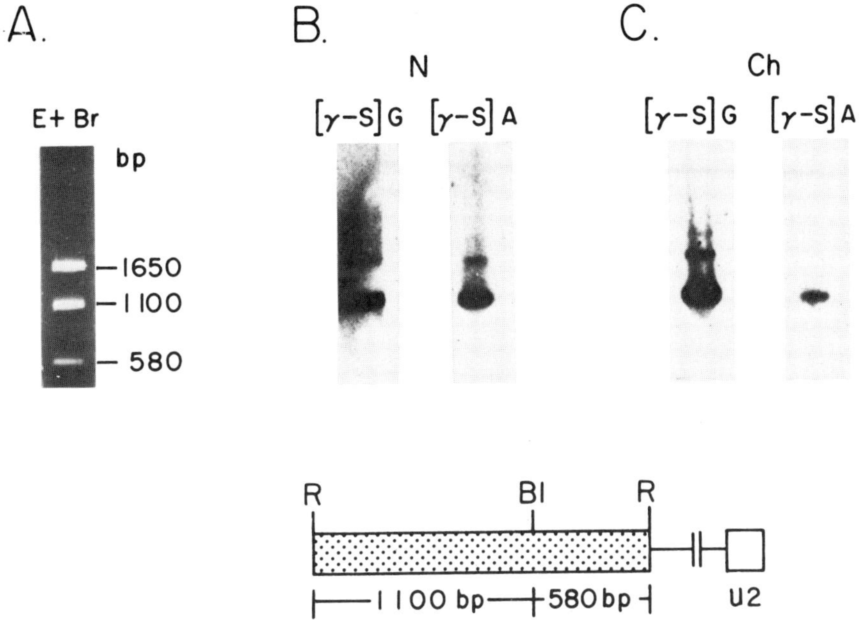

Figure 8. [γ-S]NTP initiated RNAs hybridized to the repetitive element in mR2. A) Ethidium bromide stained agarose gel containing BglII restricted Rsa2a (mR2 insert) DNA. EtBr, ethidium bromide; R, RsaI; B1, BglII; bp, base pair. B) and C) Autoradiographs of [γ-S]GTP and [γ-S]ATP initiated, ^{32}P-labeled RNA synthesized in isolated myeloma nuclei (N) or chromatin (Ch).

4. DISCUSSION

Huang and Calarco (13) provided the first evidence that IAP proteins produced in mouse pre-implantation embryos are immunologically related to IAP gene products produced in myeloma cell lines. In this study, the highest level of IAP antigens was detected on two to eight-cell embryos. IAP antigens were absent on morulae, blastocysts and inner cell masses, but detected at low levels on zygotes. Our data on IAP transcription during embryogenesis yielded similar results (Fig. 5). However, in a recent report Piko et al. (10) analyzed the synthesis of IAP related RNA (the 5.4 kb species) in early mouse embryos and concluded that IAP RNA is relatively abundant in ovarian oocytes but quantitatively reduced in the ovulated egg. The amount of IAP RNA increased one hundred times as development proceeded from one-cell stage to late blastocyst stage. Our data correspond to that of Piko et al. with regard to the onset of IAP RNA expression but differ from their determination of the time during embryogenesis when IAP transcrips disappear. The Balb/c mouse embryos used in our study showed a decrease of IAP RNA transcription at the morula-blastocyst stage. In embryos of

$CD2F_1$ mice, as studied by Piko et al., abundant IAP transcripts were detected even in late blastocysts. Other strain specific variations in IAP synthesis have been previously reported (29). It will be interesting to analyze the precise mode of IAP gene expression during early embryogenesis and to determine whether the turning off of IAP gene expression in post-implantation embryos is related to IAP gene methylation, rearrangement or chromatin structure.

The first report of IAPs in mouse tumors is now almost 30 years old (7) and yet their physiological role has not been determined. Because IAPs are a multi-gene family, any attempt to assign a general function for IAPs will likely be unsuccessful. Although IAPs are broadly grouped into two classes there exist many variant genes (containing both insertions and deletions) within each class (15). IAPs have classically been thought of as retroviral derivatives, but we believe that they should be considered more as endogenous cellular elements.

In this regard, various IAP genes have now been shown to act as insertional mutagens (presumatively via mechanisms involving reverse transcription and reintegration) in two kappa immunoglobulin and two cellular oncogene loci (30,31,32). An IAP insertion may also be responsible for the activation of the renin-2 gene in DBA/2 mice (33). Another recent result, suggesting that IAP genes may be less retroviral-like than originally thought, comes from work with an IAP complementary DNA clone (pMCT-1) isolated from a mouse colon tumor (34). When the expression of sequences homologous to pMCT-1 DNA was studied in Friend erythroleukemia cells, it was shown to be linked to the early G_1 phase of the cell cycle (35). These studies taken _in toto_ indicate that IAP genes may be a significant source of genetic variability in mice as suggested by Piko (10).

The identification and function of the repetitive DNA element contained in mR2 are currently unknown. The observation that large transcripts homologous to mR2 DNA are detected to differing extents in every cell type examined is consistent with the expression of a large number of mouse repetitive genes, e.g., B1, B2, R, MIF-1 and evolutionarily conserved sequences (36). Studies of Constantini et al. (37) and Scheller et al. (38) have demonstrated that different repeated sequences are preferentially represented in the hnRNA population of sea urchin oocytes, gastula and intestinal cells. They propose that coordinately expressed transcripts are covalently linked to a common repeat sequence and controlled as gene "batteries" or gene sets throughout development. However, it is not known whether such repeated sequences are "the cause or consequence of differentiated gene expression" (39).

The most interesting aspect of the transcription of the repetitive DNA family, represented by mR2, is its homology to a 1.4 kb RNA which is present in two myeloma and two embryonal carcinoma (EC) cell lines but absent in liver, brain and EC cells induced to differentiate. This interesting expression pattern relates the DNA family in mR2 to a select group of tumor-specific and, perhaps, embryonic genes. For example, Scott et al. (40) have isolated a cDNA clone, called Set-1, from mRNA present in SV40 transformed fibroblasts but not in the normal parental cell line. A highly repetitive element in Set-1 hybridizes to three transformation-specific RNAs (1.6 kb, 0.7 kb and 0.6 kb) found in all mouse tumors and most transformed fibroblasts examined. The unique sequences present in Set-1 hybridizes only to the 1.6 kb RNA. No adult mouse tissues contain the smaller transcripts and only normal thymus contains even traces of the 1.6 kb species (41). Furthermore, early embryos and undifferentiated EC cell lines exhibit a much larger number of Set-1 transcripts ranging in size from less than 0.7 kb to over 5 kb (42). Induction of EC cells to embryoid bodies (pluripotential core cells surrounded by a layer of endodermal cells) resulted in a dramatic decrease in the number of Set-1 related transcripts (42). The DNA sequence of the Set-1 clone revealed that the unique region codes for a class I major histocompatibility (MHC) antigen of the Qa/Tla type which has a member of the B2 repeat family (175 bp) inserted into the 3' non-coding end of the mRNA. Although the gene family represented by mR2 DNA shows some similarities to the Set-1 clone, preliminary studies assure us that they are distinctly different genes. That the mR2 DNA element is homologous to a transcript which is tumor specific and possibly embryonic indicates to us that the characterization of this gene family may lead to the elucidation of transcriptional regulatory controls which are important to normal and neoplastic development.

5. ACKNOWLEDGEMENTS

This research was supported by NIH grants R01AG04350 and 5P01AG03633. J.A.M. is supported by 2T32AG00069. R.A.M. is supported by a W. R. Grace Fellowship. We thank Dr. Jacob Mayer for performing the in situ hybridization experiment reported in this article.

6. REFERENCES

1. Teratomas and Differentiation, ed. Sherman, M. I. and Solter, D., Academic Press, 1975.

2. Teratoma Stem Cells, ed. Silver, L., Martin, G. and Strickland, S., Cold Spring Harbor Symposium, Vol. 10, 1983.

3. Brinster, R. L.,'The effect of cells transferred into mouse blastocyst on subsequent development.' J. Exp. Med., 140: 1049-1056, 1974.

4. Mintz, B. and Illmensee, K. 'Normal genetically mosaic mice produced from malignant teratocarcinoma cells.' Proc. Natl. Acad. Sci. USA, 72: 3585-3589, 1975.

5. Strickland, S. and Mahdavi, V. 'The induction of differentiation in teratocarcinoma stem cells by retinoic acid.' Cell, 15: 393-403, 1978.

6. Nicolas, J. F., Avner, R., Gaillard, J., Guenet, L., Jakob, H. and

Jacob, F. 'Cell lines derived from teratocarcinomas.' Cancer Res., 36: 4224-4231, 1976.

7. Howatson, A. F. and McCollough, E. A. 'Virus-like bodies in translatable mouse plasma cell tumour.' Nature (Lond.), 181: 1213-1214, 1958.

8. Hojman-Montes de Oca, F., Dianoux, L., Peries, J. and Emanoil-Ravicovitch, R. 'Intracisternal A particles: RNA expression and DNA methylation in murine teratocarcinoma cell lines.' J. Virol., 46: 307-310, 1983.

9. Lueders, K. K., Segal, S. and Kuff, E. L. 'RNA sequences specifically associated with mouse intracisternal A particles.' Cell, 11: 83-94, 1977.

10. Piko, L., Hammons, M. O. and Taylor, K. D. 'Amounts, synthesis, and some properties of intracisternal A particle-related RNA in early mouse embryos.' Proc. Natl. Acad. Sci. USA, 81: 488-492, 1984.

11. Wujick, K. M., Morgan, R. A. and Huang, R. C. C. 'Transcription of intracisternal A-particle genes in mouse myeloma and Ltk⁻ cells.' J. Virol., 52: 29-36, 1984.

12. Lueders, K. K. and Kuff, E. L. 'Sequences associated with intracisternal A particles are reiterated in the mouse genome.' Cell, 12: 963-972, 1977.

13. Huang, T. T. F. and Calarco, P. G. 'Immunological relatedness of intracisternal A-particles in mouse embryos and neoplastic cell lines.' J. Natl. Cancer Inst., 68: 643-648, 1982.

14. Marzluff, W. F., Murphy, E. C. and Huang, R. C. C. 'Transcription of ribonucleic acid in isolated mouse myeloma nuclei.' Biochemistry, 12: 3440-3446, 1973.

15. Ono, M., Cole, M. D., White, A. T. and Huang, R. C. C. 'Sequence organization of cloned intracisternal A particle genes.' Cell, 21: 465-473, 1980.

16. Moshier, J. A., Deutch, A. H. and Huang, R. C. C. submitted J. Biol. Chem.

17. Maniatis, T., Fritsch, E. F. and Sambrook, J. Molecular Cloning: A Laboratory Manual. Coldspring Harbor Laboratories, 1982.

18. Chirgwin, J. M., Przybyla, A. E., MacDonald, R. J. and Rutter, W. J. 'Isolation of biologically active ribonucleic acid from sources enriched in ribonuclease.' Biochemistry, 18: 5294-5299, 1979.

19. Marzluff, W. F. and Huang, R. C. C. 'Transcription of RNA in isolated nuclei.' In Transcription and Translation: A Practical Approach. ed. Hames, B. D. and Higgins, S. J., IRL Press, pp. 89-129, 1984.

20. Harrison, P. R., Conkie, D. and Paul, J. 'Localization of cellular globin messenger RNA by *in situ* hybridization to complementary DNA.' FEBS Lett., 32: 109-

21. Cole, M. D., Ono, M. and Huang, R. C. C. 'Intracisternal A-particle genes: structure of adjacent genes and mapping of the transcriptional unit.' J. Virol., **42**: 123-130, 1982.

22. Chiu, I. M., Huang, R. C. C. and Aaronson, S. A. 'Genetic relatedness between intracisternal A-particles and other major oncovirus genera.' Virus Res., in press.

23. Paterson, B. M., Segal, S., Lueders, K. K. and Kuff, E. L. 'RNA associated with murine intracisternal type A particles codes for the main particle protein.' J. Virol., **27**: 118-126, 1978.

24. Shen-Ong, G. L. and Cole, M. D. 'Differing populations of intracisternal A-particle genes in myeloma tumors and mouse subspecies.' J. Virol., **42**: 411-421, 1982.

25. Brown, S. D. M. and Dover, G. 'Organization and evolutionary progress of a dispersed repetitive family of sequences in widely separated rodent genomes.' J. Mol. Biol., **150**: 441-466, 1981.

26. Fanning, T. G. 'Characterization of a highly repetitive family of DNA sequences in the mouse.' Nucl. Acids Res., **10**: 5003-5013, 1982.

27. Gebhard, W., Meitinger, T., Hochtl, J. and Zachau, H. G. 'A new family of interspersed repetitive DNA sequences in the mouse genome.' J. Mol. Biol., **157**: 453-471, 1982.

28. Christy, R. J., Brown, A. R., Gourlie, B. B. and Huang, R. C. C. 'Nucleotide sequence of murine intracisternal A-particle gene LTRs have extensive variability within the R region.' Nucl. Acids Res., **13**: 289-302, 1985.

29. Kuff, E. L. and Fewell, J. W. 'Intracisternal A-particle gene expression in normal mouse thymus tissue: gene products and strain-related variability.' Mol. Cell Biol., **5**: 474-483, 1985.

30. Kuff, E. L., Feenstra, A., Lueders, K., Smith, L., Hawley, R., Hozumi, N. and Shulman, M. 'Intracisternal A-particle genes as movable elements in the mouse genome.' Proc. Natl. Acad. Sci. USA, **80**: 1992-1996, 1983.

31. Canaani, E., Dreazen, O., Klar, A., Rechari, G., Ram, D.,

Cohen, J. B. and Givol, D. 'Activation of the c-mos oncogene in a mouse plasmacytoma by insertion of an endogenous intracisternal A-particle genome.' Proc. Natl. Acad. Sci. USA, 80: 7118-7122, 1983.

32. Cohen, J. B., Unger, T., Rechovi, G., Canaani, E. and Givol, D. 'Rearrangement of the oncogene c-mos in mouse myeloma NSI and hybridomos.' Nature (Lond.), 300: 797-799, 1983.

33. Burt, D. W., Reith, A. D. and Brammar, W. J. 'A retroviral provirus closely associated with the Ren^{-2} gene of DBA/2 mice.' Nucl. Acids Res., 12: 8579-8593, 1984.

34. Augenlicht, L. H., Kobrin, D., Pavlovec, A. and Royston, M. E. 'Elevated expression of an endogenous retroviral long terminal repeat in a mouse colon tumor.' J. Biol. Chem., 258: 1842-1847, 1984.

35. Augenlicht, L. H. and Halsey, H. 'Expression of mouse long terminal repeat is cell cycle-linked.' Proc. Natl. Acad. Sci. USA, 82: 1946-1949, 1985.

36. Bennett, K. L., Hill, R. E., Pietras, D. F., Woodworth-Gutai, M., Haas, C. K., Houston, J. M., Heath, J. K. and Hastie, N. D. 'Most highly repeated dispersed DNA families in the mouse genome.' Mol. Cell. Biol., 4: 1561-1571, 1984.

37. Constantini, F. D., Britten, R. J. and Davidson, E. H. 'Message sequences and short repetitive sequences are interspersed in sea urchin egg poly $(A)^+$ RNAs. Nature (Lond.), 287: 111-117, 1980.

38. Scheller, R. H., Constantini, F. D., Kozlowski, M. R., Britten, R. J. and Davidson, E. H. 'Specific representation of cloned repetitive DNA sequences in sea urchin RNAs.' Cell, 15: 189-203, 1978.

39. Jelinek, W. R. 'Repetitive sequences in eukaryotic DNA and their expression.' Ann. Rev. Biochem., 51: 813-844, 1982.

40. Scott, M. R. D., Westphal, K. H. and Rigby, P. W. J. 'Activation of mouse genes in transformed cells.' Cell, 34: 557-567, 1983.

41. Brickell, P. M., Latchman, D. S., Murphy, D., Willison, K. and Rigby, P. W. J. 'Activation of a Qa/Tla class I major histocompatibility antigen gene is a general feature of oncogenesis in the mouse.' Nature (Lond.), 306: 756-760, 1983.

42. Murphy, D., Brickell, P. M., Latchman, D. S., Willison, K. and Rigby, P. W. J. 'Transcripts regulated during normal embryonic development and oncogenic transformation share a repetitive element.' Cell, 35: 865-871, 1983.

METHOTREXATE RESISTANCE, GENE AMPLIFICATION, AND SOMATIC CELL HETEROGENEITY

Robert T. Schimke
Department of Biological Sciences
Stanford University, Stanford CA, USA

Abstract: This review describes properties of the acquisition of resistance to methotrexate in cultured somatic cells as a consequence of amplification of the dihydrofolate reductase gene. We propose that gene amplification occurs following overreplication of DNA in a single cell cycle as a result of perturbation of the regulation of DNA synthesis. Cells surviving this process may contain multiple chromosomal changes. A similar phenomenon may partially underlie the progressive heterogeneity of tumor cell populations and the increase in heterogeneity of cellular responses during aging.

This paper will review studies from the author's laboratory concerning methotrexate (MTX) resistance and amplification of the dihydrofolate reductase (DHFR) gene, and will emphasize those aspects of amplification events that result in a marked heterogeneity in cell populations. The interested reader is referred to Schimke (1984, Cell 57,705-713) and Stark and Wahl (1984, Ann. Rev. Biochem. 53, 447-492) for more complete reviews of the subject of gene amplification.

Properties of DHFR Gene Amplification. Three mechanisms have been shown to result in MTX resistance: 1) altered affinity of DHFR for MTX (Haber et al, 1981, J. Biol. Chem. 256, 9501-9510); 2) altered MTX transport (Sirotnak et al, 1981, Cancer Res. 41, 4447-4452); 3) overproduction of DHFR as a result of gene amplification. These multiple modes of resistance can occur in the same cells. Resistance as a result of gene amplification has the following properties:

1. Resistance is a result of step-wise selection. We interpret this property to indicate that gene amplifications occur in small increments, thus requiring multiple step selections to obtain highly resistant cells by an amplification process. In the laboratory the use of large, single-step selections is more frequently employed to study resistance mechanisms, whereas in nature and clinically, small

B. Pullman et al. (eds.), Interrelationship Among Aging, Cancer and Differentiation, 117–120.

incremental selection processes are more common. Perhaps this is the reason why gene amplification has not been recognized previously.

2. Resistance results from overproduction of a normal protein. In the case of MTX resistance, the overproduction of DHFR basically involves a titration of cellular MTX by additional enzyme.

3. The resistance phenotype and amplified genes can be either stable or unstable. When stable, the genes reside on one or more chromosomes, often constituting expanded regions of chromosomes, so-called marker chromosomes or HSR regions (Biedler and Spengler, 1976, Science, 185-187). When unstable the genes reside on extrachromosomal, self-replicating elements called minute chromosomes. Because they lack centromeric regions, the minute chromosomes can undergo micronucleation and unequal distribution into daughter cells at mitosis, resulting in their instability (Schimke et al, 1981, Cold Spring Harbor Symp. Quant. Biol. 55, 785-797). When cells are first selected for resistance and gene amplification, the population predominately has cells with unstably amplified genes. However, when cells are maintained under selection conditions for long time periods, the cells that emerge often have stably amplified genes (Kaufman and Schimke, 1981, Mol. Cell. Biol. 1, 1069-1076). Empirically cell lines that are highly aneuploid tend to have minute chromosomes. We have proposed that the two end-results of gene amplification, HSRs or minute chromosomes, arise from the same initial event, i.e. overreplication of DNA (Mariani and Schimke, 1984, J. Biol. Chem. 259, 1901-1910) and result from differing modes of secondary recombination events (Schimke, 1984, Cell 37, 705-713). Thus if the overreplicated DNA recombines to form a circular structure (Hamkalo et al, 1985, Proc. Natl. Acad. Sci. 82, 1126-1130), it will constitute an extrachromosomal minute chromosome. If, as a rarer event, the overreplicated DNA recombines into the chromosome, this would result in amplified genes at the site of the resident (non-amplified) gene. Cell populations, therefore, can be vastly heterogeneous with respect to both the number and localization of amplified genes.

4. The same MTX resistance phenotype can result from amplification of two, different DNA sequences. In Leishmania Beverley et al (1984, Cell 38, 431-439) have reported MTX resistance resulting from amplification of different DNA sequences. Such variability in molecular events resulting in the same phenotype constitutes an additional form of cell heterogeneity.

5. MTX resistance and DHFR gene amplification occurs in cells from patients treated with MTX (see Schimke, 1984, Cell 37, 307-313 for ref.).

6. The spontaneous frequency of DHFR gene amplification in cultured somatic cells (Chinese Hamster Ovary Cells) is extremely high, and of

the order of a doubling of gene copy number in 1 x 10^{-3} cells per cell generation (Johnston et al, 1983, Proc. Natl. Acad. Sci. 80, 3711-3715).

7. The frequency of DHFR gene amplification can be increased dramatically by a variety of agents that reversibly inhibit DNA synthesis, hydroxyurea (Brown et al, 1983, Mol. Cell. Biol. 3, 1097-1107), MTX, UV light, and the carcinogen N-acetoxy N-acetylaminoflourene (Tlsty et al, 1984, Mol. Cell. Biol. 4. 1050-1056).

On the Mechanism of Gene Amplification. Recent studies from our laboratory have concentrated on the mechanism whereby hydroxyurea (as well as another inhibitor of DNA replication, aphidicolin) enhances the frequency of MTX resistance by virtue of gene amplification. Mariani and Schimke (J. Biol. Chem. 259. 1901-1910) concluded that transient (6 hr) inhibition of DNA synthesis, upon resumption of DNA synthesis, results in rereplication (overreplication) of DNA in a subset of the cells in a single S-phase. This effect was observed only in cells that had progressed into the S-phase prior to inhibition of DNA synthesis. This overreplication process involves not only the DHFR gene (which is replicated early in S-phase), but a large amount of the DNA replicated prior to the inhibition. More recently we have found (manuscript in preparation) that a critical variable in generating cells with additional DNA/cell following recovery from inhibition of DNA synthesis is the duration of the inhibition, i.e. the longer the duration of inhibition, the greater is the degree of overreplication.

We propose that the initial event in selective gene amplification, i.e. DHFR genes, is a non-specific overreplication of DNA. Inasmuch as DHFR is replicated in the first hour of the S-phase, it will have a high probability of being involved in an overreplication process. Various forms of recombination are proposed to occur subsequently to generate chromosomal or extrachromosomal genes. Only the rare cell with a productive recombination to generate a cell with increased DHFR enzyme activity will be selected subsequently. We wish to emphasize that the majority of the overreplicated DNA is unstable, and will be lost rapidly in progeny that are not under selective pressure. Thus the selectivity comes from the selection process, not from the initial amplification process.

Overreplication of DNA and the Generation of Various Chromosomal Aberrations. We are currently exploring the hypothesis that overreplication of DNA in a single cell cycle generates free-ended, double-stranded DNA that is highly recombinogenic (Schimke, 1984, Cell, 37, 705-713). As long as such overreplicated DNA does not "invade" the sister chromatids, the structure of the newly replicated sister chromatids is intact. However, if such recombination events occur, it can result in inversions, sister chromatid exchanges, normal chromosomes with extrachromosomal DNA, or varying degrees of polyploidation and fragmentation of chromosomes. Hill and Schimke (Cancer Res. in press) have subjected mouse lymphoma cells to a 6-hour treatment with hydroxyurea and have analyzed such cells for DNA content and a spectrum of chromosomal aberrations as an immediate consequence of such treatment, i.e., in the first M-phase following recovery from inhibition with hydroxyurea. Cells subjected to such treatment contain a subset of cells with increased DNA/cell, and 42% of metaphase spreads of this subset of cells contain gross chromosomal aberrations of the type described above. Cells with normal DNA/cell content show virtually no chromosomal aberrations. Similar results are obtained with cells treated with aphidicolin or subjected to UV irradiation (Sherwood and Tlsty, manuscript in preparation). Such data lead us to suggest that major rearrangements-alterations in somatic cell genomes can occur as a consequence of altered regulation of DNA synthesis such that overreplication-recombination occurs in a single cell cycle.

Concluding Remarks. That amplification-rearrangements events are common can be attested to readily from both the current understanding of the structure of genomes, and the increasing numbers of reports of selective amplifications and amplifications of "oncogenes" in directly isolated human tumors (see Stark and Wahl, Schimke, op. cit.). One can, in fact, reasonably defend the hypothesis that the most frequent type of genomic alteration occurring in nature is that of amplification (see Schmookler-Reis, this volume for the presence of amplified circular DNA sequences in human lymphocytes). Our formulation of the mechanism of gene amplification, i.e., overreplication-recombination, raises, perhaps, a fundamental question concerning DNA replication that has not previously been ennunciated. What determines the number and timing of initiations of replication on complex chromosomes containing multiple origins of replication? Our understanding of DNA replication comes from circular DNA structures (E. coli and SV40) studied under laboratory conditions where there are no perturbations of the organism (cells). Such systems may not provide an adequate basis for understanding replication in complex chromosomes, and in particular under the conditions where organisms or cells of organisms are subjected to a variety of environmental and drug (carcinogen)-induced stresses. The question posed is not that related to transition from resting to dividing cells, or that related to progression from G1 to S phase of the cell cycle. We suggest that the control of number-timing of replication initiations can be altered, and a consequence of this alteration occasional overreplication-recombination events occur. To the extent that such a process occurs, extensive somatic cell "mutation" may result, leading to heterogeneous cell populations in viable progeny of such recombinations, extensive cell death (as a result of chromosomal fragmentation), and an occasional recombination event resulting in overexpression of certain genes as a result of gene amplification or recombinational activation of genes, the consequence of which is overcoming normal growth constraints, i.e., cancer.

MACROMOLECULAR CORRELATES OF CELLULAR SENESCENCE AND CANCER

Robert J. Shmookler Reis, Arun Srivastava,
and Samuel Goldstein
Departments of Medicine and Biochemistry
University of Arkansas for Medical Sciences and
Geriatric Research Education and Clinical Center,
V.A. Medical Center, Little Rock, AR 72205, USA

ABSTRACT

Human diploid fibroblasts undergo a passage-dependent amplification of the H-ras proto-oncogene, accompanied by a coordinate increase in H-ras mRNA level and in the 21 kd H-ras protein. Other genes, including α-hCG and N-myc, are not amplified in these cells. Methylation of cytosines in the γ-globin and α-hCG gene regions, assessed in fibroblast mass cultures and in derived clonal lineages, declined in a site-specific and clone-specific pattern. The consequence of both processes is increasing heterogeneity of the cell population, in its DNA structure and potential for gene expression. Such heterogeneity may underlie the age-dependent incidence of clonal-origin diseases such as cancer.

INTRODUCTION

The incidences of many human diseases show a striking correlation with age: e.g., most cancers, atherosclerosis, and a variety of auto-immune conditions increase dramatically toward the end of the human lifespan.[1] Such diseases also have another feature in common--they arise clonally from rare deviant cells which appear increasingly with age, even in healthy normal individuals. Each of us consists of at least 10^{13} cells, and during the course of a lifetime, our bodies go through over 10^{16} cell divisions.[2] Only a very small fraction of those cells will ever give rise to benign neoplasms, and a still smaller fraction will in some individuals produce malignant tumors. Similarly, relatively few cells initiate atherosclerotic plaques, and only a small fraction of lymphocytes will be stimulated to produce auto-immune antibodies. Thus, the vast majority of cells remain normal throughout our lifespans, and show no outward signs of aging. This is not to say that most cells in our bodies do not age, but by many criteria they need not show a severe impairment of function unless stressed severely.

B. Pullman et al. (eds.), Interrelationship Among Aging, Cancer and Differentiation, 121–132.

Much of the morbidity and mortality associated with senescence may therefore be ascribed, plausibly, to increasing heterogeneity of the cell population as we age. The emergence of a wide range of deviant cellular phenotypes would in aggregate diminish optimal tissue function, and most importantly would increase the probability of pathological foci arising. Is there evidence of an age-dependent increase in phenotypic variance? The coefficient of variation for any parameter is almost certain to increase with age, whether the underlying physiological or biochemical variable increases, decreases or remains constant in average value.[3] As an example at the cellular level, it has been known for some years that the scatter of interdivision times increases during serial passage of diploid fibroblasts.[4] While the outcome of that process (measured by the endpoint, failure of self-replacement) is tightly synchronized at a characteristic limit to population doublings,[5] the underlying broadening of the distribution of intermitotic intervals occurs progressively and is one of the best objective criteria of cellular senescence.[6]

Cancer cells pose a curious contrast to the aging cell population from which they arise. They are characterized by an unlimited replicative potential, and a markedly decreased requirement for growth factors.[2,5] They often show enhanced expression for a variety of proto-oncogenes,[7] many of which turn out to be growth factors, growth factor receptors, or receptor-associated proteins.[8,9] This is commonly, but not necessarily, associated with amplification of the proto-oncogene DNA,[9,10] and may also involve their hypomethylation, since both benign and malignant tumors were recently shown to have lost methylations at a variety of other gene loci.[11] The aging fibroblast, on the other hand, as its frequency of cell replication declines, shows an increased requirement for growth factors.[12,13] The number of receptors for glucocorticoids, such as hydrocortisone, declines in senescent cells,[12] while the epidermal growth factor receptors (corresponding to the oncogene erb-B) remain constant in number but diminish strikingly in autophosphorylating activity.[13] If gene amplification and hypomethylation allow cancer cells to augment gene expression at the level of RNA transcription, and thus ultimately at the protein level, it is of interest to know whether aging cells undergo similar changes, or indeed retain the ability to utilize these mechanisms. In fact, our evidence strongly supports a replication-dependent increase in clonal diversification, resulting from both gene amplification and hypomethylation, leading to increasing heterogeneity of aging somatic cells in their genetic potential and expression.

GENE AMPLIFICATION

Prokaryote gene duplications occur at high frequency (10^{-4}-10^{-5}) relative to other mutations, and are unstable unless maintained under selective pressure (reviewed in ref. 14). That DNA sequence amplifications have also occurred in eukaryotes (and have been fixed by selection), on numerous occasions in their evolutionary history, is indicated by the existence of small clusters of specialized, related genes (e.g., globins and actins), moderate-size families of highly-conserved, functionally important DNA sequences (e.g., histone genes and ribosomal cistrons), and very large arrays of highly repetitive sequences (e.g., the hRI family in humans). It has recently become apparent that gene duplications for selectable markers can occur at remarkably high frequency (10^{-3}-10^{-4}) in eukaryote cells, allowing proportionately higher maximal protein synthesis levels for the enzymes encoded.[15-18]

Proto-oncogenes are the normal cellular homologues of retroviral transforming genes, many of which function as mitogens, mitogen receptors and co-factors, in the regulation of cell division.[8,19-21] Amplification of proto-oncogene DNA has been observed in a variety of neoplasias, and has been proposed as one of several independent steps in their etiology.[9,10] We asked whether similar amplification occurs during cellular senescence of normal diploid cells, and surprisingly we found amplification of the c-Ha-ras-1 proto-oncogene (H-ras) in all seven fibroblast strains examined during <u>in vitro</u> aging.[22] This passage-dependent amplification ranged up to four-fold over spans of 18-36 population doublings, and appeared in fibroblasts taken from donors of widely differing chronological ages (Figure 1). H-ras copy number correlated directly with the percent of replicative lifespan completed (PLC) <u>in vitro</u> (Figure 2). There was no effect on copy number due to donor age <u>in vivo</u> (Figures 1 and 2), apart from associated variation in the PLC. The N-myc oncogene was also examined, since it has been shown to amplify in advanced (but not early) stage neuroblastomas, and in derived cell lines.[23] No N-myc DNA amplification occurred during serial passage of several diploid fibroblast strains tested (Figure 3). Moreover, rehybridization of the same filters to other genes, unrelated to oncogenes (e.g., α-hCG, β- and γ-globins), also demonstrated no age-dependent amplification (data for α-hCG are plotted in Figure 2).

Using a variety of restriction enzymes and probing with several H-ras sub-fragments derived from a 6.6 kb genomic clone (pEJ 6.6), we have defined a narrow amplification gradient, extending only a few kilobases beyond the four <u>ras</u> exons.[22] Thus, the amplified region of DNA is not as extensive as those observed for other genes amplified developmentally or under strong selective pressure.[17,18,24,25] In view of this, and the low copy number achieved during H-ras amplification, we would not expect substantial production of double-minute chromosomes or homogeneously staining regions (HSRs), which frequently accompany "large-scale" DNA amplifications.[14,23] Consistent with this expectation, we have seen no HSRs and only an occasional putative double-minute chromosome in late-passage fibroblasts (data not shown).

The passage-dependent amplification of H-ras DNA is associated with a corresponding increase in mRNA levels corresponding to this gene (bands at 1.2 kb in Figure 4), again approximately four-fold.

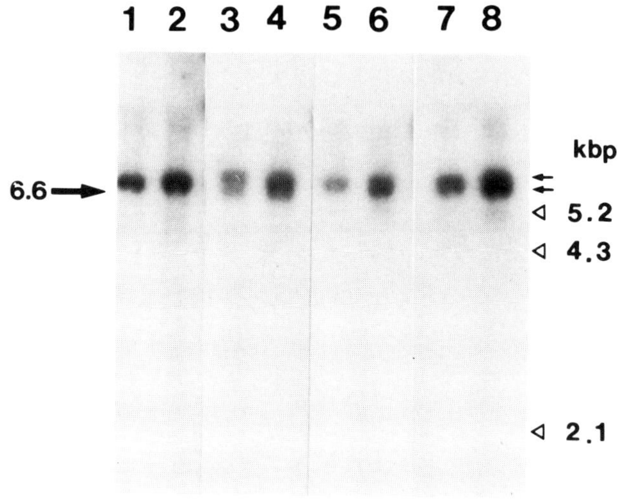

Figure 1: Southern blot analysis of c-Ha-ras-1 gene sequences in genomic DNA of fibroblasts at early and late passage. DNA samples were prepared from four fibroblast strains at early and late passage levels: HSC172 (fetal lung) at 25 MPD (lane 1) and 45 MPD (lane 2); A2 (forearm skin of an 11-year old male) at 22 MPD (lane 3) and 42 MPD (lane 4); J069 (forearm skin of a 67-year-old male) at 16 MPD (lane 5) and 37 MPD (lane 6); and J088 (forearm skin of a 76-year old female) at 21 MPD (lane 7) and 39 MPD (lane 8) were digested to completion with the BamHI restriction endonuclease. Cleaved DNA samples were electrophoresed in 1% horizontal agarose slab gels and transferred to nitrocellulose filters, Southern blotted, and probed with the c-Ha-ras-1 (pEJ6.6) probe. Reprinted, with permission, from ref. 22.

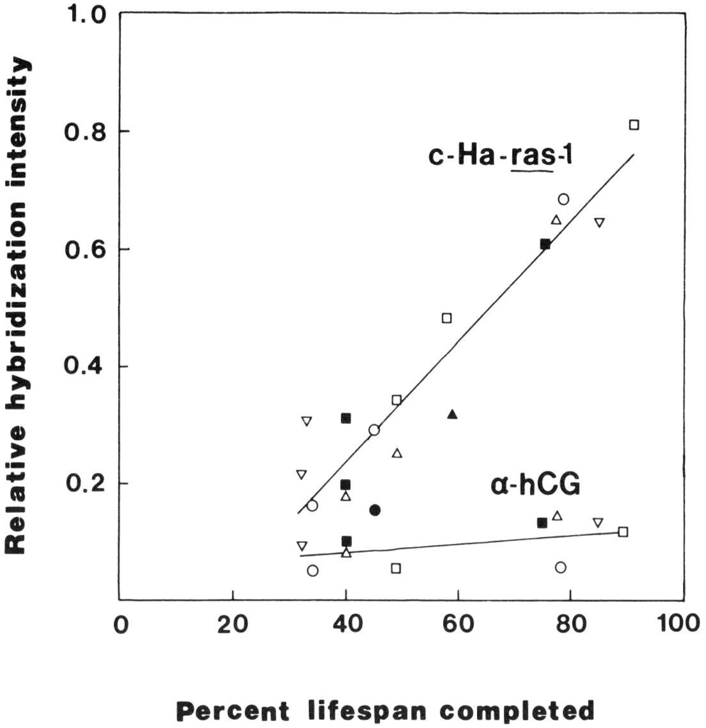

Figure 2: Intensity of hybridization to c-Ha-ras-1 and α-hCG DNA probes as a function of percent life span completed for several fibroblast strains. All autoradiographs were scanned using a Cliniscan microdensitometer. Integrated areas under band peaks (corrected for DNA load in some instances) were plotted against the percentage of maximal <u>in vitro</u> life span completed at the time of DNA preparation. The maximal population doubling levels obtained for these fibroblast cultures were: HSC172 (■), 62 MPD; MRC 5 (●), 64 MPD; WI38 (▲), 58 MPD, A2 (△), 55 MPD, A25 (▽), 55 MPD; J069 (○), 47 MPD; and J088 (□), 43 MPD. The correlation coefficient for c-Ha-ras-1 data points was $r = 0.937$, with a 95% confidence interval, $0.893 \leq r \leq 0.963$. The statistical significance of a non-zero correlation, $p < 10^{-4}$, was calculated from a linear regression analysis. Reprinted, with permission, from ref. 22.

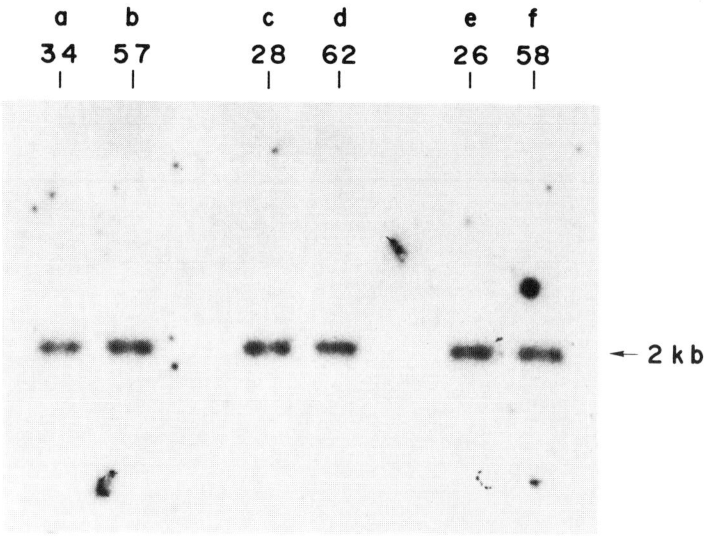

Figure 3: Southern-blot analysis of N-myc copy number in the DNA of three fibroblast strains at early and late passage. DNAs were prepared and analyzed as described in the legend to Figure 1. Fibroblast DNA preparations are from a fetal lung strain, WI-38 (lanes a, b), and from strains obtained by forearm skin biopsy, of an 11-year-old male (strain A2, lanes c,d) and of a 76-year-old female (strain J088, lanes e,f), at the MPD levels indicated.

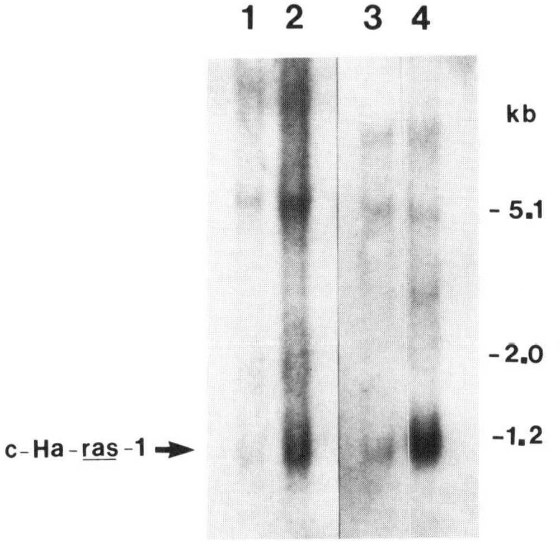

Figure 4: Analysis of RNA transcripts specified by the c-Ha-ras-1 gene on Northern blots from fibroblast strains at early and late passage. The RNA samples were from strains A2 at 18 and 54 MPD (lanes 1 and 2) and J069 at 18 and 37 MPD (lanes 3 and 4). 4 μg samples from each cell strain were fractionated on 1.5% agarose slab gels containing 6% formaldehyde and transferred to nitrocellulose filters and hybridized with the c-Ha-ras-1 probe. Reprinted, with permission, from ref. 22.

Interestingly, the several fibroblast strains examined differed in their steady-state levels of ras initial transcripts (bands at 5.1 kb) and putative processing intermediates (bands at ≈1.8 kb, lane 2, and at ≈3 kb, lane 4). This suggests that there are inter-individual differences in rates of RNA processing, which differ at specific introns and apparently also with passage level (Figure 4). The p21 ras protein level has also been estimated by electrophoretic analysis of immunoprecipitates obtained with a ras-specific monoclonal antibody, following in vivo radiolabelling of cell proteins with ^{35}S-methionine. Radioactive label in the 21 kd ras protein band again increased with fibroblast passage level, although not to as great an extent as ras gene number and mRNA level.[22]

We interpret these data as evidence of cellular selection within a population of diploid fibroblasts undergoing in vitro senescence, favoring the progeny of cells which have spontaneously duplicated the H-ras gene region. Such cells are evidently able to transcribe proportionately more H-ras mRNA, and thus to synthesize more of the ras protein product, than cells which have not amplified H-ras. This may serve to facilitate entry into the S-phase of the cell cycle, which becomes progressively blocked (at the Gl/S interface) in senescing fibroblasts.[6,26] In contrast, the N-myc gene is reported not to be expressed in fibroblasts,[27] thus precluding any selective advantage to occasional cells in which it becomes amplified, consistent with our observation of constant N-myc copy number during serial passage.

Amplification of the H-ras proto-oncogene occurs consistently in a variety of fibroblast strains, which nonetheless remain strikingly refractory to transformation. This clearly demonstrates that H-ras amplification and overexpression are not sufficient for oncogenesis, in accord with other reports indicating that multiple steps are required to create the oncogenic phenotype.[7,9,28]

DNA METHYLATION

The only post-synthetic DNA base modification known to occur in vertebrate cells is methylation of cytosine (C) to 5-methyl-cytosine (mC), occurring almost exclusively within the dinucleotide CpG.[29] The extent of methylation in specific loci is most easily assessed by the use of an isoschizomeric pair of restriction enzymes, e.g., MspI and HpaII, with a common cleavage site containing CpG, where only one of the enzymes is inhibited from cutting at sites modified to mCpG.[30,31] The initial expression of specific genes during development appears to coincide in most instances with the heritable loss of cytosine methylations in and around those genes (reviewed in refs. 32 and 33), with a few notable exceptions.[34] A causal link between hypomethylation and gene expression is most strongly implied by the correlation observed upon transfection of cells with "mosaic" plasmids constructed from unmethylated and in vitro-methylated DNA fragments, where methylation at the 5' end of a mammalian gene, but not elsewhere in the gene or in the plasmid vector, inhibits transcription of that gene.[35]

Tandemly Repeated Sequences

Three discrete classes of highly repetitive DNA which contain MspI restriction sites at intervals of 45, 110, and 175 base pairs, respectively, were largely resistant to cleavage by HpaII enzyme, indicating a consistently high level of methylation at the 3' cytosine of -CCGG- loci during fibroblast aging in mass culture, and in several clones examined at late passage.[36] Moreover, subsets of the hRI centromeric family of repetitive DNA which are cleaved by MspI restriction enzyme, were equally resistant (≃ 95%) to HpaII cleavage, implying extensive CpG methylation, in DNA from both early- and late-passage fibroblasts.[37]

Gene Sequences

Figure 5 illustrates the different clone-specific patterns of DNA methylation seen in specific gene DNA regions, including those for α-hCG, β- and γ-globins, and β- and γ-actins.[38] These data indicate a striking degree of inter-clonal heterogeneity, particularly in the α-hCG (Figure 5A) and γ-globin (Figure 5B) sequences which are not normally expressed in diploid fibroblasts. Intra-clonal variations were also evident in most instances, implying that heterogeneity of DNA methylation in and around specific genes arises de novo within pure fibroblast clones, primarily by loss of methylations. This is supported by observations of Wolf and Migeon[39] who examined mass cultures and clones of human fibroblasts for methylation of gamma-globin genes as well as 28 kb pairs of X-chromosome DNA with two specific cloned probes for the human X-chromosome. Our recent studies of the α-hCG gene region have confirmed that methylations are lost progressively within a cell lineage, at sites which vary between different subclones derived from the same fibroblast clone.[40] However, only the progeny subclones of certain "low fidelity" clones show this progressive hypomethylation, while other clones faithfully maintain essentially complete α-hCG methylation in all subclones and at all passage levels.[40] Initial studies of RNA transcripts from the α-hCG gene indicate increased expression of an incompletely processed intermediate, in a clone which undergoes hypomethylation around the α-hCG locus, compared to a high-methylation clone (unpublished data).

Thus, when individual populations of relatively homogeneous cells are studied, DNA methylation patterns in and around endogenous gene loci appear to be inherited somewhat unstably, and this may be associated with random gene derepression in some cell lineages during cellular aging. Tandemly-repetitive sequences, in contrast, show relatively faithful maintenance of their high CpG methylation levels. This suggests either the existence of a distinctive maintenance methylase specific for such repetitive sequences, and/or a strong selective pressure for high methylation of centromeric DNA.

A strikingly similar set of observations has recently been made for a variety of benign and malignant human colon neoplasms, each

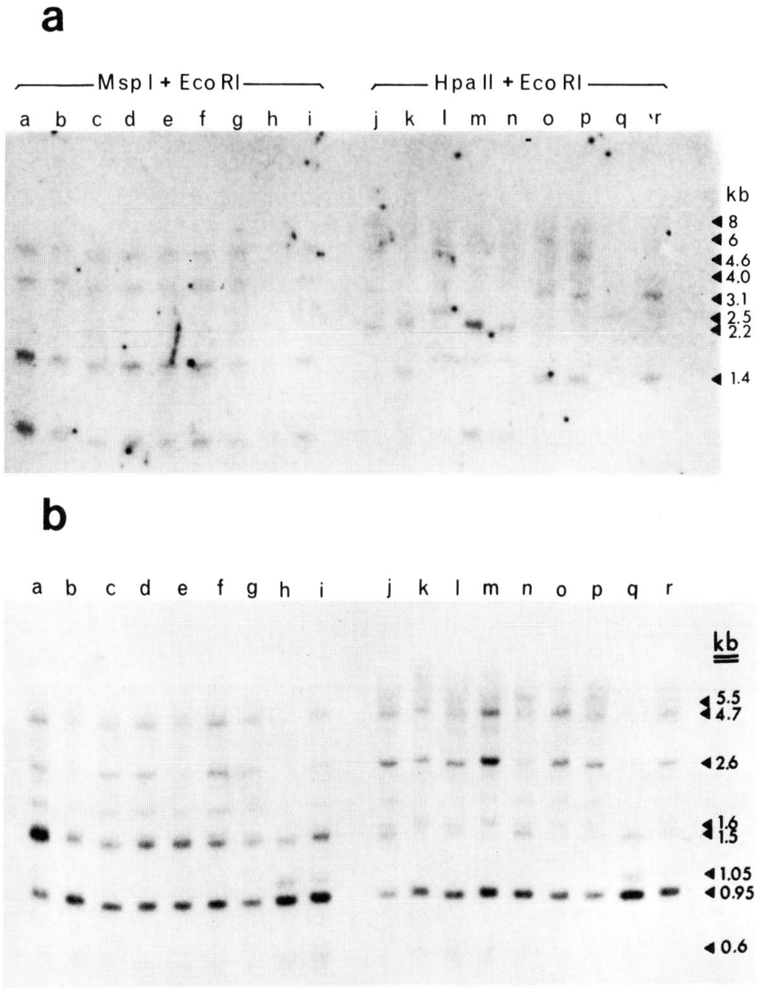

Figure 5: Variation of DNA methylation patterns in a human fibroblast mass culture and derived clones, for the α-hCG and γ-globin gene regions. Autoradiographs are shown of Southern-transferred DNA fragments hybridizing to ^{32}P-cDNA probes for (A) α-hCG, and (B) γ-globin genes. DNA samples were obtained from A2 diploid human fibroblasts at 28 mean population doublings (a,j) and eight clones isolated from that strain; clone 1 (b/k), clone 2 (c/l), clone 3 (d/m), clone 4 (e/n), clone 5 (f/o), clone 6 (g/p), clone 7 (h/q), and clone 8 (i/r). All DNA loads were 6 μg, except clone 7 which was 3 μg. DNA samples were digested with MspI and EcoRI (a-i) or HpaII and EcoRI (j-r). Fragment sizes in kilobase pairs (kb) were determined from PM2/HindIII and Lambda/HindIII markers. Reprinted, with permission, from Shmookler Reis and Goldstein (38[A] and 36[B]).

presumably having a clonal origin.[11] In all 23 tumors examined, hypomethylation was observed in four gene regions, including γ-globins and α-hCG, but not for the hRI family of tandemly repeated sequences. Patterns of gene-specific hypomethylation observed in benign and malignant tumors[11] could, in the light of our data, merely reflect their clonal origins, since normal control tissues would be polyclonal in nature and thus might mask clone-specific hypomethylation patterns by "averaging" them. However, the degree of hypomethylation in neoplasms appeared too great to support such an explanation.[11] If, instead, extensive hypomethylation is a common precondition for neoplasia and malignancy, perhaps associated with the widespread gene derepression which is observed in many neoplastic cells,[41,42] then the clonal heterogeneity we have observed for hypomethylation during cellular senescence, would create an increasing likelihood of oncogenic transformation.

CONCLUSIONS

Even in the absence of overt pathology, normal aging is accompanied by an inexorable increase in clonal lesions: benign neoplasias, atherosclerotic plaques, and auto-immune antibodies. A certain fraction of these will become clinically important, and of course will contribute to age-dependent morbidity and mortality. These may reflect a clonal diversification in gene structure and expression, which increases during in vitro senescence, where cell lineages can be analyzed, and presumably also in somatic tissues in vivo.

While we have not yet completed a clonal analysis of ras gene amplification, our data[22] and those of others,[14-17] are consistent with the random spontaneous amplification of 0.01-0.1% of each cell's DNA complement. Although only a small number of examples have been studied thus far, transcriptional enhancement in each case appears to be coupled to DNA amplification, and is thus commensurately augmented for affected loci.[14-17,22,23] Sporadic hypomethylation of genes, during repeated cell division, is now well-documented[11,36,38,40] and is apparently also accompanied by increased transcription. Thus the patterns of genetic expression in individual somatic cells may become quite heterogeneous, and increasingly diverged from the tissue-specific norm, with the result that the "signal-to-noise ratio" for programmed expression of each cell will be eroded. This could be a common process in senescence, underlying both the age-dependent decline in fitness observed under stressful conditions, and the age-dependent increase in diseases of clonal origin.

Acknowledgements

We are grateful for the technical assistance of Kathryn Rowley and the secretarial support of Joyce Stevens and Marijo Nelsen. Supported by grants (AG-03314 and AG-03787) from the National Institutes of Health, and from the Veterans Administration.

REFERENCES

1. Upton, A.C.: 1977, "Pathobiology," in Finch, C.E. & Hayflick, L., eds., Handbook of the Biology of Aging. New York, Van Nostrand Reinhold Co., pp. 513-535.
2. Alberts, B., Bray, D., Lewis, J., Raff, M., Roberts, K., and Watson, J.D.: 1983, Molecular Biology of the Cell. New York, Garland Publishing Co., pp. 612, 620, 940.
3. Shmookler Reis, R.J.: 1981, Mech. Ageing and Devel. 17, pp. 311-320; see also Finch and Hayflick (ref. 1), figures on pp. 15, 18, 26, 79, 284, 385, etc.
4. Absher, P.M., Absher, R.G., and Barnes, W.D.: 1974, Exp. Cell Res. 88, pp. 95-104.
5. Hayflick, L., and Moorehead, P.: 1961, Exp. Cell Res. 25, pp. 585-610.
6. Cristofalo, V.J., and Sharf, B.B.: 1973, Exp. Cell. Res. 76, pp. 419-427.
7. Slamon, D.J., DeKernion, J.B., Verma, I.M., and Cline, M.J.: 1984, Science 224, pp. 256-262.
8. Cooper, G.M., 1982, Science, 218, 801-806.
9. Land, H., Parada, L.F., and Weinberg, R.A., 1983, Science 222, pp. 771-778.
10. Alitalo, K., Schwab, M., Lin, C.C., Varmus, H.E., and Bishop, J.M.: 1983, Proc. Natl. Acad. Sci. U.S.A. 80, pp. 1707-1711.
11. Goelz, S.E., Vogelstein, B., Hamilton, S-R., and Feinberg, A.P.: 1985, Science 228, 187-190.
12. Rosner, B.A., and Cristofalo, V.J.: 1984, Endocrinology 108, pp. 1965-1971.
13. Carlin, C.R., Phillips, P.D., Knowles, B.B., and Cristofalo, V.J.: 1983, Nature 306, 617-619.
14. Schimke, R.T.: 1982, in Schimke, R.T., ed., Gene Amplification. New York, Cold Spring Harbor Press, pp. 1-6, 317-333.
15. Mariani, B.D., and Schimke, R.T.: 1984, J. Biol. Chem. 259, pp. 1901-1910.
16. Johnston, R.N., Beverly, S.M., and Schimke, R.T.: 1983, Proc. Natl. Acad. Sci. U.S.A. 80, pp. 3711-3715.
17. Beach, L.R., and Palmiter, R.D.: 1981, Proc. Natl. Acad. Sci. U.S.A. 78, pp. 2110-2114.
18. Spradling, A.C.: 1982, in Schimke, R.T., ed., Gene Amplification. New York, Cold Spring Harbor Press, pp. 121-127.
19. Weinberg, R.A.: 1980, Cell 22, pp. 643-644.
20. Bishop, J.M.: 1982, Mol. Cell. Biochem. 23, pp. 515-524.
21. Kelly, K., Cochran, B.H., Stiles, C.D., and Leder, P.: 1983, Cell 35, pp. 603-610.
22. Srivastava, A., Norris, J., Shmookler Reis, R.J., and Goldstein, S.: 1985, J. Biol. Chem., in press.
23. Brodeur, G.M., Seeger, R.C., Schwab, M., Varmus, H.E., and Bishop, J.M.: 1984, Science 224, pp. 1121-1124.
24. Bostock, C.J., and Tyler-Smith, C.: 1982, in Schimke, R.T., Gene Amplification. New York, Cold Spring Harbor Press, pp. 15-21.

25. Roberts, J.M., Buck, L.B., and Axel, R.: 1983, Cell 33, pp. 53-63.
26. Olashaw, N.E., Kress, E.D., and Cristofalo, V.J.: 1983, Exp. Cell Res. 149, pp. 547-554.
27. Eva, A., et al.: 1982, Nature 295, pp. 116-119.
28. Land, H., Parada, L.F., and Weinberg, R.A.: 1983, Nature 304, pp. 596-602.
29. Vanyushin, B.F., Tkacheva, S.G., and Belozersky, A.N.: 1970, Nature 225, pp. 948-951.
30. Singer, J., Roberts-Ems, J., and Riggs, A.D.: 1979, Science 203, pp. 1019-1020.
31. Waalwijk, C., and Flavell, R.A.: 1978, Nucl. Acids Res. 5, pp. 3231-3236.
32. Ehrlich, M., and Wang, R.Y.H.: 1981, Science 212, pp. 1350-1357.
33. Razin, A., and Riggs, A.D.: 1980, Science 210, pp. 604-610.
34. Bird, A.P.: 1984, Nature 307, pp. 503-504.
35. Busslinger, M., Hurst, J., and Flavell, R.A.: 1983, Cell 34, pp. 197-206.
36. Shmookler Reis, R.J., and Goldstein, S.: 1982, Proc. Natl. Acad. Sci. U.S.A. 79, pp. 3949-3953.
37. Shmookler Reis, R.J., Srivastava, A., Beranek, D.T., and Goldstein, S.: 1985, J. Mol. Biol., in press.
38. Shmookler Reis, R.J., and Goldstein, S.: 1982, Nucl. Acids Res. 10, pp. 4293-4304.
39. Wolf, S.F., and Migeon, B.R.: 1982, Nature 295, pp. 667-671.
40. Goldstein, S., and Shmookler Reis, R.J.: 1985, submitted.
41. Rosen, S.W., Weintraub, B.D., and Aaronson, S.A.: 1980, J. Clin. Endorinol. 50, pp. 834-841.
42. Milsted, A., Day, D.L., and Cox, C.P.: 1982, J. Cell. Physiol. 113, pp. 420-426.

DOMINANCE OF IN VITRO SENESCENCE IN SOMATIC CELL HYBRIDIZATION AND BIOCHEMICAL EXPERIMENTS

James R. Smith and Olivia M. Pereira-Smith
Department of Virology and Epidemiology, Baylor College of Medicine, Houston, Texas 77030

ABSTRACT

Studies reported here have led us to conclude that both senescent and quiescent cells produce a protein(s) inhibitor that blocks the initiation of DNA synthesis. In quiescent cells the inhibitor is rapidly lost after the addition of growth factors, whereas in senescent cells the inhibitor is produced even in the presence of serum growth factors. In a further series of experiments we have found that the phenotype of finite proliferative lifespan is dominant in somatic cell hybrids between normal human cells and immortal cells and, in addition, that fusion of certain immortal cell lines with other immortal cell lines yield hybrids with finite proliferative lifespan. These results suggest that in vitro cellular senescence results from a genetic program that causes a potent inhibitor of DNA synthesis to be produced at the end of the in vitro lifespan, and that immortal cells arise as a result of recessive changes in this program.

The limited capacity for division exhibited by normal animal cells in culture is well documented and a striking contrast to the unlimited division potential of abnormal tumor-derived or virus- or carcinogen-transformed cells. In 1965 Hayflick (1) proposed that the limited division of normal cells was a manifestation of aging at the cellular level, and we have studied normal human cells, particularly fibroblasts, as a model for in-vitro aging. Although our studies have been aimed primarily at understanding the phenomenon of cellular aging, they are of relevance to the fields of cancer and differentiation. Cancerous cells in culture have somehow lost the ability to limit cell division and do not exhibit a finite in-vitro lifespan. Our cell hybrid studies have demonstrated that the phenotype of immortality is recessive and senescence is dominant. Therefore it is obvious that an understanding of the mechanisms that limit proliferation in normal cells would lead to an understanding of the processes that result in these immortal, abnormal cells. Cellular differentiation is very often accompanied by cessation of cell division. From our studies we have concluded that

B. Pullman et al. (eds.), Interrelationship Among Aging, Cancer and Differentiation, 133–141.

normal cells in culture stop dividing as the end result of a genetic program that causes the cells to produce a new gene product which blocks the initiation of DNA synthesis. In this sense, then cellular senescence is a form of cellular differentiation.

Exp.	Senescent cell	Young cell	Young cytoplast x young cell	Senescent cytoplast x young cell
I	0.4* (2/529)	66* (209/317)	ND	33* (108/325)
II	1.3 (7/547)	55 (119/217)	65* (79/121)	30 (90/305)
III	1.9 (1/53)	89 (222/248)	78 (50/64)	51 (113/223)
IV	4.9 (8/163)	62 (187/301)	71 (213/300)	33 (87/262)
V	14 (36/266)	68 (168/248)	64 (110/171)	42 (76/182)

* Percent labeled nuclei (No. labeled nuclei/No. of cells scored)

TABLE 1. INHIBITION OF DNA SYNTHESIS BY SENESCENT CYTOPLASTS

For the past several years we have studied DNA synthesis and proliferation of various cell hybrids in an attempt to learn more about these processes. Norwood and coworkers first demonstrated in 1974 that senescent human cells could inhibit DNA synthesis in cells capable of proliferation following fusion (2). More recently, Burmer et al. (3) and we (4) have independently reported that senescent cytoplasts were as efficient as whole cells in inhibiting DNA synthesis in cells capable of proliferation (Table 1). Norwood and coworkers had determined that cycloheximide treatment of senescent cells at the time of fusion eliminated the inhibitory effect (5), indicating that a protein(s) was involved in the DNA synthesis inhibition observed. We found that treatment of senescent cytoplasts with 5 ug/ml cycloheximide or 10 ug/ml puromycin for as little as 2 hours prior to fusion also eliminated the inhibitory effect of the cytoplasts (Table 2) (6). We studied the kinetics of recovery of inhibitory activity in senescent cytoplasts incubated for various times after treatment with these protein synthesis inhibitors and before fusion to proliferating cells. Cycloheximide treated cytoplasts regained their inhibitory activity within 2-3 hours after removal of the cycloheximide, and detectable levels of inhibition were observed within 1 hour of removal of cycloheximide (Table 2). Puromycin treated cytoplasts did not regain inhibitory activity. The differential effect of these protein synthesis inhibitors can be explained by the difference in the mechanisms by which they act. Cycloheximide inhibits translocation and release of tRNA and prevents GTP dependent breakdown of polyribosomes, whereas puromycin causes premature release of the polypeptide chain from polysomes resulting in their breakdown. These results were consistent with the hypothesis that the loss of proliferative potential in normal animal cells results from the expression of a new gene or a change in the expression of a gene product that is involved in the regulation of cell cycle. The gene product which blocks initiation of DNA synthesis appear to be a protein(s) present in the senescent cytoplast.

Time of treatment	Cycloheximide (5 ug/ml)	Puromycin (10 ug/ml)
	Percent Inhibition	
0	38	35
1	29	18
2	2	15
3	2	3
Time of Recovery		
0	5	5
1	10	10
2	25	8
3	30	5

TABLE 2. EFFECT OF CYCLOHEXIMIDE AND PUROMYCIN ON INHIBITOR ACTIVITY

Our next goal was to determine the localization of the inhibitory proteins(s), i.e. cytosol versus cell membrane. We found that treatment of senescent cytoplasts with 0.025% twice crystallized trypsin for 1 minute at 4°C or growth of senescent cells on coverslips that had been coated with fibronectin (FN) prior to enucleation and fusion, eliminated inhibitory activity (Table 3) (7). (We had coated coverslips with either FN or poly-L-lysine in an attempt to increase cell cell adhesion during enucleation). If the cytoplasts were allowed to recover from trypsin treatment, inhibitory activity was regained with kinetics similar to recovery from cycloheximide treatment. Since trypsin and FN, agents that affect surface membranes, cause loss of the inhibitory activity, indications were that the protein inhibitor(s) was present in the surface membranes of senescent cells (Table 4).

Experimental condition	Percent inhibition		
	Experiment		
	1	2	3
Untreated	59	40	42
Trypsin	6	2	4
Poly-L-lysine	52	39	30
FN	0	4	2

TABLE 3. EFFECT OF TRYPSIN AND FIBRONECTIN (FN) ON INHIBITORY ACTIVITY OF SENESCENT CYTOPLASTS

Recovery time (hour)	Percent inhibition		
	Senescent cytoplasts		Quiescent cytoplasts
	Experiment		Experiment
	1	2	1
0	8	6	0
1	0	7	0
2	23	17	9
4	46	27	24
Untreated control	45	33	35

TABLE 4. RECOVERY OF INHIBITOR ACTIVITY IN SENESCENT AND QUIESCENT CYTOPLASTS FOLLOWING TRYPSIN TREATMENT

It was of interest to us that Norwood's group had determined that young cells made quiescent by maintenance in low serum for 2-4 weeks exhibited inhibitory activity similar to senescent cells (8). We extended their observations and found that cytoplasts derived from quiescent cells also possessed the ability to inhibit DNA synthesis initiation in proliferating cells. We found that the quiescent cell inhibitor was eliminated by trypsin treatment and growth of the cells on FN, but not by cycloheximide treatment (Table 5). The inhibitory activity was regained if recovery from trypsin treatment was permitted and the kinetics of recovery were similar to that occurring in senescent cells (Table 4).

The fact that cycloheximide treatment of quiescent cells (9) or cytoplasts (7) prior to fusion did not eliminate their inhibitory activity raised the question of whether proteins were involved in the inhibition of DNA synthesis mediated by quiescent cells and cytoplasts. We therefore treated quiescent cytoplasts with trypsin and allowed them to recover in the presence of cycloheximide. The inhibitory activity was not regained, indicating that protein synthesis must occur for the inhibitory effect and that the inhibitory activity is either a protein(s) or mediated by a protein(s). Since senescent cells are maintained in medium containing 10% fetal bovine serum (FBS) compared with quiescent cells that are in 0.5% FBS, we tested the effect of cycloheximide treatemnt of senescent cytoplasts derived from senescent cells that had been maintained on 0.5% FBS for 2 weeks. Cycloheximide did not reverse the inhibitory activity in cytoplasts derived from these senescent cells.

These results indicated that protein inhibitors of DNA synthesis are produced by senescent and quiescent cells, that they are associated with surface membranes and stablized by low serum.

To confirm the membrane association of the inhibitor proteins, we isolated surface membrane enriched fractions from quiescent cells and found that fusion of 100 ug protein equivalent of surface membranes with proliferating cells resulted in DNA synthesis inhibition (Table

6) Proteins extracted from these isolated membranes with 30 mM octyl-glucoside, a dialyzable detergent, for 30 minutes also inhibited DNA synthesis in cells exposed to the proteins for 24 hours in the presence of 10% FBS (Table 7).

Exp.	Experimental condition	Labeled Nuclei: Arrested cytoplasts x young cells	Labeled Nuclei: Unfused young cells	Percent inhibition**
1	Untreated	62[a] (90/146)[b]	77 (82/106)	20
	Cycloheximide	61 (42/69)	78 (78/100)	22
	Trypsin	69 (73/106)	72 (80/111)	4
	FN	77 (77/100)	72 (78/109)	-7 (stimulation)
2	Untreated	42 (19/45)	72 (180/249)	42
	Cycloheximide	38 (15/39)	70 (97/139)	46
	Trypsin	71 (41/58)	70 (73/104)	-1
	FN	79 (39/49)	70 (49/113)	-12
3	Untreated	43 (21/49)	65 (91/140)	34
	Cycloheximide	46 (46/101)	66 (66/100)	30
	Trypsin	68 (38/56)	66 (188/285)	-3
4	Untreated	50 (68/137)	73 (205/280)	32
	FN	73 (107/146)	73 (168/231)	0

* Quiescent cells were maintained on 0.5% serum for 1 week (Exp. 1) and 2 weeks (Exp. 2-4).

$$\text{**Percent inhibition} = 1 - \left(\frac{\text{\% labeled nuclei in cybrids}}{\text{\% labeled nuclei in unfused young cells}} \right) \times 100$$

[a]Numbers represent percent labeled nuclei.
[b]Numbers in parentheses are actual counts.

TABLE 5. DNA SYNTHESIS INHIBITORY ACTIVITY OF QUIESCENT* CYTOPLASTS AND EFFECT OF VARIOUS TREATMENTS ON THE ACTIVITY

Since the method of membrane isolation required large numbers of cells and the yield of membranes was variable from preparation to preparation, we have more recently extracted proteins directly from quiescent cell monolayers. The extraction procedure does not lyse nuclei, which appear intact when observed by a phase contrast microscope. We obtained inhibition with 80 ug cell protein extracted in this way as compared with 10 ug protein extracted from isolated membranes (Table 8). We have worked with two other preparations of quiescent protein and obtained variable inhibitory activity with the same protein concentration. The reason for this is the variability in amount of protein extracted at each time. We are presently testing lower concentrations of octylglucoside to determine one that will extract the inhibitory proteins without the variation in total protein extracted. The results obtained with preparation 3 indicate that the activity of the senescent protein is equivalent to that of the quiescent protein, but experiments must be repeated to confirm this.

Exp.	Membranes fused to young cells	Percent labeled nuclei	Inhibition (Percent of control)
1	Control (unfused)	88	
	Young	88	0
	Quiescent	70	20
2	Control	88	
	Young	89	-1
	Quiescent	67	24
3	Control	90	
	Young	86	4
	Quiescent	69	24

TABLE 6. DNA SYNTHESIS INHBITION BY SURFACE MEMBRANES ISOLATED FROM QUIESCENT CELLS

Exp.	Cell membranes used for protein extraction	Percent labeled nuclei	Inhibition (Percent of control)
1	Control	58	
	Young	57	0
	Quiescent	27	53
2	Control	56	
	Young	56	0
	Quiescent	38	32
3	Control	78	
	Young	78	0
	Quiescent	47	44

TABLE 7. INHIBITION OF DNA SYNTHESIS BY PROTEINS EXTRACTED FROM ISOLATED SURFACE MEMBRANES

Our aim now is to determine whether the senescent and quiescent inhibitors are the same or different proteins. If they are the same, we have to determine why they are expressed constitutively in senescent and not quiescent cells. If they are distinctly different proteins, we must determine whether they are regulated by similar or different mechanisms and if they interact in any way.

Cell extract (ug/ml)	Percent Inhibition		
	Preparation No.		
Quiescent cell	1	2	3
25	-	12	0
50	-	12	5
80	40 ± 7	-	25
100	-	60	39
Senescent cell			
40			7
80			20
160			55

TABLE 8. INHIBITION BY CELL EXTRACTS

While trying to determine the actual mechanisms involved in normal growth control, we have also attempted to gain an idea of the degree of complexity of the mechanisms by studying the long-term proliferation of various human somatic cell hybrids. We had previously found that fusion of normal cells with a variety of immortal cell lines yielded hybrids that ceased division after as many as 70 population doublings (PD) in vitro (10). This demonstrated that limited growth was a dominant phenotype in hybrids and, conversely, that the phenotype of immortality was recessive. This led us to hypothesize that immortal cells arise as a result of recessive changes in the growth control mechanisms of the normal cell. If more than one such change could result in an immortal cell, since immortality is recessive, fusion of various immortal cell lines with each other would allow for complementation of the different changes and yield hybrids having limited division. We therefore fused a variety of immortal cell lines with each other. In some cases only immortal hybrids were obtained, indicating that immortalization in the particular parents involved had occurred by the same processes. However, in other fusions we obtained hybrids that ceased division after as many as 65 PD (10), indicating that since different recessive changes had led to immortalization of the parent lines involved, complementation of these changes in the hybrids resulted in limited proliferation. To date, we have separated 15 different cell lines into 3 complementation groups (Table 9). Cell lines have been assigned to group C on the basis that they complement cell lines in both groups A and B. It remains to be determined if the lines in this group C indeed belong in but one group, or whether they fall into a number of other groups. The data so far indicate that 1) SV40 virus induces the same recessive changes in human cells of varied origin, to cause immortalization. This is based on the observation that all hybrid clones obtained from fusions of SV40 transformed fibroblasts with each other or with SV40 transformed keratinocyte or amnion cells, are immortal. 2) Different DNA tumor viruses immortalize human cells by different processes, since the adenovirus transformed line (293) complements SV40 transformed cells (639). 3) Cell type (epithelial versus fibroblast), embryonal layer of origin (ecto, meso, endodermal) and type of tumor (lung versus skin, carcinoma versus sarcoma) do not affect complementation group assignment. 4) Cell lines that contain an active Ki-ras oncogene always fall into to Group C indicating that activation of this oncogene may be involved in immortalization. However, it is also possible that the presence of the oncogene could be merely coincidental.

		Fusions				Group Assigned
		A		B	C	
		639	1080	HeLa	143B	
639 fibroblast		+	+	-	-	A
VA13 fibroblast		+	ND	ND	ND	A
847 fibroblast	(SV)	+	"	"	"	A
wtB keratinocyte		+	"	"	"	A
A268IV amnion		+	"	-	"	A
293 kidney	(Ad)	-	ND	ND	ND	B/C
HT1080 *N-ras fibro		+	+	-	-	A
108021 fibro	(S)	+	+	ND	ND	A
143B *K-ras osteo		-	-	-	"	C
HeLa cervical		-	-	+	-	B
A1698 *K-ras bladder	(C)	-	ND	-	"	C
5637 bladder		ND	"	-	"	A/C
J82 bladder		-	"	+	"	B
A2182 lung *K-ras		-	"	-	"	C
T98G glioblastoma		-	ND	+	ND	B

(Table 9 continued)

+ = all immortal hybrids
SV = SV40 transformed
Ad = adenovirus transformed
S = sarcoma
C = carcinoma
* = activated ras oncogene

TABLE 9. COMPLEMENTATION GROUP ASSIGNMENT

These results are encouraging because we have found a limited number of complementation groups. We therefore know that it is feasible to determine exactly what changes lead to immortality. If we had found many complementation groups, identifying the numerous events leading to immortality would have been a much more difficult task.

REFERENCES

1. Hayflick,L.: 1965, Exp. Cell Res. 37, pp. 614-636.
2. Norwood,T.H., Pendergrass,W.R., Sprague,C.A.,and Martin,G.M.: 1974, Proc. Natl. Acad. Sci. USA 71, pp. 223-236.
3. Burmer,G.C., Motulsky,H., Ziegler,C.J., and Norwood,T.H.: 1983, Exp. Cell Res. 145, pp. 79-84.
4. Drescher-Lincoln,C.K., and Smith,J.R.: 1983, Exp. Cell Res. 144, pp.455-462.
5. Burmer,G.C., Zeigler,C.J., and Norwood,T.H.: 1982, J. Cell Biol. 94, pp. 187-192.
6. Drescher-Lincoln,C.K., and Smith, J.R.: 1984, Exp. Cell Res. 153, pp. 208-217.
7. Pereira-Smith,O., Fisher,S., and Smith,J.R.: 1985, Exp. Cell Res. (in press).
8. Rabinovitch,P.S., and Norwood,T.H.: 1980, Exp. Cell Res. 130, pp. 101-109.
9. Burmer,G.C., Rabinovitch,P.S., an Norwood,T.H.: 1984, J. Cell Physiol. 118, pp. 97-103.
10. Pereira-Smith,O.M., and Smith,J.R.: 1983, Science 221, pp. 964-966.

STUDIES ON GENE STRUCTURE AND FUNCTION IN AGING: COLLAGEN TYPES I AND II AND THE ALBUMIN GENES.

D. Gershon*, K. Kohno**, G.R. Martin** and Y. Yamada**
*Dept. of Biology, Technion-Israel Institute of Technology, Haifa, Israel and ** Laboratory of Developmental Biology and Anomalies, National Institute of Dental Research, NIH, Bethesda, MD 20205.

ABSTRACT. The possible occurrence of age-related alterations in gene structure and methylation was investigated in liver DNA of young and old rats. The DNA was digested with PstI, EcoRI, BamHI and HindIII. Southern blot analysis and hybridization with probes for the collagen type I, collagen type II and albumin genes revealed no age-associated changes in structure. Digestion with HpaII and MspI showed the same degree of low methylation in the albumin gene in young and old rats. The collagen type I gene showed a considerable reduction in methylation in the old rat as revealed by higher sensitivity to HpaII and MspI restriction enzymes.

INTRODUCTION.

There are several lines of suggestive evidence that alterations in gene expression may be a contributing factor to the deterioration in the adaptive capacity of senescent organisms. For instance, Bezooijen and and Knook (1) provided evidence that albumin synthesis declined in hepatocytes derived from middle-aged and old rats up to the ages of 24-27 months. However, in animals aged 30 months or over they detected a considerable increase in the rate of albumin synthesis in hepatocytes. This unexplained increase could possibly be due to defective control of the albumin gene expression. Adelman (2) and Jacobus and Gershon (3) found that in many cases old animals had altered response to metabolic and hormonal stimuli as reflected in the patterns of synthesis of inducible enzymes such as glucokinase, tyrosine aminotransferase and ornithine decarboxylase. These altered patterns can be attributed, at least in part, to alterations in the control of the expression of the pertinent genes. The response of peritoneal macrophages from old animals to specific stimuli, such as opsonized zymosan, which act through defined receptors is considerably altered (Lavie and Gershon, in preparation). This alteration is manifested in considerably reduced levels of production of superoxide.

The production of α_2-globulin declines in livers of old rats as a result of a gradual decrease in transcription of the gene encoding this

B. Pullman et al. (eds.), Interrelationship Among Aging, Cancer and Differentiation, 143–148.

protein (4). Amplification of a DNA sequence situated in the human genome between Alu-repeat clusters occurs in lymphocytes of senescent but not young humans (5,6). Such amplification was not found in other human tissue (5). Earlier work showed that the frequency of sister chromatid exchange decreased with age in mammalian cells both in vitro and in vivo (7). Klass et al. (8) have described in the nematode Caenorhabditis elegans an age-dependent exponential increase in the proportion of methylated cytosine in the DNA. Furthermore, an age-associated increase in single strand breaks and reduced in vitro transcriptional capacity of C.elegans DNA was also reported (8).

Most of the above mentioned studies were conducted at the whole genome level. Studies of individual genes may yield more revealing information regarding the role of the modulation of gene expression in aging. It was, therefore, decided to conduct a series of studies employing restriction enzyme cleavage and Southern blot analysis in DNA of young and old rats with probes derived from the albumin, collagen type I and collagen type II genes. The former two are normally expressed in the adult liver while the latter, being cartilage specific, is not.

MATERIALS AND METHODS.

Young (5-8 months) and old (25-28 months) Wistar and Fisher 344 rats were used in this study. The probe for the type I collagen gene (designated α1R1) was a 1 Kb c-DNA fragment from the 3' region of the gene (9). The probe for the rat collagen type II gene was a 1.2 Kb genomic clone from the promoter region of the gene (10). It was prepared and processed as described elsewhere (10). The albumin gene probe was a 600 base PstI fragment derived from c-DNA clone RSA 57 (11). The RSA 57 clone covers exons 4-7 of the albumin gene (11,12).

High molecular weight DNA was prepared according to Blin and Stafford (13). DNA electrophoresis in agarose gels, transfer to nitrocellulose paper and hybridization with nick translated probes was carried out according to Maniatis et al.(14).

RESULTS.

DNA from livers of young and old rats was cleaved with restriction enzymes PstI, EcoRI and BamHI. It was then subjected to Southern blot analysis following hybridization with the probes derived from collagen type I, collagen type II and albumin genes. This analysis revealed no differences in hybridization patterns between DNA of young or old animals (figure 1a-c).

The extent of methylation of -C-C-G-G- sites in the 3' region of the collagen type I and in the region of methylation sites 2,3 and 4 (12) of the albumin gene was estimated in DNA derived from young and old animals. This was achieved by determination of the sizes of the fragments which hybridized with appropriate probes after digestion with the methylation specific restriction enzymes HpaII and MspI. As can be seen in fig. 2a the collagen type I gene, at least in the 3' region, contains

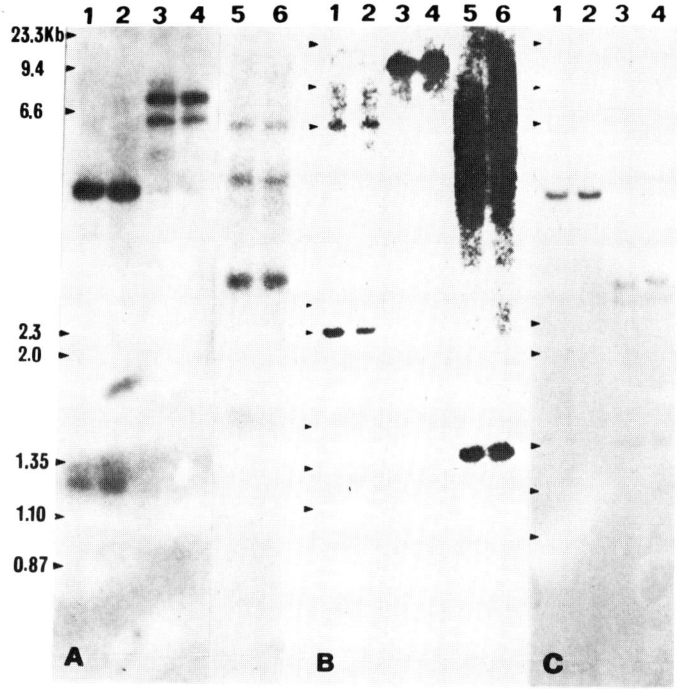

Fig. 1. Restriction analysis of young and old rat liver DNA with probes from collagen type I, collagen type II and albumin genes. High molecular weight liver DNA was prepared according to Blin and Stafford (13). The DNA of 3 animals of each age was pooled. Southern blot and hybridization with nick-translated probes were carried out according to Maniatis et al (14). a) 1 Kb c-DNA fragment from the 3' region of the rat collagen type I gene (9) was hybridized to young and old DNA digested by PstI (lanes 1,2), EcoRI (lanes 3,4) and BamHI (lanes 5,6). b) 1.2 Kb genome fragment from the promoter region of the rat collagen type II gene (10) was hybridized to young and old rat liver DNA digested by PstI (lanes 1,2), EcoRI (lanes 3,4) and BamHI (lanes 5,6). c) 0.6 Kb PstI fragment of c-DNA clone RSA 57 (11) of the rat albumin gene hybridized to young and old rat liver DNA cleaved by PstI (lanes 1,2) and EcoRI (lanes 3,4).

considerable methylation of the inner cytosine of -C-C-G-G- sequences. This is indicated from the poorer digestion with HpaII as compared to that obtained with MspI. More interestingly, the larger HpaII fragments (7-14 Kb) and MspI fragments (1.5-2.4 Kb) which are seen in DNA from young liver, are digested into smaller fragments in "old" DNA (fig.2a). This is particularly obvious in the MspI digest (fig.2a, lanes 3,4) in which a 2.1 Kb fraction disappears and a 1.6 Kb fraction diminishes in size in the "young" preparation while the "old" preparation contains additional 1.2 and 0.5 Kb fragments. It should be noted that the "young" DNA preparation was not, for any trivial reasons, less sensitive to digestion with HpaII and MspI than the "old" preparation, because it was

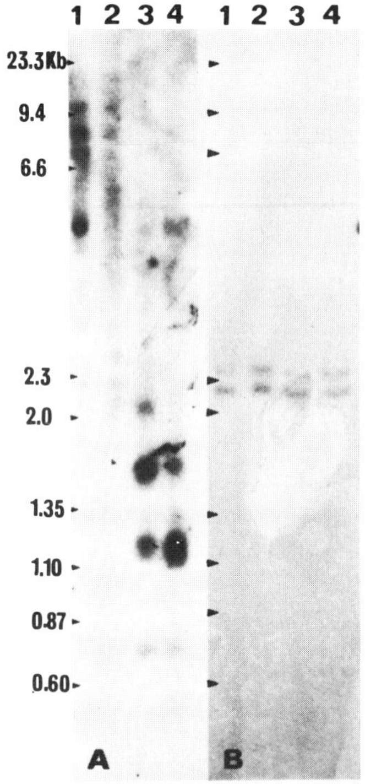

Fig. 2. Methylation state of rat collagen type I and albumin genes in young and old rat liver DNA. High molecular weight rat liver DNA was prepared according to Blin and Stafford (13). Southern blot and hybridization with the appropriate probes were carried out according to Maniatis et al. (14). a) 1 Kb c-DNA fragment from the 3' region of the collagen type I gene hybridized to DNA from young and old livers digested with HpaII (lanes 1,2) and MspI (lanes 3,4). b) 0.6 Kb PstI fragment of c-DNA clone RSA 57 (11) of the rat albumin gene hybridized to liver DNA from young and old rats digested with HpaII (lanes 1,2) and MspI (lanes 3,4).

cleaved very well in the albumin gene with the same enzymes (fig. 2b). These findings mean that in the DNA of old liver the 3' portion of the collagen type I gene is notably less methylated than it is in the DNA of young liver. On the other hand, the portion of the albumin which was studied by us, appears to be either considerably undermethylated or unmethylated to the same degree in "young" and "old" DNA. This conclusion is derived from the fact that the restriction fragments obtained with HpaII and MspI appear to be identical in their electrophoretic pattern in "young" and "old" preparations (fig. 2b).

DISCUSSION.

The results presented above indicate that no gross rearrangements appear with age in various regions of three separate genes in the liver. This conclusion is based on an analysis which utilized a few restriction enzymes possessing different specificities. Preliminary studies also indicate no age-related differences in the myc gene. Similar analysis of several more genes, including oncogenes which are prone to rearrangements, should be pursued before final conclusions are drawn regarding the possibility that structural changes in genes play a role in aging.

It is interesting that the collagen type I gene becomes undermethylated with age in the 3' region. it is not clear whether methylation in the 3' region plays a role in the control of gene expression (15,16). It is clear, however, that the collagen genes are expressed with a relatively high degree of methylation. McKeon et al. (17) and Fernandez et al. (18) have demonstrated that in chick embryo the methylation pattern of two regions of the pro $\alpha 2$(I) collagen gene was identical in several tissues regardless of whether or not this gene was expressed in them. The latter authors (18) have found, however, that the pro $\alpha 1$(II) collagen gene was undermethylated in chondrocytes as compared to other tissues in which the gene was not expressed. Yet in chondrocytes dedifferentiated by treatment with either 5-bromodeoxyuridine or retinoic acid in which collagen $\alpha 2$(II) synthesis ceases, there was no increase in methylation.

The fact that we obtained enhanced cleavage with MspI in the collagen type I gene requires special attention: MspI does not cleave in the -C-C-G-G- sequence if the external cytosine is methylated (19). For instance, it was claimed that unique MspI sites in A_γ and G_γ human globulin genes, as well as the chick collagen 2(I) in sperm cells, are methylated in the external cytosine of a CCGG sequence (19,17). Keshet and Cedar (20) have shown, however, that in such instances where MspI does not cleave at this sequence, which has an internal methylated cytosine, it is not usually due to methylation of the external cytosine but due to special flanking sequences which are G-C rich. It is thus most probable that MspI recognition is affected by DNA secondary structure. In our case the "old" DNA was probably cleaved at a specific site in which either an external cytosine is demethylated or, more likely, the secondary structure is altered to allow for MspI recognition. The latter possibility may be of general interest in aging research and should be further investigated.

Our finding that the state of very low methylation of the albumin gene does not change with age is also of interest. As mentioned in the Introduction, Bezooijen and Knook reported a considerable increase in albumin synthesis in rats 30 months or older following a decline in animals up to age 27 months. If the degree of methylation is directly related to gene expression, one would thus expect increased methylation of this gene in livers of 24-27 month old rats, and possibly demethylation in animals over the age of 30 months. We studied animals of up to 27 months of age and did not observe increased methylation in the albumin gene. Further analysis of specific sites of methylation, especially in the promoter region of the various genes in different tissues, is required before a clearer picture emerges with regard to the control of gene activity in old animals. This will have to be combined with estimation of the amount of specific mRNAs present in cells of animals of various ages. Another possibility should be explored that, like a considerable number of other enzymes (21), the methylases involved in the methylation of deoxycytidine decline in activity with age. Such a decline may explain some of our findings on the lower state of methylation in the collagen type I gene and the lack of increased methylation in the albumin gene in the liver of aging rats.

REFERENCES.

1) C.F.A. van Bezooijen and D.L. Knook, in Liver and Ageing, D. Platt ed., Schattauer verlag, Stuttgart, pp. 227-235 (1977).
2) R.C. Adelman, in Handbook of the Biology of Aging, C.E. Finch and L. Hayflick eds., Van Nostrand Reinhold Co., New York, pp. 63-72 (1977).
3) S. Jacobus and D. Gershon, Mech. Age. Dev. 12, 311 (1980).
4) A. Roy, T.S. Nath, N.M. Motwani and B. Chatterjee, J. Biol. Chem. 258, 10123 (1983).
5) R.J. Shmookler-Reis, C.K. Lumpkin Jr., K.T. Riabowol and S. Goldstein, in Cold Spring Harbor Symposia on Quantitative Biology, C.S. Harbor Laboratory Vol. XLVII, pp. 1135-1139 (1983).
6) R.J. Shmookler-Reis, C.K. Lumpkin, Jr., J.R. McGill, K.T.Riabowol and S. Goldstein, Nature 301, 394-396 (1983).
7) E.L. Schneider, D. Kram, Y. Nakanishi, R.E. Monticone, R.R. Tice, B.A. Gilman and M. Nieder, Mech. Age. Dev. 9, 303-310 (1979).
8) M. Klass, P.N. Nguyen and A. Decharvigny, Mech. Age. Dev. 22, 253 (1983).
9) C. Genovese, D. Rose and B. Kream, Biochem. 23, 6210 (1984).
10) K.Kohno, M. Sullivan and Y. Yamada, J. Biol. Chem. (1985, in press).
11) T.D. Sargent, J.R. Wu, J.M. Sala-Trepat, R.B. Wallace, A.A. Reyes and J. Bonner, Proc. Natl. Acad. Sci. USA 76, 3256 (1979).
12) M. Vedel, M. Gomez-Garcia and M.Sala-Trepat, Nucl. Acid. Res. 11, 4335 (1983).
13) N. Blin and D.W. Stafford, Nucl. Acid Res. 3, 2303 (1976).
14) T. Maniatis, E. Fritsch and J. Sambrook, in Molecular Cloning, Cold Spring Harbor Laboratory, 1982, pp. 382-389.
15) W. Doerfler, Ann. Rev. Biochem. 52, 93 (1983).
16) A. Razin and M. Szyf, Biochim. Biophys. Acta 782, 331 (1984).
17) C. McKeon, H. Ohkubo, I. Pastan and B. de Crombrugghe, Cell 29, 203 (1982).
18) M.P. Fernandez, M.F. Young and M.E. Sobel, J. Biol. Chem. 260, 2374 (1985).
19) L.H.T. van der Ploeg, J. Graffen and R.A. Flavell, Nucl. Acid. Res. 8, 4563 (1980).
20) E. Keshet and H. Cedar, Nucl. Acid Res. 11, 3571 (1983).
21) D. Gershon, Mech. Age. Dev. 9, 189 (1979).

RETROVIRUS-LIKE GENE FAMILIES IN NORMAL CELLS: POTENTIAL FOR AFFECTING CELLULAR GENE EXPRESSION

E. Keshet, A. Itin and G. Rotman
Dept. of Virology
The Hebrew University-Hadassah Medical School
P.O. Box 1172
Jerusalem, Israel

ABSTRACT. DNAs of normal animal cells contain gene families that are distinguished by many properties which are also shared with retroviruses and transposable elements, but are, nevertheless, unrelated to retroviruses (i.e. there is no nucleic acid homology with known viruses). It has recently been shown that cellular retrovirus-like elements may occasionally act as natural mutagens. In order to assess the potential of retrovirus-like elements to affect expession of cellular genes it is necessary to expose all resident retrovirus-like families, study their involvement in DNA-rearrangements and analyze their control elements. We addressed these issues in respect to a specific 'retrovirus-like' family, designated VL30, of which over one hundred copies are dispersed throughout the mouse genome. We have shown that VL30 elements contribute to the overall genomic fluidity through participation in recombinations. These recombinations may juxtapose cellular sequences next to VL30 LTRs and may also create new LTR structures from the pool of cellular LTR units. Certain resident LTR units possess transcriptional promotor and enhancer activities that are more efficient than those of a strongly transforming retrovirus. The notion that additional retrovirus-like families still await exposure has been demonstrated here by the exposure of a novel retrovirus-like gene family.

INTRODUCTION

Chromosomes of normal, uninfected animal cells contain nucleotide sequences closely related to those of infectious retroviruses. These genetic elements, designated 'endogenous proviruses', are inherited as stable Mendelian genes, are evolutionarily unstable, and vary greatly in expression (1). Endogenous proviruses with known exogenous

B. Pullman et al. (eds.), Interrelationship Among Aging, Cancer and Differentiation, 149–161.

counterpart account, however, only for a small fraction of the cellular 'retrovirus-like' genetic information. Additional DNA elements reside in normal cells that do not share nucleotide sequences with known retroviruses but, nevertheless possess distinct retrovirus-like properties. Notable are the striking similarities in the structural organizaiton, which are suggestive of similar strategies for gene expression, reverse transcription and integration (2). The lack of cross-reactivity with known retroviruses hinders the exposure of retrovirus-like gene families. It is thus likely that the overall cellular repertoire of retrovirus-like genetic elements has been exposed only parially. The genetic origins of retrovirus-like elements are not known. They may either represent descendents of once-competent, unidentified retroviruses that have been inserted into germ cell DNAs, or, alternatively, they may have evolved through recombinations between cellular sequences (3). A great deal of consideration has recently been given to the possibility that cellular retrovirus-like elements, irrespective of their inherited role (if any), may act as natural mutagens. That is, the can affect the activity of cellular genes in a non-programmed fashion. This thesis stems from two features of retrovirus proviruses that are apparently shared by resident retrovirus-like element. Firstly, the striking resemblance to eukaryotic transposable elements and, secondly, the high degree of fluidity of their retroviral genomes. Comparative structural analysis of eukaryotic transposable elements and proviruses revealed that the two types of genetic elements share the same sequence arrangements, particularly of the LTR units and putative primer binding sites used during reverse transcription (4-6). Moreover, the two types of elements may also share a considerable sequence homology (7). Functionally, there are examples for the re-insertions of both endogenous proviruses (8) and retrovirus-like DNA units (9) into new cellular loci, in the absence of exogenous infection. In general, the tendency of retrovirus-like elements to self-amplification while retaining their dispersed distribution pattern is likely to result in the occupancy of new cellular loci. It has been shown that retrovirus-like elements may affect cellular genes adjacent to the site of insertion either negatively, i.e. insertional inactivation (9), or positively, through the insertion of transcriptional promotor or enhancer functions. Activation of an oncogene by insertion of a virus-like DNA element has recently been shown (10). Fluidity of retroviral genomes has been shown to occur through frequent recombinations with cellular retrovirus-related genes and, at a much lower frequency, with non-viral cellular genes. Well-known examples are the emergence of novel pathogens through recombination with endogenous retrovirus-related sequences (11, 12) and the transduction of cellular oncogenes onto retroviral genomes (13), respectively.

We have examined the possible contribution of retrovirus-like elements, which are natural residents in animal cells, to the overall genomic fluidity, and their potential role in affecting expression of cellular genes. Evaluation of these putative roles requires the exposure of the entire cellular pool of retrovirus-like families and study of the involvement of each family in DNA rearrangements, such as transpositions and recombinations, that may also lead to juxtapositioning of the LTRs next to otherwise remote genes. Also, it is necessary to characterize the efficiency and specificity of the control elements present in the LTR units. We have addressed these issues with respect to a particular murine multigene family designated VL30 (14), a representative of a dispersed retrovirus-like family (100-200 copies per haploid genome) with no apparent homology to known retroviruses. These elements constitutively transcribe 30S RNA molecules carrying a cis-acting packaging signal recognized by c-type retroviruses (14-16). We have previously shown that VL30 elements possess all structural hallmarks in common with proviruses and transposable elements (17-19) and that different species of the genus _mus_ and different strains of _mus musculus_ are highly variable in respect to VL30 copy number and distribution patterns (20, 21). In this study we will focus on VL30-mediated recombinations as an additional mean for creating new genomic linkages and on analysis of the transcriptional promotor and enhancer activities possessed by diverse VL30-associated LTRs. We shall also describe the exposure of a novel retrovirus-like gene family.

EXPERIMENTAL PROCEDURES

30S RNA was obtained from purified virions released by Balb/c mouse cells chronically infected with MuLV. Virion preparations were composed of roughly equimolar amounts of MuLV RNA and VL30 RNA encapsidated in MuLV virions. Complementary DNA synthesized from this RNA was used as the hybridization probe for screening mouse genomic library. The mouse library was constructed from Balb/c embryonic DNA, partially digested with EcoRI and digested with Charon 4A lambda DNA vector. Colonies that gave positive signals in an _in situ_ plaque hybridization assay were selected and the VL30 DNA clones were unambiguously identified as those recombinant phages detecting 30S RNA in northern blots prepared from polyadenylated RNA derived from uninfected Balb/c cells (17).

Clones containing complete VL30 DNA units were identified through heteroduplexing independently cloned VL30 DNAs with each other. Only clones containing the complete 5.2-5.4 kb long element (and mouse flanking sequences at both sides) were selected for further studies.

Physical maps of restriction enzyme cleavage sites were obtained for some representative VL30 clones. The approximate locations of the LTRs were initially determined by heteroduplexing followed by fine physical mapping with restriction endonucleases and detection of symmetrical cuts, and finally by sequencing.

To aid analysis of VL30 sequences that are present in molecular contexts other than 'standard' VL30 units, a battery of subgenomic VL30 probes was obtained by subcloning relatively short VL30 segments in either plasmid or M13 phage vectors. Colonies with incomplete representation of VL30 sequences were identified by screening the genomic library with multiple subgenomic VL30 probes.

In order to indentify genomic linkages with defined non-VL30 sequences, phage replica were obtained on multiple nitrocellulose filters, challenged with the respective hybridization probes and screened for overlapping hybridization signals. A VL30 LTR unit was also subcloned in the VX miniplasmid and homologous recombination was used to facilitate isolation of LTR-containing genomic DNA fragments.

Analysis of LTR transcriptional activity was carried out by constructing recombinant plasmids containing different LTR fragments linked to an assay gene (CAT) devoid of its own promotor and enhancer and measuring CAT activity following transfection of DNAs onto mouse cells.

RESULTS

Our approach in analyzing the genomic fluidity of VL30 elements was to carry out a comparative structural analysis of several members of the VL30 family randomly cloned from normal mouse DNA. The rationale of this approach was that the current status of the genome arrangement of individual VL30 units is the consequence of DNA rearrangements that have occurred in the past and accummulated in the germ line. We further reasoned that comparison of diverse VL30 units will shed light on the mechanism(s) responsible for their diversification. Since gross rearrangements of VL30 elements were sought, we heteroduplexed a 'prototypic' VL30 DNA unit with each newly isolated VL30 DNA. This procedure enabled visualization of relatively large (>100 bp) segments on non-homologous DNA that were incorporated into VL30 sequences (See Fig. 1).

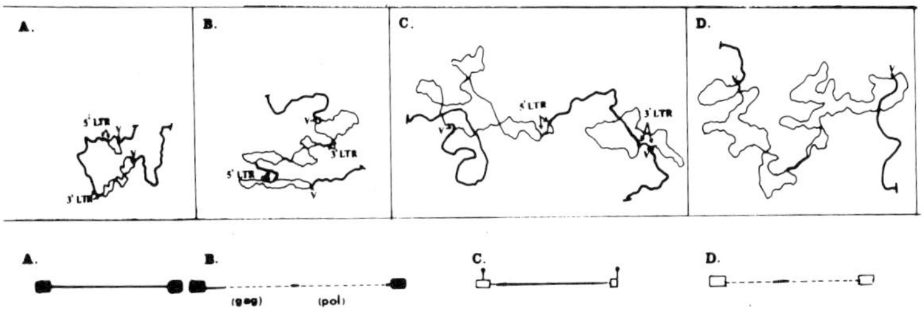

Figure 1. VL30 DNA recombinants visualized by heteroduplexing. Top figures show tracing of typical heteroduplexes and bottom figures show schematic organization of DNA elements. Filled boxes and solid lines symbolize VL30 LTRs and internal sequences, respectively. Open boxes and broken lines symbolize non-VL30 LTRs and internal sequences, respectively.

When two prototypic VL30 DNA units were heteroduplexed with each each other, an uninterrupted duplex of approximately 5.4 kbp was observed (Fig. 1A). This double stranded region represents complete VL30 units bounded by roughly 600 bp long terminal repeats (LTR) and flanked by unrelated cellular sequences. In contrast, when the same prototypic VL30 DNA clone was heteroduplexed with a putative VL30 recombinant, duplexed regions were interrupted by large segments of non-homologous DNA (Fig. 1B). Further analysis by blot hybridization has shown that in the particular clone shown in Fig. 1B both VL30 LTR are present, but that the majority of VL30 internal sequences were replaced by other DNA sequences. The non-VL30 sequences were subsequently identified as MuLV related gag and pol sequences (22). This particular clone was initially cloned by a linkage-selection procedure, in which library phages were challenged-with both VL30 and MuLV specific probes, and those recombinant clones that hybridized with both probes were selected. This approach was based on the assumption that since co-packaging of heterologous RNA species is thought to be the first step in retrovirus-mediated recombination (23), recombination with packageable RNA molecules should occur at high frequencies . Another type of VL30 recombinant is illustrated in Fig. 1C. This recombinant is reciprocal in its sequence arrangement to the one shown in 1B, namely, VL30 LTRs were replaced by non-homologous LTR units whereas the majority of internal sequences were indistinguishable from those present in the prototype VL30 unit. Hybridization experiments suggested that the acquired LTR units are distinct from LTR sequences of known retrovirus-related elements. Subsequent sequencing of the LTR substantiated this conclusion. We suspected, therefore, that this DNA represnets a recombinant between VL30 and a yet unrecognized retrovirus-like family. In order to test this possibility we challenged the genomic library with a specific subgenomic probe derived from the non-VL30 portion. Several DNA elements were cloned and were subsequently shown to constitute a novel retrovirus-like gene family. A heteroduplex between a representative of this family, designated B-3, and a prototype VL30 DNA is shown in Fig. 1D. With the exception of a short cross-reactive segment, these two genomes seem to be unrelated. Preliminary sequence data have indicated that this new element, apart from its unique LTR units, is also unique in respect to its primer binding site sequence (PBS). Members of this family contain the sequence (5') TGGAGGTTCCACTGAGAT which corresponds to the 3' end of tRNAgln.

Points of recombination grossly suggested by the heteroduplex analysis could also be demonstrated at the resolution level of sequence determination. Experimentally determined sequences were subjected to dot-matrix comparison in order to identify putative recombination points. An example is shown in Fig. 2.

As can be seen, VL30 DNA (represented by VL-5) and a member of the new retrovirus-like family described above (represented by VL-C3) have unrelated nucleotide sequences downstream of the 5' LTR unit. On the other hand, an apparent recombinant VL30 clone, designated VL-9, shares almost identical sequences with VL-C3, starting 60 nucleotides downstream from its 5' LTR unit, thereby suggesting that this point in VL-9 is a recombination point.

We next examined the possibility that recombinations may also occur within LTR units. The sequences of LTR units of several randomly selected VL30 clones was determined and compared also with a recently published sequence of a VL30 cDNA clone (24). An example of a computer-aided dot matrix comparison is shown in Fig. 3A.

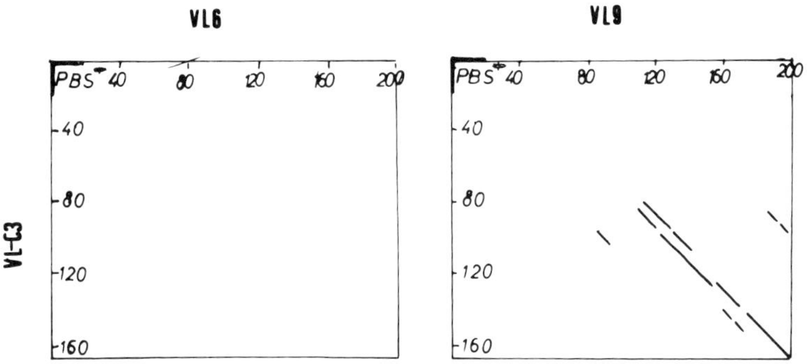

Figure 2. A putative recombination point between VL30 and an unrelated retrovirus-like DNA element visualized by sequence comparison. Sequences were compared by dot matrix analysis at a stringency level of 10 out of 12 matches. Reions shown are downstream from the 5' LTR (starting with pBS^-).

Clearly, LTRs are very different from each other. For each pair of LTRs compared, both stretches of highly homologous sequences and stretches of poorly homologous sequences were detected. The pattern of distribution of homologous LTR sequences is shown in Fig. 3A, and as drawn schematically in Fig. 3B, is consistent with the contention of recombinant LTR units. As shown in Fig. 3B, the sharp points of sequence divergence often coincided with U3-R and R-U5 junctions. From a pair of different VL30 LTRs (VL11 and VL3) subclones were isolated that represent two overlapping but non cross-reactive U3 domains and two overlapping non cross-reactive U5 domains. With the aid of these probes we could demonstrate that a single LTR is composed of two subsets that are of different evolutionary origins (data not shown). These findings are consistent with the thesis that LTRs might be assembled, presumably through recombinations, from the pool of retrovirus-like elements present in normal cells.

Retroviral LTRs possess efficient transcriptional promotors and enhancers. A fundamental question is to what extent the LTR elements abundantly present in normal cells share these capacities of infectious retroviruses. A study of the transcriptional elements present in the population of VL30-associated LTRs was therefore undertaken. Chimeric DNAs were constructed in which either complete or disected LTRs were linked to a test gene devoid of a promotor (or

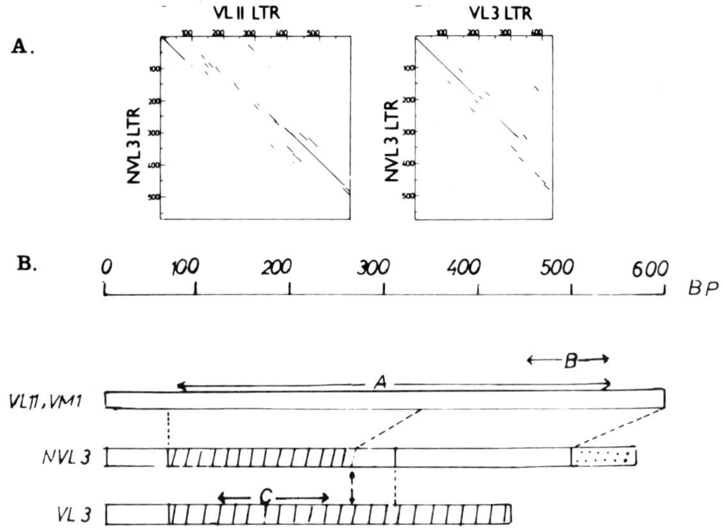

Figure 3. Diversity of VL30 LTR sequences. A. Examples of comparisons of LTR sequences by dot matrix. B. Schematic representation of cross-homologies in certain VL30 LTRs. Homologous DNA blocks were indicated by the use of the same signs. Vertical arrows represent U3-R junctions. VL11, VMi and VL3 and VL30 DNA elements randomly selected form the genomic library. NVL-3 is a VL30 cDNA clone analyzed by Norton et al (24).

with a promotor but lacking an enhancer). Following transfection onto mouse cells, transient activity of the test gene (chloramphenicol acetyl transferase) was measured. In the first series of constructs, whole LTR units were ligated upstream from a promotorless CAT gene. In these constructs, CAT activity is totally dependent upon the presence of a functional transcriptional promotor in the LTR.

As can be seen in Fig. 4, certain VL30 LTRs efficiently promote CAT activity (provided that the LTR was inserted in the same transcriptional polarity as the test gene). S-1 mapping was used to confirm that CAT transcripts were indeed initiated within the LTRs (data not shown). A striking observation was that certain VL30 LTRs are 5-6 times more effective in mouse cells than the LTR of Harvey-MSV, a strongly transforming murine retrovirus. Noteworthy is also the observation that diverse VL30-associated LTRs possess variable transcriptional capacities. Moreover, efficient promotor/enhancer activity could be correlated with the presence of specific U3 sequences.

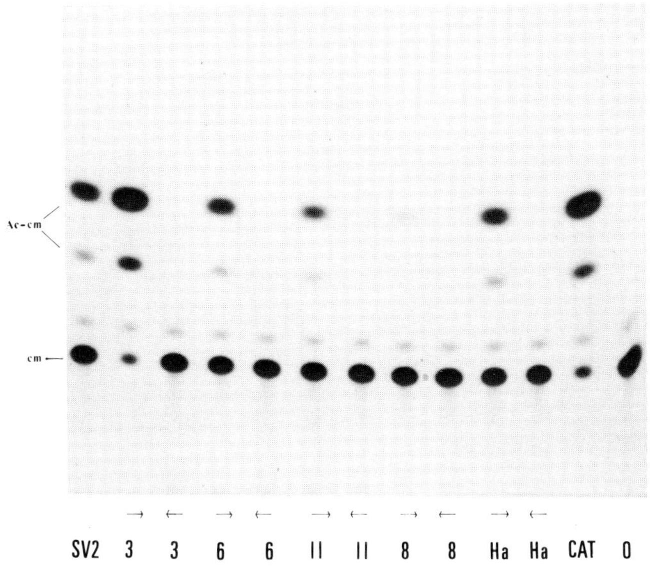

Figure 4. Promotor activity of VL30 LTRs. Promotor activity was analyzed by ligating LTR elements upstream from the coding region of chloramphenicol acetyl transferase (CAT) gene in plasmid pSVO-CAT (25, transfection onto mouse L-cells, and measuring transient enzymatic activity of CAT. Numbers refer to VL30 LTR used, and arrows indicate the orientation of the LTR in respect to CAT. Ha - LTR of Ha-MSV; SV2 - pSV2-CAT (25); CAT - commercial enzyme; O - mock transfection.

The registered CAT activity shown above is necessarily driven by the LTR promotor but may also be augmented by the presence of a transcriptional enhancer in the LTR. We wished, therefore, to determine whether a VL30 LTR contains a functional enhancer. To this end, different plasmid constructs were made, in which VL3 LTR was inserted in both orientations upstream or downstream from a CAT gene promoted by the SV40 early promotor but lacking a transcriptional enhancer.

As can be seen in Fig. 5, VL3 LTR augmented CAT activity even when inserted in an orientation where its own transcriptional polarity is opposite to the CAT gene and even when inserted downstream from the CAT gene. Moreover, an orientation-independent augmentation of enzymatic activity was also exerted by a U3 specific fragment derived from VL3 LTR. These results have indicated that certain VL30-associated LTRs contain both functional promotor and enhancer, the combined activities of which exceed those displayed by a strongly transforming retrovirus.

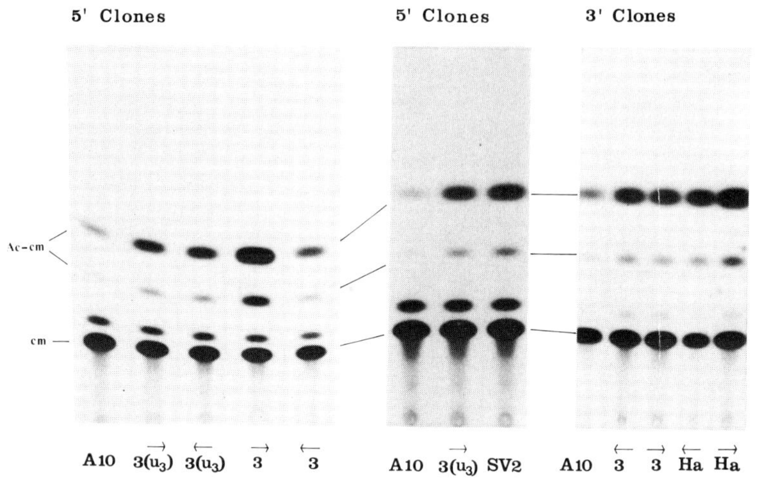

Figure 5. Enhancer activity of VL30 LTRs. LTR sequences were inserted into the plasmid pA10-CAT, in which activity is promoted by the early SV40 promotor. CAT activity of pA10-CAT served as reference for non-enhanced activity. Symbols are as in Fig. 4. 3(U3) refers to a fragment of VL3 LTR that contains U3 sequences only without the promoter region. 5' and 3' clones refer to site of insertion in respect to CAT.

DISCUSSION

In this study we have demonstrated a high level of fluidity in the genome organization of a particular cellular retrovirus-like gene family. Diversification of several members of this family has been shown to result primarily from participation in recombination and the stable deposition of the recombinant elements in the genome. As the apparent recombinant DNA elements were isolated from a normal mouse embryonic gene library, these DNA rearrangements should be viewed as a contribution to the fluidity of the mouse genome, in general. The retrovirus-like family described here is distinguished among other known retrovirus-like families in the sense that it furnishes both embryonic and differentiated cells with abundant RNA transcripts that contain a functional cis-acting packaging signal. These features of VL30 genes are likely to be responsible for the frequent recombination with other packageable RNA transcripts, as retrovirus-mediated recombinations are thought to neccessitate co-packaging of heterologous RNAs (23). Not surprisingly, the recombining partners identified in this study are also retrovirus-like elements that presumably possess packaging signals. More significant, however, in terms of the potential effects on cellular phenotype are recombination with non-viral related cellular sequences. These events, albeit occurring at a much lower frequency, are precedented by the transduction of oncogenes onto retroviral genomes. Interestingly, two oncogenic retroviruses - the Kirstein and Harvey strains of Mouse Sarcoma Virus - which acquired ras sequences from the rat genome also acquired rat VL30 sequences (26). Since these two independent isolates were selected only on the basis of ras transduction, the concomitant recombination with VL30 sequences suggests a possible role for VL30 in facilitating the transduction of the oncogene. It is possible that genetic interaction between VL30 and ras sequences renders the latter packageable and, hence, available for further recombinations. It can be further speculated that recombination of VL30 with cellular genes can generate a potential for transmissibility of cellular genes, in general.

Insertion of retroviral LTRs via viral infection may result in altering the normal regulation of genes adjacent to the insertion site. For resident LTRs to exert similar effects they must participate in a DNA rearrangement that brings otherwise distant cellular sequences to their proximity. In addition to transposition of whole retrovirus-like elements, recombinations may also create novel genomic linkages of LTR units. Insertion of retroviral LTRs has been shown to alter expression of a nearby gene by either promotor insertion or enhancer insertion mechanisms. Studies reported here clearly demonstrate that cellular LTRs possess both promotor and enhancer activities that can activate linked cellular genes to higher levels than those exerted by the LTR of a strongly transforming retrovirus (like Ha-MSV). Linking the LTR of Ha-MSV to a normal cellular oncogene led to its activation to levels sufficient to trigger cellular transformation (27).

Recombination between retrovirus-like elements may also occur within LTR sequences, thereby creating LTR units with altered sequence arrangements. These structural changes may also be accompanied by alteration in function, e.g., in promotor or enhancer activities. Thus, also the reservoir of cellular LTRs should also be viewed as a potentially fluid population. In order to assess the overall potential of retrovirus-like elements to act as mutagens we must recognize all retrovirus-like gene families, study the frequencies of their DNA rearrangements and the regulatory capacities of their LTR. These parameters may be different for different families. The rather incidental encounter of the new retrovirus-like gene family reported here attests to the possibility that additional gene families still await exposure.

REFERENCE

1. Coffin I.M. In: The molecular biology of tumor viruses, part III, RNA tumor viruses (Eds. R.A. Weiss, N. Teich, H.E. Varmus and J.M. Coffin). Cold Spring Harbor Laboratory, 1982, pp. 1109-1203.

2. Temin, H.M. Cell 21: 599-600, 1980.

3. Temin, H.M. Persp. Biol. Med. 14: 11-26, 1970.

4. Eibel, H., Gafner J., Statz A. and Philippsen P. Cold Spring Harbor Symp. Quant. Biol. 45: 609-617, 1980.

5. Elder, R.T., Loh E.Y. and Davis R. W. Proc. Natl. Acad. Sci. USA 80: 2432-2436, 1983.

6. Flavell A. and Ish-Horowitz D. Nature 292: 591-595, 1981.

7. Kugimiya W., Ikenga H. and Saigo A. Proc. Natl. Acad. Sci. USA 80: 3193-3197, 1983.

8. Jenkins N.A., Copeland N.G., Taylor B.a. and Lee B.K. Nature 293: 370-374, 1981.

9. Kuff E.L., Feenstra A., Lueders K.K., SMith R., Hawley N., Hozumi N. and Shulman M. Proc. Natl. Acad. Sci. USA 80: 1992-1996, 1983.

10. Rehavi, G., Givol D. and Canaani E. Nature 300: 607-610, 1982.

11. Chattopadhyay S.K., Cloyd M.W., Linemeyer D.L., Lander, M.R, Rands E. and Lowy D.R. Nature 295: 25-31, 1982.

12. Herr W., and Gilbert W. J. Virol. 50: 155-162, 1984.

13. Bishop J.M. and Varmus H. In: The Molecular Biology of Tumor Viruses Part III, Eds. R.A. Weiss, N. ,Teich H.E. Varmus and J.M. Coffin. Cold Spring Harbor Laboratory 1982, pp. 999-1108.

14. Besmer P., Olshevsky U., Baltimore D. and Fan H. J. Virol. 29: 1168-1176, 1979.

15. Hawk R.S., Troxler D.H., Lowy D., Duesberg P.H., and Scolnick E.M. J. Virol. 25: 115-123, 1978.

16. Sherwin S.A., Rapp U.R., Benveniste R.E., Sen A. and Todaro G.I. J. Virol. 26: 257-264, 1978.

17. Keshet E., Shaul Y., Kaminchik J. and Aviv H. Cell 20: 431-439, 1981.

18. Keshet E. and Shaul Y. Nature 289: 83-85, 1981.

19. Itin A. and Keshet E. J. Virol. 77: 656-659, 1983.

20. Keshet E. and Itin A. J. Virol. 43: 50-58, 1982.

21. Itin A., Rotman G. and Keshet E. Virology 127: 374-384.

22. Itin A. and Keshet E. J. Virol. 47: 174-184.

23. Coffin J.M. J. Gen. Virol. 42: 1-26, 1979.

24. Norton J., Connor J. and Avery R. Nucleic Acid Res. 12: 3345-3460, 1984.

25. Gorman C., Moffat L. and Howard B. Molec. Cell Biol. 2: 1044-1051, 1982.

26. Ellis R.W., Defeo D., Maryak J.M. Young H.A., Shik T.Y., Chang E.H., Lowy D.R. and Scolnick E.M. J. Virol. 36: 408-420, 1980.

27. Chang E., Furth M., Scolnick E. and Lowy D. Nature 297: 479-483, 1982.

The Conformational Effects of UV Induced Damage on DNA

David A. Pearlman, Stephen R. Holbrook and Sung-Hou Kim

Chemical Biodynamics Division, Lawrence Berkeley Laboratory &
Department of Chemistry
University of California, Berkeley, CA 94720

ABSTRACT

Energy minimization techniques are used in conjunction with the results of small molecule crystallographic studies on relevant compounds to propose structural models for photodamaged DNAs. Specifically, we present models both for a DNA molecule containing a psoralen photo-crosslink and for a DNA molecule containing a thymine photodimer. In both models, significant distortions of the nucleic acid helix are observed, including kinking and unwinding at the damage site and numerous changes in the backbone torsion angles relative to their standard conformations.

1. INTRODUCTION

With the advent of more reliable methods of structure prediction in the past few years (1), the rational approach to drug design has come into favor. In the incarnation of this approach relevent to our work, the effects of particular drug molecules on intrinsic cellular structures is gleaned from energy minimization calculations, allowing otherwise unavailable insights into the conformational effects of the drugs on their targets. Clearly, this methodology has great potential usefulness in the study of chemical and physical damage of DNA and in the interaction of DNA with both carcinogens and drugs. A knowledge of the actual structure-activity relationships of carcinogens and anti-cancer agents allows the design of potentially much more effective drugs (2), which can then be tested in clinical settings.

Recently, we have become interested in the conformational effects of UV radiation induced DNA damage. In particular, we have studied both the structural deformations brought about by the formation of thymine dimers, and the changes induced by the formation of psoralen crosslinks, in otherwise "healthy" DNA helices (3). The former type of DNA damage is often associated with high cell mutation rates (4) and when left unrepaired, as in the genetic disorder xeroderma pigmentosa can lead to various skin cancers. On the other hand, the latter DNA aberration has

B. Pullman et al. (eds.), Interrelationship Among Aging, Cancer and Differentiation, 163–171.

been effectively utilized in the treatment of various skin diseases, such as psoriasis and vitiligo, and in cancer chemotherapy (5). A greater understanding of the conformational consequences produced by thymine dimer formation may be of considerable value in understanding the basis of its biological effects and of repair mechanisms. Knowledge of the structure of psoralen crosslinked DNA, could lead to design of even more effective crosslinking reagents. In both cases, the knowledge gained is potentially valuable in understanding the possible interactions of DNA with other drugs, mutagens, and carcinogens.

2. METHODS

Our study has entailed combining molecular model building and energy minimization with X-ray crystallographic structures of model compounds. Direct experimental investigations on the systems of interest, i.e. the lesions discussed above, have not yet been been performed, due both to our present inability to isolate sufficient amounts of the necessary molecular species and to the instability of these systems to such

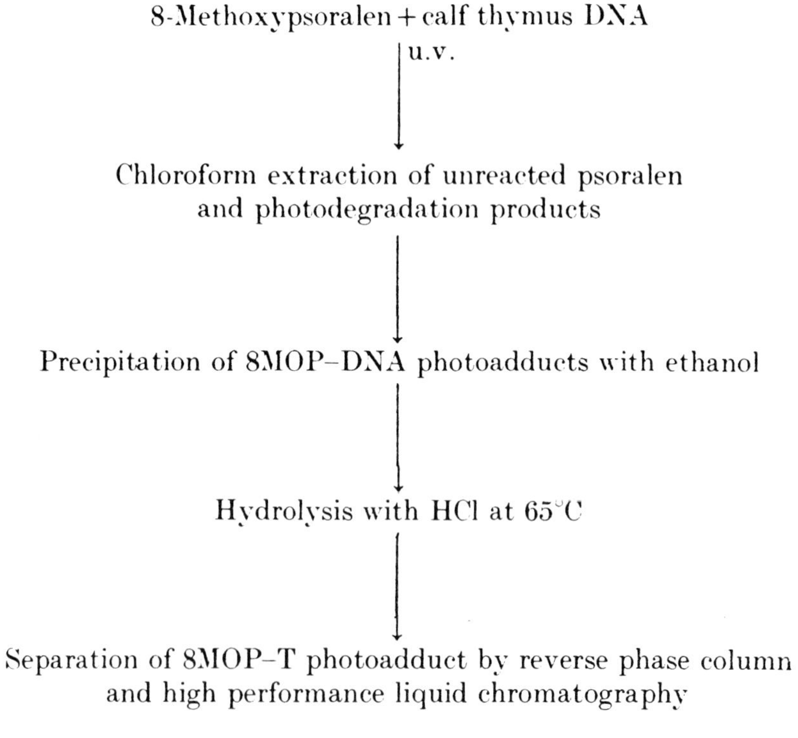

Figure 1: Procedure followed to synthesize and isolate the 8-methoxypsoralen-thymine monadduct from calf thymus DNA reacted with 8-methoxypsoralen. The crystal structure of this monoadduct served as the starting model for the psoralen-DNA lesion site.

experimental probes as X-rays. Thus, for both types of damage, we started with the crystallographically determined geometry of a small molecule complex which could serve as a model for the geometry at the site of damage.

2.1. The Psoralen Crosslink.

The starting model for DNA incorporating a psoralen crosslink was the crystal structure of a cis-syn monoadduct of psoralen isolated from calf-thymus DNA (6). Figure 1 shows the scheme used to isolate the psoralen-thymine monoadduct from a mixture of DNA and psoralen which had been irradiated by UV light. A schematic view of the monoadduct is shown in figure 2b, and the crystallographically determined structure is shown in figure 3a. Since the system we were interested in was the diadduct

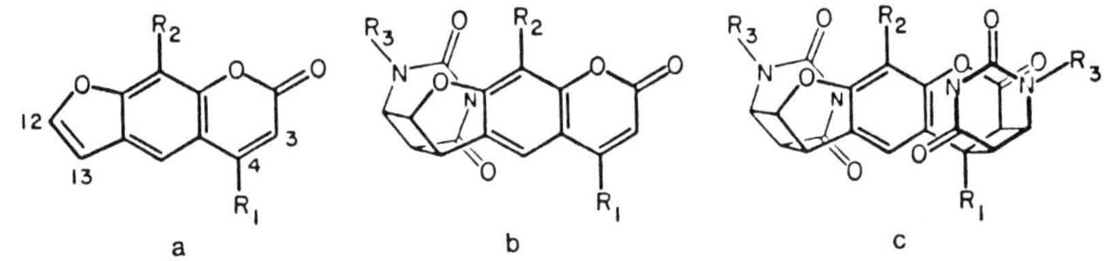

Figure 2: a) A psoralen molecule with the crosslink sites (C3, C4, C12, and C13) labeled; b) A monoadduct of thymine with psoralen having the cis-syn configuration; c) A diadduct of thymine with psoralen having the cis-syn configuration at each linkage. For our work, 8-methoxypsoralen, where R_1=H and R_2=OCH_3, was used. The thymine bases are attached to the furanose rings of the nucleic acid at R_3.

(crosslink) of thymines on opposite strands to psoralen (see figure 2c), it was necessary for us to fit a second thymine base to psoralen by assuming that the geometries about the crosslink regions joining the thymine bases were the same. Once we had derived a model for the diadduct system, we were ready to incorporate this into a DNA oligonucleotide. This was done by least squares fitting a thymine base at the end of a double helical B-form DNA trimer to each of the crosslinked thymines. The resulting structure had a break in each nucleic acid backbone strand. To close these gaps and produce a stereochemically acceptable structure, torsion angle manipulation was performed using the energy minimization package AMBER (7). Subsequently, complete energy minimization was performed on the crosslinked hexamer. A classical empirical energy function (1) was used:

$$E_{total} = \sum K_r(R-R_0)^2 + \sum K_\theta(\theta-\theta_0)^2 + \sum V_n/2\,[1-\cos(n\phi-\gamma)] + \\ + \sum_{i<j} [A_{ij}/R_{ij}^{12} - B_{ij}/R_{ij}^{6} + q_i q_j/\varepsilon R_{ij}] +$$

$$+ \sum_{i<j} [C_{ij}/R_{ij}^{12} - D_{ij}/R_{ij}^{10}] .$$

The summations represent the contributions of bond stretching, angle bending, torsion angle twisting, non-bonded atom pair interactions and hydrogen bonding, respectively. Parameters used in this function are from Weiner et al. (1) and Rao et al. (8). When the hexamer structure had relaxed into an energy minimum, we added two additional base pairs to each end of the structure, and energy minimization was again carried out. The extra base pairs were added to insure that end effects were minimal near the intercalation site, and to help in defining helical deformation parameters such as the kink angle, etc. The final resulting structure was

```
5' CGCAT-ATGCG 3'
        ᒪP┐
3' GCGTA-TACGC 5' ,
```

where P represents the crosslinked psoralen. A canonical B-DNA helix (9) of the same sequence as that above was also subjected to energy minimzation (in two steps, just as for the psoralen model), for purposes of comparison.

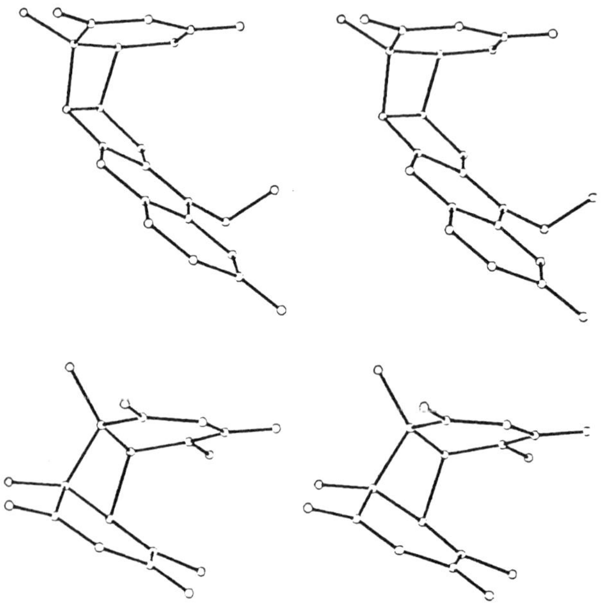

Figure 3: The crystallographically determined geometries of the models used for the starting (before minimization) local geometries at the lesion sites: a) The psoralen monoadduct; b) The thymine dimer.

2.2. The Thymine Dimer

For our model thymine dimer lesion site, we used the crystallographically determined geometry of Camerman and Camerman (10). Unlike that of the psoralen crosslink, the crystallographic model for the thymine dimer structure was not obtained from a DNA mixture, but rather from a solution of thymine bases which had been irradiated under UV light. In a thymine dimer, two crosslinks are formed between a pair of thymine bases, covalently fusing the bases through a cyclobutyl ring. The experimentally determined geometry is shown in figure 3b. A thymine base at the end of a double-stranded B-DNA trimer was least squares fit to each of the crosslinked thymines, in analogy to the procedure adopted for generating the psoralen crosslinked structure. Torsion angle manipulation closed the gaps in the nucleic acid backbone strands, which was followed by full energy minimization on the resulting hexamer using the AMBER program and the same parameters as for the psoralen crosslink structure. Two additional base pairs were added to each end of the minimized hexamer and energy minimization was repeated. The final, energy minimized structure was

3' CGCGT=TCGCG 5'
5' GCGCA-AGCGC 3' ,

where T=T denotes the thymine dimer. As for the psoralen crosslink minimization, a B-form canonical DNA helix of the same same sequence as that of the thymine dimer damaged structure was also minimized, for comparative purposes.

3. RESULTS

3.1. The Psoralen Crosslink

The psoralen crosslink is found to cause significant disruptions in DNA helical geometry. These can be clearly seen in figure 4c, which is a space-filling representation of the psoralen crosslink-DNA model, and in figure 5c, which shows a simplified ribbon drawing of the model. The kink in the helix axis caused by the psoralen crosslink is calculated to be 46.5°. In addition, a large amount of unwinding occurs concommitantly: 87.7° over the entire structure. In fact, the unwinding is so severe at the crosslink site that the helix actually winds 13° in a left-handed sense at this step. This contrasts with a 45° right-handed step in the minimized B-form structure. The large amount of unwinding results from physical constraints on the cis-syn crosslink geometry itself. Perhaps more suprisingly, the base steps sandwiching the crosslink site both also exhibit unwinding (by 8-9° relative to the minimized B-form, rather than compensating the central disruption by overwinding.

On a local conformational level, numerous changes are observed in the torsion angles of the crosslinked thymine residues and in the two nucleotides on the 3' ends of these thymines (11). The sugar puckers in

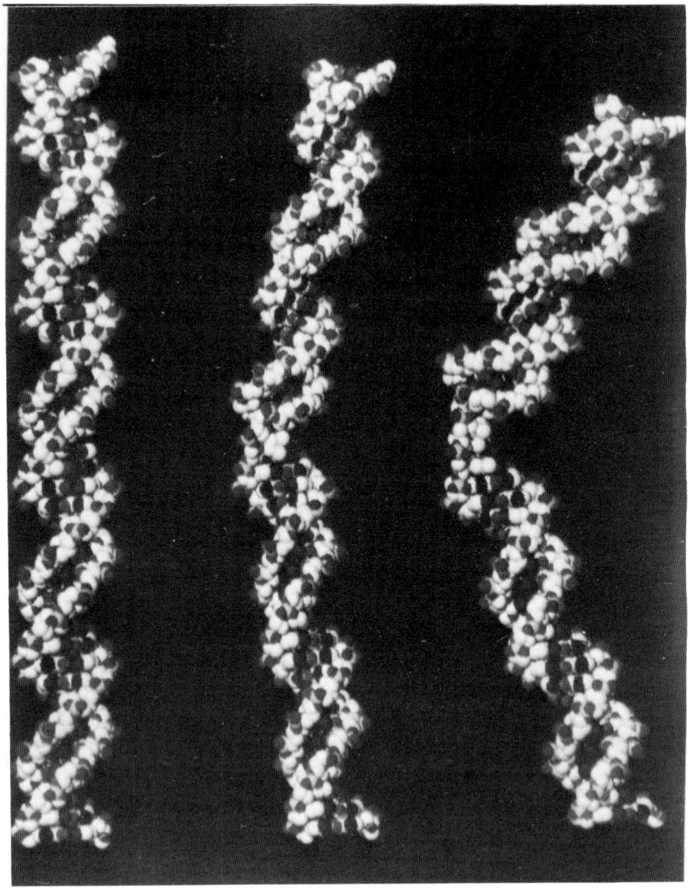

Figure 4: Space-filling representations of the final energy minimized structures considered in this work, generated using computer graphics. a) B-form DNA; b) Double-stranded DNA containing a thymine dimer; c) Double-stranded DNA containing a psoralen crosslink.

these regions are seen to be particularly prone to conformational change. While the the Watson-Crick hydrogen bonds from the crosslinked thymines to their adenine partners are maintained, there are significantly distorted, while the base pairs formed by adjacent residues are normal. The disruptions are most clearly visible in the angles of the hydrogen bonds, which are 10-19° smaller than those for minimized B-form DNA.

3.2. The Thymine Dimer

The changes brought about by the thymine dimer, while large, are not as great as those forced by the psoralen crosslink. This can be seen in the

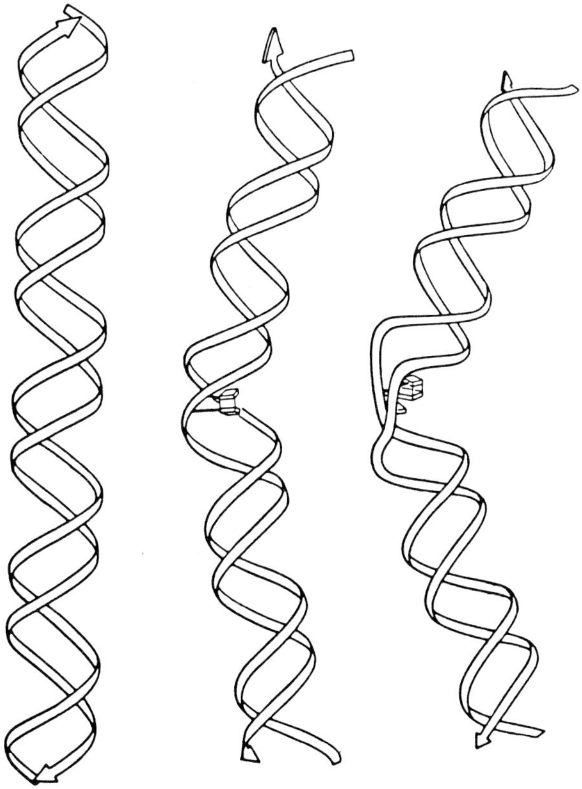

Figure 5: Simplified ribbon drawings of the final energy minimized structures considered in this work. a) B-form DNA; b) Double-stranded DNA containing a thymine dimer; c) Double-stranded DNA containing a psoralen crosslink.

space-filling model and ribbon drawing (figures 4b and 5b, respectively). Specifically, the kink induced by the thymine dimer is 27°, about 3/5 that seen in the psoralen crosslinked DNA, while the total amount of unwinding effected by the dimer is only 20°, less than 1/4 that calculated for the psoralen crossslinked structure. A considerable portion of the unwinding induced at the dimer step is compensated for by overwinding at adjacent steps, but a net unwinding results from the geometries at more distant base steps.

In contrast to the psoralen crosslinked helix, where torsion angle changes are symmetrically distributed about the lesion, here more changes occur in the strand opposite the thymine dimer (11). This is a result of the fact that the bases of the thymine dimer open at an angle which forces the opposing strand to stretch more than the strand incorporating the dimer. Significant conformational rearrangement occurs out to the first base pair past the thymine on the 5' end, and out to the second base pair from the dimer on the 3' end. This is consistent with NMR data suggesting greater distortions on the 3' end of the dimer (12). As for the psoralen structure, the sugar puckers appear to be the structural parameters most sensitive to major conformational change.

The gross overall conformational changes calculated for the thymine dimer damaged helix in our study contrast with the recent energy minimization findings of Rao et al. (8). These researchers, using the same force field, predicted that the incorporation of a thymine dimer causes essentially no kinking or unwinding of the helical axis. The probable reason for the difference between our results and theirs lies in the respective starting models used for minimization. The resulting structures obtainded from macromolecular minimizations are highly dependent on the input geometries. Thus it was likely that since we started with a kinked and unwound model (the monomer crystal structure), our final structure would also be kinked and unwound, while Rao et al., whose initial model was neither kinked nor unwound (B-DNA structure), would not predict such changes. In light of experimental studies which indicate topological unwinding, we feel our results are a more likely representation of physical reality.

4. DISCUSSION

The results obtained here clearly indicate that the UV induced lesions studied induce gross conformational changes in the DNA helices incorporating them. The large magnitudes of these changes make them excellent candidates for recognition sites by cellular repair enzymes (13). In addition, the deleterious effects such lesions have on a cell's well being are reconcilable with these major alterations.

One interesting possibility suggested by the results is that the modes of action of the lesion may not be strictly local. Perhaps the large kinks bring within interaction range portions of DNA helices which are ususlly far from one-another. It is conceivable that repair enzymes interacting with the DNA may require such changes for some aspects of recognition.

At any rate, the results here are qualitatively consistent with existing experimental evidence that both photodimer formation and psoralen crosslink formation force great changes in the conformation of DNA (14-19). As we have noted elsewhere (3), no direct experimental data indicating the exact conformational effects of these lesions is currently available. Hopefully, the models proposed here will help in designing future experiments which will in fact be able to quantify and further characterize such structural distortions and their biological effects.

5. ACKNOWLEGEMENTS

We thank David A. Pirkle, who helped with various aspects of this work. We also acknowledge the generosity of Professor Peter Kollman of the University of California, San Francisco, California who provided us with his program AMBER with a preprint of his group's work on thymine photodimers, and Dr. Doug Ohlendorf of Genex Corporation, Gaithersburg, Maryland for generating the computer graphics photographs of the space filling models presented here. This work has been supported by grants from NIH (GM31616, GM29287), NSF (PCM8019468) and the Department of Energy.

6. REFERENCES

1. Weiner, S. J., Kollman, P. A., Case, D. A., Singh, U. C., Ghio, C., Alagona, G., Profeta, S., Jr. & Weiner, P. (1984) J. Am. Chem. Soc. **106**, 765-784.
2. Kopka, M.L., Yoon, C., Goodsell, D., Pjura, P. and Dickerson, R.E. (1985) Proc. Natl. Acad. Sci. USA **82**, 1376-1380.
3. Pearlman, D. A., Holbrook, S. R., Pirkle, D. H. & Kim, S.-H. (1985) Science **227**, 1304-1308.
4. Wang, S. Y., Ed. (1976) Photochemistry and Photobiology of Nucleic Acids, Vol. I, Academic Press, New York.
5. Hearst, J. E. (1981) Ann. Rev. Biophys. and Bioengineer. **10**, 69-86.
6. Peckler, S., Graves, B., Kanne, D., Rapoport, H., Hearst, J. E. & Kim, S.-H. (1982) J. Mol. Biol. **162**, 157-172.
7. Weiner, P. K. & Kollman, P. K. (1981) J. Comp. Chem. **2**, 287-303.
8. Rao, S. N., Keepers, J. W. & Kollman, P. (1984) Nucl. Acids Res. **12**, 4789-4807.
9. Arnott, S. & Hukins, D. W. L. (1972) Biochem. Biophys. Res. Commun. **47**, 1504-1509.
10. Camerman, N. & Camerman, A. (1970) J. Am. Chem. Soc. **92**, 2523-2527.
11. Pearlman, D. A. & Kim, S.-H. (1985) Proceedings of the International Symposium on Biomolecular Structure and Interactions (in press).
12. Hruska, F. E., Wood, D. J., Ogilvie, K. K. & Charlton, J. L. (1975) Can. J. Chem. **53**, 1193-1203.
13. Haynes, R. H. (1964) Photochem. and Photobiol. **3**, 429-450.
14. Denhart, D. T. & Kato, A. C. (1973) J. Mol. Biol. **77**, 479-494.
15. Legerski, R. J., Grey, H. B., Jr. & Robberson, D. L. (1977) J. Biol. Chem. **252**, 8740-8746.
16. Wang, T.-M. & McLaren, A. D. (1972) Biophysik **8**, 237-244.
17. Wiesehahn, G. & Hearst, J. E. (1978) Proc. Natl. Acad. Sci. USA **75**, 2703-2707.
18. Ciarrocchi, G. & Pedrini, A. M. (1982) J. Mol. Biol. **155**, 177-183.
19. Chatterjee, P. K. (1981) Ph.D. Thesis, Columbia University, New York.

DNA SUPERCOILING AND GENE EXPRESSION

James C. Wang
Department of Biochemistry and Molecular Biology, Harvard University, Cambridge, Massachusetts, U.S.A.

The supercoiling of intracellular DNA and its influence on gene expression is briefly reviewed. In prokaryotes such as Escherichia coli, the degree of supercoiling is maintained by the actions of at least two DNA topoisomerases: an ATP-dependent enzyme DNA gyrase which reduces the linking number (negative supercoiling) and an ATP-independent enzyme DNA topoisomerase I which relaxes negatively supercoiled DNA. The expression of different genes is affected differently by supercoiling of the DNA; as the DNA becomes more negatively supercoiled, the expression of some genes is enhanced, while that of others is depressed or relatively unaffected. In eukaryotes, the existence of a gyrase-like enzyme that actively supercoils intracellular DNA is uncertain. Indirect evidence suggests, however, that DNA supercoiling may also occur and may have strong effects on gene expression. Some possibilities are raised that DNA supercoiling does not necessarily require a gyrase-type activity; in particular, the possibility that the transcription process itself in eukaryotes may cause supercoiling is discussed.

DNA SUPERCOILING AND GENE EXPRESSION

Supercoiled DNA

Because of the double-helix structure of DNA, a double-stranded circular DNA is composed of two topologically linked single-stranded rings. The order of linkage between such an intertwined pair of circular strands is defined by the linking number α, which is the total number of turns one strand makes around the other when the molecule is laid on a flat surface. The quantity α is a topological invariant so long as the continuity of the strands is maintained. If the strands are broken and then rejoined to permit the DNA to assume its most stable structure, the linking number α° of this relaxed DNA is determined by its structure; it has been shown that for a duplex ring of a typical nucleotide sequence N base pairs (bp) in length, α° is very close to $N/10.5$ (1-6).

B. Pullman et al. (eds.), Interrelationship Among Aging, Cancer and Differentiation, 173–181.

Vinograd and his associates found twenty years ago that the linking number α of the animal virus polyoma DNA is significantly smaller than α°, the linking number of the same DNA in its most stable form (7). When viewed in the electron microscope, the DNA has a twisted appearance similar to that of a torsionally unbalanced rope (8; see also ref. 9), the DNA is said to be "supercoiled" or "superhelical." Because the linking difference ($\alpha - \alpha^\circ$) is negative, polyoma DNA is "negatively" supercoiled. A large number of negatively supercoiled DNAs have since been found from both prokaryotic and eukaryotic sources (reviewed in ref. 10).

The Origin of Supercoiling

Several hypotheses were proposed for the origin of supercoiling. The intracellular DNA, for example, could have a helical pitch different from that of purified DNA (11-13). Another postulate was that the nicked molecule was partially unwound at the time of closure (7), due to the existence of a replication or transcription "bubble" in the DNA for example. Vinograd et al. pointed out a third possibility that because of the higher energy state of a supercoiled DNA, it would not be formed from a nicked molecule unless "the latter were supercoiled as a result of an interaction with an organizer, such as viral core protein, or with an as yet unknown core substance upon which duplex DNA is wound as it is synthesized in the cell" (13; see also ref. 7).

A few years after the discovery of supercoiled DNA and encouraged by the finding that the helical pitch of DNA is dependent on counterions and temperature (12), the author proceeded to test whether such dependences might explain the negative supercoiling of DNA from natural sources (14). A number of viral and plasmid DNAs of different sizes were isolated from *E. coli* under various conditions and their linking differences ($\alpha - \alpha^\circ$), then called the number of superhelical turns, were determined. The data obtained suggest that whereas ($\alpha - \alpha^\circ$) is sensitive to the size of the duplex DNA ring, the variability in ($\alpha - \alpha^\circ$) *per unit length* of the DNA is much less. Furthermore, several factors are *not* important for the linking number deficiency of intracellular DNA. These include replication, transcription and the presence of proteins coded by the episomal DNAs; similarly, the temperature and counterion effects appear to be small relative to the magnitude of ($\alpha - \alpha^\circ$) of the isolated DNAs.

Whereas this early peek into the origin of DNA supercoiling has not been too revealing, an observation made during that expedition turns out to be significant: It was found that in *E. coli* cell extracts there is an activity capable of relaxing negatively supercoiled DNA (14). Subsequent identification of this activity led to the discovery of a new type of enzyme, now termed DNA topoisomerase (15; reviewed in 16; 17).

Two major findings regarding the origin of supercoiling were made in the mid-seventies. In eukaryotes, the nucleosome has been postulated (and later confirmed) to be the structural unit of the bulk of DNA, in which the DNA is wrapped around an octameric histone core in a left-handed way (reviewed in 18; 19). A circular duplex DNA organized around nucleosomes has a lower linking number when it is relaxed by cycles of breakage and rejoining; experimentally, it appears that each nucleosome reduces α° by about 1 (20; for a discussion, see 21). For negatively supercoiled DNAs from eukaryotic sources, such as polyoma and simian virus (SV) 40 DNA, the observed linking differences for the purified DNAs can be attributed entirely to the lowering of the values of α° *in vivo* due to the coiling of the DNAs round the nucleosomes.

In prokaryotes, an enzyme DNA gyrase was discovered in 1976 (22). Gyrase is the first known DNA topoisomerase that can break and rejoin a pair of DNA strands in concert. Topoisomerases that reversibly break and rejoin a pair of DNA strands in concert are classified as type II topoisomerases; type I topoisomerases break and rejoin DNA one strand at a time. Furthermore, gyrase has a DNA-dependent ATPase activity and it utilizes ATP hydrolysis to vectorially transport a DNA segment through the transient gate it generates in the DNA (reviewed in 17; 23; 24; see also 25). The enzyme can actively reduce the linking number α to as much as 10% below α°.

In Prokaryotic Cells the Degree of Supercoiling is Regulated by the Balancing Actions of DNA Topoisomerases

In the bacterium *E. coli*, three DNA topoisomerases are known. *E. coli* DNA topoisomerase I was discovered in 1971 as the "ω-protein" (15). DNA topoisomerase II is gyrase and DNA topoisomerase III was reported in 1983 (26). Both DNA topoisomerase I and III are type I topoisomerases; they change DNA topology by breaking and rejoining one DNA strand at a time, and they do not require a cofactor. Whereas DNA gyrase catalyzes ATP hydrolysis-coupled negative supercoiling of DNA, DNA topoisomerase I and III relax negatively supercoiled DNA toward a state of lower free energy.

A number of experiments have shown that the degree of supercoiling of intracellular *E. coli* DNA is dependent on the levels of gyrase and DNA topoisomerase I. DNA isolated from cells in which gyrase action is blocked by drugs or mutations shows an increase in its linking number, whereas DNA from cells with reduced levels of DNA topoisomerase I has a lower linking number (reviewed in 17; 24).

Additional Evidence that Intracellular DNA in Prokaryotes is Negatively Supercoiled

The dependence of the linking number of an intracellular DNA ring on the levels of DNA gyrase and DNA topoisomerase I provide strong evidence that intracellular DNA is in a higher free energy state in terms of its linking number; the linking number of intracellular DNA is lower than that of the most stable structure in the cellular milieu, and this deficiency is maintained by a gyrase catalyzed active process, which is in turn countered by the relaxation action of DNA topoisomerase I (and likely topoisomerase III as well). This picture is supported by two additional experiments. In one, the binding of psoralen to intact and gamma-ray severed intracellular DNA was examined by photo-crosslinking of this intercalative agent to DNA. It was found that the intact DNA binds more psoralen than broken DNA, a result consistent with intracellular DNA being negatively supercoiled (27). In another, the linking numbers of a pair of nearly identical plasmids, one containing a stretch of alternating CG sequence and the other not, are compared. Although in wild-type E. coli strains such a pair of plasmids have nearly identical specific linking differences (28-30), in a strain in which the topA gene encoding DNA topoisomerase I is expressed from a regulated lac promoter the specific linking difference of the CG-containing plasmid is considerably lower when the topA gene is switched off (30). It is known that switching off topA increases the negative superhelicity of intracellular DNA, and that in vitro a stretch of alternating CG several dozen bp in length flips to a left-handed helical conformation when the negative specific linking difference $[-(\alpha - \alpha^{\circ})/\alpha^{\circ}]$ is greater than 0.04 (in a dilute aqueous buffer containing 100 mM monovalent cation with or without mM amounts of Mg(II) ions [31]). Thus the experiments with the CG-containing plasmid suggest that the "effective" negative superhelicity is lower than 0.04 under normal physiological conditions but higher than 0.04 when topA is switched off. Various estimates place the "effective" specific linking difference of intracellular DNA in the range -0.02 to -0.04. It has also been shown that when protein synthesis is blocked by chloramphenicol in E. coli, an alternating CG-containing plasmid has a lower linking number than its control without the CG sequence (28). This observation is consistent with the inference drawn above, although extrapolation of the chloramphenicol result to normal physiological conditions is less certain.

In Prokaryotes Gene Expression is Affected by DNA Supercoiling

In 1963, a recessive suppressor gene, initially termed su leu 500 and later renamed supX, was identified in the bacterium Salmonella typhimuriam (32). Mutations in the gene suppress the leucine auxotrophy imposed by the leu 500 mutation, which is found to be a promoter mutation (33, and references therein). In addition to suppressing the leu 500 auxotrophy, supX⁻ mutations are pleiotropic and affect the expression of a number of genes. The supX loci in both S. typhimuriam and E. coli are close to cysB. For almost a decade genetic studies of supX and biochemical studies of DNA topoisomerase I proceeded along parallel tracks. In 1980, however, the structural gene topA encoding the topoisomerase was identified and its proximity to cysB on the E. coli chromosome suggests that topA and supX might be the same gene (34-36). Furthermore, a number of amber and deletion mutants of supX exhibit no or reduced level of antigenic determinants that can be recognized by rabbit antibodies specific to the topoisomerase (37).

To rule out the possibility that supX might code an activator for regulating a nearby topA structural gene, nonsense mutations in E. coli supX gene carried on an F' factor in a strain of S. typhimuriam carrying a supX deletion were selected for suppressing the leu 500 mutation. The cysB-topA region of F' was then cloned and the resulting plasmids were used to transform a ΔtopA E. coli strain. Three such transformed strains carrying cloned DNA from episomes carrying different nonsense mutant supX alleles all showed the presence of antigenic determinants of DNA topoisomerase I which are lacking in the untransformed ΔtopA parent. No or little active DNA topoisomerase I can be detected in these transformed cells, however. Furthermore, it is shown that in the transformed cells the 100,000 dalton topoisomerase I is missing but in each of the three strains a new smaller plasmid-coded peptide appears, which presumably represents fragments of the enzyme resulting from translation termination at the supX nonsense codons (38). These experiments show conclusively that supX and topA are identical.

The identity of supX and topA provides a molecular interpretation of how mutations in the gene suppress leu 500 auxotrophy. Namely, the reduction of topoisomerase I activity increases the negative superhelicity of intracellular DNA, which in turn affects transcription from the leu 500 promoter. In vitro, DNA supercoiling has been known to affect transcription, although the situation in vivo might be more complex, especially when regulatory elements are involved (for discussions see 39, 40).

In Eukaryote Cells There is No Direct Evidence that Intracellular DNA is Negatively Supercoiled

Although duplex DNA rings purified from eukaryotic sources are negatively supercoiled (reviewed in ref. 10), this deficiency in linkage can be attributed to the reduction in the linking number of a DNA when it is relaxed in the form of a string of nucleosomes, as mentioned earlier. No gyrase activity has been found in eukaryotes. The ATP-dependent type II topoisomerase found in all eukaryotes, at least in the purified form, can not catalyze the supercoiling of DNA like bacterial gyrase (16, 17). In addition, with eukaryotic cells, no difference in binding was observed with intact and gamma-ray severed DNA (27). Thus the strong evidence in favor of negative supercoiling of intracellular DNA in prokaryotes lacks its equivalent in eukaryotes .

Indirect Evidence Suggests, However, that Intracellular DNA in Eukaryotes Might be Negatively Supercoiled as Well

On the other hand, several recent findings are suggestive that DNA supercoiling might be of paramount importance in eukaryotes as well. Weintraub et al. found that nuclease-sensitive regions of the chromosome, which correlated with activated genes, are sensitive in vitro to agents that are specific to single-stranded DNA if the DNA is negatively supercoiled (41 and references therein). Harland et al. (42) reported that genes cloned on plasmids are transcribed when injected into oocyte nuclei if and only if the circularity of the DNA is maintained. Luchnik et al. (43) reported that a small fraction of intracellular simian virus 40 (SV40) DNA appears to be under torsional stress, and that only this subpopulation appears to be transcriptionally active. Worcel et al. studied transcription and nucleosome assembly of a plasmid carrying a 5S ribosomal RNA gene after its injection into Xenopus laevis oocyte nuclei, and concluded that nucleosome assembly and expression of the 5S gene require the DNA being in a supercoiled state maintained by the two known eukaryotic topoisomerases (44, 45).

The Possibility of DNA Supercoiling by Mechanisms Differing from that of Bacterial Gyrase

Extensive studies on how bacterial gyrase supercoils DNA have been carried out in a number of laboratories (reviewed in 16, 17, 23). A more recent model, which incorporates structural features deduced from electron microscopic examination of the enzyme and its complexes with DNA, will be reported elsewhere (25).

As mentioned earlier, purified eukaryotic DNA topoisomerase II does not catalyze the supercoiling reaction. The possibility that the enzyme might nevertheless catalyze the supercoiling reaction in vivo has been widely implied, and factors that have been invoked to account for the difference in vitro and in vivo include the presence of another subunit in vivo, the association of the enzyme with other cellular entities which might impose a directionality in its transport of one DNA segment across another, and the association of the enzyme with specific DNA sequences in vivo. At this time, it remains an open question whether there is a eukaryotic topoisomerase which actively supercoils DNA in vivo.

It is equally plausible that DNA supercoiling can be achieved by mechanisms that are entirely different from the one described for gyrase. Indeed, the earlier models proposed for gyrase provide such examples (46, 47). Basically, these earlier models assume two distinct DNA binding sites on a gyrase. One site is bound to a fixed locus on a DNA. At the other site the DNA can move relative to the enzyme. Although the earlier models proposed for gyrase are incorrect for this particular enzyme, as a general alternative way of supercoiling DNA these models might be valid. Such a model has been invoked for the plausible formation of a transient supercoiled loop when a type I restriction enzyme tracks along DNA (48).

It is of particular interest regarding gene expression that the transcription process itself might cause DNA supercoiling. As the polymerase moves along the DNA, the DNA rotates relative to the enzyme. The topological problem that comes with rotation and the possible need of a topoisomerase have been pointed out (49-51). A situation of particular interest is one in which the DNA is bound to another protein X distal to the polymerase but interacts with it to form a DNA loop in between. In this situation, transcription would lead to the supercoiling of the loop. Protein X does not have to interact with the polymerase directly; if both are embedded in a cellular structure (the "nuclear matrix" for example), the same topology would result. A number of transcriptional factors that interact with sequences distal to the site of RNA synthesis are known, and their possible interaction with the polymerase has been pointed out (see for examples, 52 and 53 and references therein). Furthermore, it has been reported that the sites of transcription are located in the "nuclear matrix" (54-56). These observations suggest that the possibility of transcriptionally activated DNA supercoiling deserves further experimentation.

ACKNOWLEDGEMENTS

Work of this laboratory on DNA topology and DNA topoisomerases has been supported by grants from the U.S. Public Health Service, the National Science Foundation, and the American Cancer Society.

REFERENCES

1. Wang, J.C., 1979, Cold Spring Harbor Symp. Quant. Biol. 43, pp. 29-33.
2. Wang, J.C., 1979, Proc. Natl. Acad. Sci. USA 76, pp. 200-203.
3. Peck, L.J. and Wang, J.C., 1981, Nature 292, pp. 375-378.
4. Strauss, F., Gaillard, C. and Prunell, A., 1981, Eur. J. Biochem. 118, pp. 215-222.
5. Rhodes, D. and Klug, A., 1980, Nature 286, pp. 573-578.
6. Rhodes, D. and Klug, A., 1981, Nature 292, pp. 378-380.
7. Vinograd, J., Lebowitz, J., Radloff, R., Watson, R. and Laipis, P., 1965, Proc. Natl. Acad. Sci. USA 53, pp. 1104-1111.
8. Weil, R. and Vinograd, J., 1963, Proc. Natl. Acad. Sci. USA 50, pp. 730-738.
9. Wang, J.C., 1980, Trends in Biochem. Sci. Aug., pp. 219-221.
10. Bauer, W.R., 1978, Ann. Rev. Biophys. Bioeng. 7, pp. 287-313.
11. Crawford, L.V. and Waring, M.J., 1967, J. Mol. Biol. 25, pp. 23-30.
12. Wang, J.C., Baumgarten, D. and Olivera, B.M., 1967, Proc. Natl. Acad. Sci. USA 58, pp. 1852-1858.
13. Vinograd, J., Lebowitz, J. and Watson, R., 1968, J. Mol. Biol. 33, pp. 173-197.
14. Wang, J.C., 1969, J. Mol. Biol. 43, pp. 263-272.
15. Wang, J.C., 1971, J. Mol. Biol. 55, pp. 523-533.
16. Gellert, M., 1981, Ann. Rev. Biochem. 50, pp. 879-910.
17. Wang, J.C., 1985, Ann. Rev. Biochem. 54, pp. 665-697.
18. Kornberg, R., 1977, Ann. Rev. Biochem. 46, pp. 931-1154.
19. McGee, J.D. and Felsenfeld, G., 1980, Ann. Rev. Biochem. 49, pp. 1115-1156.
20. Germond, J.E., Hirt, B., Oudet, P., Gross-Bellard, M. and Chambon, P., 1975, Proc. Natl. Acad. Sci. USA 72, pp. 1843-1847.
21. Wang, J.C., 1982, Cell 29, pp. 724-726.
22. Gellert, M., Mizuuchi, K., O'Dea, M.H. and Nash, H.A., 1976, Proc. Natl. Acad. Sci. USA 73, pp. 3872-3876.

23. Cozzarelli, N.R., 1980, Science 207, pp. 953-960.
24. Gellert, M., 1981, Ann. Rev. Biochem. 50, pp. 879-910.
25. Kirchhausen, T., Wang, J.C. and Harrison, S.C., 1985, Cell, in press.
26. Dean, F., Krasnow, M.A., Otter, R., Matzuk, M.M., Spengler, S.J., et al., 1982, Cold Spring Harbor Symp. Quant. Biol. 47, pp. 769-777.
27. Sinden, R.R., Carlson, J.O. and Pettijohn, D.E., 1980, Cell 21, pp. 773-783.
28. Haniford, D.B. and Pulleyblank, E.D., 1983, J. biomolec. Struct. Dynam. 1, pp. 593-609.
29. Gellert, M. and Felsenfeld, G., personal communication.
30. Snyder, L. and Wang, J.C., unpublished.
31. Peck, L.J. and Wang, J.C., 1983, Proc. Natl. Acad. Sci. USA 80, pp. 6206-6210.
32. Mukai, F.H. and Margolin, P., 1963, Proc. Natl. Acad. Sci. USA 50, pp. 140-148.
33. Overbye, K.M., Basu, S.K. and Margolin, P., 1983, Cold Spring Harbor Symp. Quant. Biol. 47, pp. 785-791.
34. Sternglanz, R., DiNardo, S., Wang, J.C., Nishimura, Y. and Hirota, Y., 1980, in "Mechanistic Studies of DNA Replication and Genetic Recombination" (ed., Alberts, B.M.), New York: Academic, pp. 833-837.
35. Sternglanz, R., DiNardo, Voelkel, K.A., Nishimura, Y., Hirota, Y., Becherer, K., Zumstein, L. and Wang, J.C., 1981, Proc. Natl. Acad. Sci. USA 78, pp. 2747-2751.
36. Trucksis, M. and Depew, R.E., 1981, Proc. Natl. Acad. Sci. USA 78, pp. 2164-2168.
37. Trucksis, M., Golub, E.I., Zabel, D.J. and Depew, R.E., 1981, J. Bacteriol. 147, pp. 679-681.
38. Margolin, P., Zumstein, L., Sternglanz, R. and Wang, J.C., 1985, Proc. Natl. Acad. Sci. USA, in press.
39. Wang, J.C., 1982, in "Promoters, Structure and Function" (eds., Rodriguez, R.L. and Chamberlin, M.J.), New York: Praeger, pp. 229-241.
40. Wang, J.C., 1983, in "Genetic Rearrangement: Biological Consequences of DNA Structure and Genome Arrangement" (eds., Chater, K.F., Cullis, C.A., Hopwood, D.A., Johnston, A.W.B. and Woolhouse, H.W.), London: Croom Helm Ltd., pp. 1-26.
41. Kohwi-Sigematzu, T., Gelinas, R. and Weintraub, H., 1983, Proc. Natl. Acad. Sci. USA 80, pp. 4389-4393 and references therein.
42. Harland, R.M., Weintraub, H. and McKnight, S.L., 1983, Nature 302, pp. 38-43.
43. Luchnik, A.N., Bakayev, V.V., Zbarsky, I.B. and Georgiev, G.P., 1982, EMBO Journal 1, pp. 1353-1358.
44. Ryoji, M. and Worcel, A., 1984, Cell 37, pp. 21-32.
45. Glikin, G.C., Ruberti, I. and Worcel, A., 1984, Cell 37, pp. 33-41.
46. Liu, L.F. and Wang, J.C., 1978, Proc. Natl. Acad. Sci. USA 75, pp. 2098-2102.
47. Gellert, M., Mizuuchi, K., O'Dea, M.H., Ohmori, H. and Tomizawa, J., 1979, Cold Spring Harbor Symp.Quant. Biol. 47, pp. 35-40.

48. Yuan, R., Hamilton, D.L. and Burckhardt, J., 1980, Cell 20, pp. 237-244.

49. Maaloe, O. and Kjeldgaard, N.O., 1966, "Control of Macromolecular Synthesis" (New York: Benjamin).

50. Wang, J.C., 1973, in "DNA Synthesis in Vitro" (eds., Wells, R.D. and Inman, R.B.), Baltimore: University Park Press, pp. 163-174.

51. Gamper, H.B. and Hearst, J.E., 1982, Cell 29, pp. 81-90.

52. Parker, C.S. and Topol, J., 1984, Cell 36, pp. 357-369.

53. Parker, C.S. and Topol, J., 1984, Cell 37, pp. 273-283.

54. Robinson, S.I., Nelkin, B.D. and Vogelstein, B., 1982, Cell 28, pp. 99-106.

55. Cook, P.R., Lang, L., Hayday, A., Lania, L., Fried, M., Chiswell, D.J. and Wyke, J.A., 1982, EMBO Journal 1, pp. 447-452.

56. Abulafia, R., Ben-Ze'ev, A., Hay, N. and Aloni, Y., 1984, J. Mol. Biol. 172, pp. 467-487.

INTERMEDIATES IN TRANSCRIPTION INITIATION AND PROPAGATION

David C. Straney and Donald M. Crothers
Departments of Molecular Biophysics & Biochemistry, and Chemistry, Yale University, New Haven, CT.

Control of gene expression is a subject central to an understanding of the biological basis for differentiation, cancer and aging. Nearly all cells contain a full complement of the genetic information characteristic of their species, but each cell expresses only a fraction of the RNA sequences encoded in its DNA. The program of gene expression characteristic of differentiated cells is clearly altered upon transformation. A recent example is the reported loss of expression of the major histocompatibility complex class I gene in a malignant cell line (Tanaka et al., 1985); restoration of an expressed copy of that gene renders the cell line less oncogenetic, probably by virtue of the body's ability to mount an immunological defense. Altered gene expression seems likely to be important in the aging process as well, although genes specifically related to normal aging seem not yet to have been found.

The work in our laboratory has focused on the mechanism of transcription initiation in a simple model system, the *E. coli* promoter. Expression of this classic operon (Beckwith & Zipser, 1970; Reznikoff & Abelson, 1978) is modulated by both a repressor and a cAMP binding gene activating protein, CAP or CRP.

The central actor in gene activation is RNA polymerase, which is capable of great biochemical versatility, encompassing recognition of sequence elements centered around -10 and -35 in the promoter, initiation of RNA chains, faithful copying of a DNA sequence, and recognition of chain termination signals. The enzyme is not a small one, with $\alpha_2\beta\beta'\sigma$ subunit structure and a molecular weight of over 400,000. It can be expected that a process so vital as transcription initiation and propagation will be highly evolved, and even intricate. However, the intermediate states in this complex process are transient under normal conditions, posing a challenge to experimentalists who seek to trap and characterize them.

The existing view of the mechanism of transcription initiation is summarized by the simple scheme

$$R + P \rightleftarrows RP_{closed} \rightleftarrows RP_{open} \rightarrow \text{initiation} \qquad (1)$$

B. Pullman et al. (eds.), Interrelationship Among Aging, Cancer and Differentiation, 183–186.

proposed originally by Chamberlin (1974) and characterized quantitatively by the kinetic experiments of McClure and his colleagues (1980). In the "closed" complex, RNA polymerase (R) interacts in a rapidly reversible manner with promoter DNA (P) without producing a major conformational change. The closed complex isomerizes to the "open" form in a process that is characterized by opening of about 10 base pairs in the region around -5 (Siebenlist, 1979), and which can be rate-limiting for transcription. RNA chains are initiated in the open complex, but some of these are released when only a few nucleotides long in a process called abortive initiation (McClure et al., 1978). It has generally been recognized that scheme (1) is too simple to account for all the observations, but there has been less agreement on specific additional intermediates.

A technique which we have found useful for characterizing protein-DNA interaction is electrophoresis of the complex on non-denaturing polyacrylamide gels (Fried & Crothers, 1984a,b; Wu & Crothers, 1984). The virtue of this method is that complex mixtures containing several species can be resolved into components which migrate as separate bands on the gel. Furthermore, the mobility of the complex is dependent on its conformation, a fact we were able to exploit to provide evidence implying bending of DNA by the gene activating protein CAP (Wu & Crothers, 1984).

More recently, we have extended this approach to examination of intermediates in transcription initiation from the lac promoter. The results clearly indicate a greater degree of complexity than is contained in the simple scheme (1). Species observed include: 1) a tentative "closed" complex, which is present only at low temperatures (5°-$16^{o}C$) and is in rapid equilibrium with free DNA; 2) two "open" complexes, characterized by their stability to competitor DNA, which are in rapid equilibrium with each other. The ratio of the two open complexes O_u and O_l is temperature dependent, with the upper gel band (O_u) dominant at higher temperature; 3) two "initiated" complexes which are formed after addition of ribonucleotides. These complexes are high-salt-resistant, lack the σ subunit, and contain nascent RNA chains of different lengths. The smallest RNA transcript bound in these σ-free stable complexes is an 11-mer; an 8-mer abortive initiation product seen in solution is not stably bound; 4) an RNA-polymerase-RNA complex which also appears after transcription has started. This complex contains core polymerase and the 67 bp run-off transcript but does not contain any DNA.

Experiments designed to study RNA polymerase-DNA interactions in the gel complexes include DNAse I protection, methylation interference/protection, and single strand-specific cleavage. The two open complexes produce identical DNAse I footprints and similar methylation interference/protection (one base shows different behavior), indicating that both complexes are at the same binding site. The two open complexes, however, differ in the amount of unwinding in the -13 to +3 region, as assayed by probes for single stranded DNA. A functional assay, transcription within the gel slices containing the open complexes, indicates that the open complexes also differ in their ability to escape from the abortive initiation cycle into productive transcription. The upper gel band O_u has a lower total RNA synthesis rate but a greater relative production of the stably bound 11-mer compared to the abortive 8-mer product. Dominance of O_u at higher temperature may reflect an adaptive mechanism to counter the observed tendency of abortive initiation to increase with increasing temperature. The results are summarized in the scheme shown in Figure 1.

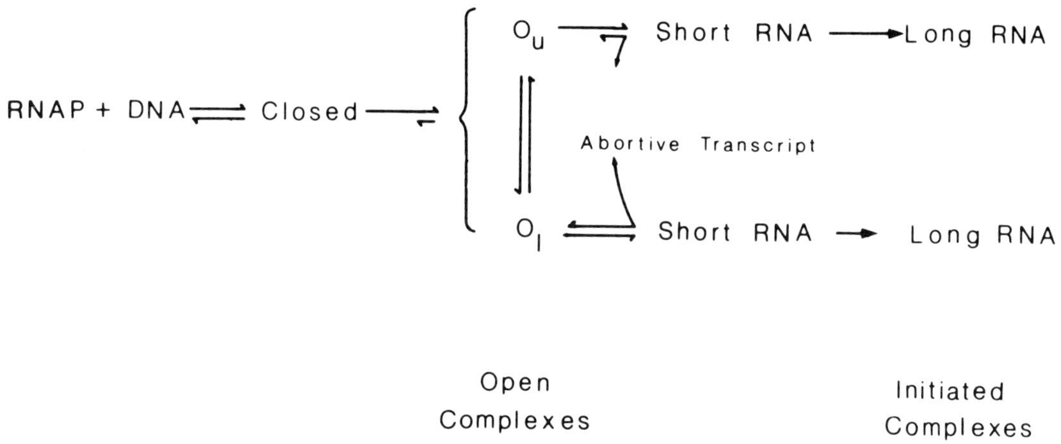

Figure 1. Summary of the intermediates in transcription initiation.

DNAse I footprinting of the initiated complex containing only the first stable transcript (the 11-mer) shows a contraction from the open complex footprints by loss of approximately 25 bp of contacts in the -35 region, and loss of the σ subunit. This drastic structural change, and the measured apparent activation energy for abortive release of nascent RNA transcripts, lead us to propose a model which features an energetically stressed intermediate in the translocation of polymerase away from its contact in the -35 region. The model focuses on mutually competitive interactions of polymerase with the -35 region or with short transcripts in the 0 to +10 region; the stability of the RNA-DNA hybrid double helix formed by the nascent transcripts can also be expected to affect the relative importance of abortive initiation.

REFERENCES

Beckwith, J. & Zipser, D. (1970) The Lactose Operon, Cold Spring Harbor Laboratory, Cold Spring Harbor, NY.

Chamberlin, M.J. (1974) Ann. Rev. Biochem. **43**, 721-775.

Fried, M. & Crothers, D.M. (1984a) J. Mol. Biol. **172**, 241-262.

Fried, M. & Crothers, D.M. (1984b) J. Mol. Biol. 172, 263-282.

McClure, W.R., Chech, C.L. & Johnson, D.E. (1978) J. Biol. Chem. 253, 8941-8948.

McClure, W.R. (1980) Proc. Natl. Acad. Sci. USA 77, 5634-5638.

Reznikoff, W.S. & Abelson, J. (1978) in The Operon, ed. J.H. Miller & W.S. Reznikoff, p. 221-243, Cold Spring Harbor Laboratory, Cold Spring Harbor, NY.

Siebenlist, U. (1979) Nature 279, 651-652.

Tanaka, K., Isselbacher, K.J., Khoury, G. & Jay, G. (1985) Science 228, 26-30.

Wu, H.M. & Crothers, D.M. (1984) Nature 308, 509-513.

New Carbohydrate Binding Proteins (Lectins) in Human Cancer Cells and their Possible Role in Cell Differentiation and Metastasation

FRIEDRICH CRAMER and HANS-JOACHIM GABIUS
Max-Planck-Institut für experimentelle Medizin, Abteilung Chemie, Hermann-Rein-Straße 3, D-3400 Göttingen, FRG

ABSTRACT

Various types of human cancer cells express a highly complex pattern of lectins which is characteristic for the particular type of cell and also for its state of development. These lectins are involved in heterotypic and homotypic cell aggregation. Tumors can be characterized via their typical lectins histochemically with fluorescent sugar derivatives. A lectin-directed chemotherapy seems possible. It is proposed that metastasation and mitotic stimulation occur via glycoconjugate-lectin interaction.

1. Introduction

What are lectins? Where do they occur? Lectins are carbohydrate-binding proteins without enzymatic activity which were discovered in plant seeds and were thought to be peculiarities of the plant kingdom until about a decade ago. They have proved to be very useful in characterizing glycoproteins and glycosylated cell membrane compounds, since they have a very high affinity and specificity to particular and highly complex oligosaccharides. The definition of a lectin also includes, that it agglutinates other cells e.g. bacteria or blood cells, which carry glycoconjugates at their surfaces. Thus, they expose at least two carbohydrate binding sites. In contrast to agglutinating antibodies they are, however, not regulated or produced on demand but constitutive substances of the lectin producing organism or cell. The amino acid sequences and the threedimensional structures of some of the lectins have been elucidated. The molecular weights normally range between 10^4 and 10^5 dalton. Some, like concanavalin A, are made up of identical subunits (1). Because of their affinity to red blood cells some are hemolytic like ricine.

In recent years it became evident that lectins also occur in higher organisms and play a very important role there (2). Vertebrate lectins seem to be developmentally regulated (3). The functions of these surface lectins in higher cells are not yet completely understood. They most

B. Pullman et al. (eds.), Interrelationship Among Aging, Cancer and Differentiation, 187–205.

likely play a leading role in cell-cell recognition during which complex carbohydrates on the surface of some cells interact with the corresponding lectins of other cells. Carbohydrate-lectin interaction seems to serve as a basis for immunogenic recognition and non-immunogenic phagocytosis which is involved in macrophage killing and in other immune reactions requiring cell-cell and cell-molecule interactions. The important role of lectin-carbohydrate interactions in recognitive processes is emphasized by the evidence for a crucial involvement of a lectin in the primary step of sperm attachment to the mammalian egg (5) and to the sea urchin egg (6).

2. Lectin pattern of normal cells

Membrane lectins can be characterized in various ways by their specific recognition of membrane glycoconjugates. The first mammalian membrane lectin from hepatocytes was isolated in 1974 (7). The possible role of some of these lectins in normal and malignant cells was discussed recently (8). Most of the work so far was done by characterization of haemaglutaniation activity in extracts and by histochemical methods using neoglycoproteins. Lectins have not been isolated as substances biochemically from malignant cells until now.

3. Methods

We have developed a method to isolate lectins via affinity columns. A set of sepharose columns derivatized with specific sugars and glycoproteins is used to bind the proteins with sugar affinity to these columns. The lectins can later be eluted with the particular sugar. Thus in a one step procedure the chromatographically pure lectin is obtained, which can be characterized in its molecular weight by acrylamide gel electrophoresis (9). The isolation scheme is shown in figure 1.

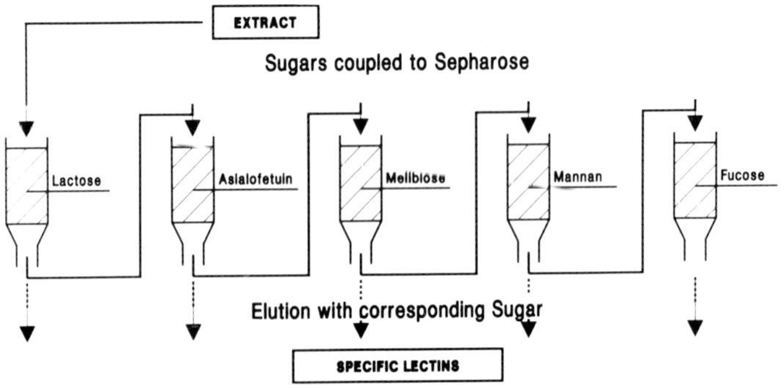

Fig. 1 Scheme for isolation of lectins with affinity columns

Usually we subdivide the lectins in groups. First the cell homogenate is extracted with buffer without detergent, which yields the soluble cytoplasmic and/or extracellular lectins. Afterwards, the homogenate is extracted with detergent and buffer which yields the membrane bound lectins. Both groups can be subdivided in calcium requiring and calcium independent lectins. Calcium requiring lectins are eluted from the affinity column by EDTA, non-calcium requiring lectins are eluted with the appropriate sugar. After elution from the affinity column the lectins are electrophoretically pure. A typical example is shown in figure 2.

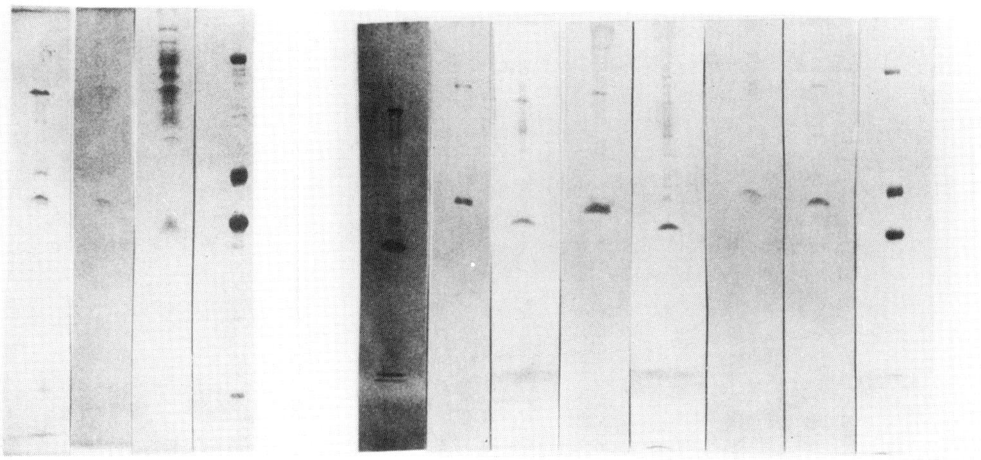

Fig. 2 Typical PAGE-pattern of lectins from sugar-affinity columns (from (13))

The lectin pattern is highly specific for particular tissues. Thus we could find that in bovine pancreas lectins for ß-galactosides are found of molecular weight 16, 35 and 64 kDa and mannan binding proteins of 37, 47 and 94 kDa and fucose specific lectins of 34, 62 and 70 kDa.

4. Why is the lectin-sugar interaction so specific?

The lectin-sugar interaction is extremely specific and very tight, in this respect comparable to the antigen-antibody-interaction. In fact the recognition of a sugar by a protein is ideal from the molecular point of view. Sugars and also more complex carbohydrates normally have a rather rigid and unique structure and a great number of places for hydrogen bonding. Therefore, oligosaccharides exposing their face to proteins can be fixed in a complex network of hydrogen bonds with extremely high specificity. Recently the structure of an arabinose binding protein has been solved which indicates this hydrogen bonding fixation very clearly (10), (Fig. 3). Within the system of oligo-saccharides containing rare sugars and having the possibility of α- and ß-configuration an immense amount of structural information can be stored and subsequently easily be read off by a recognition protein. Apparently nature is making use of this possibility in cell-cell recognition. It is certainly not by chance, that also the epitopes in the immune system are often and with preference complex carbohydrates.

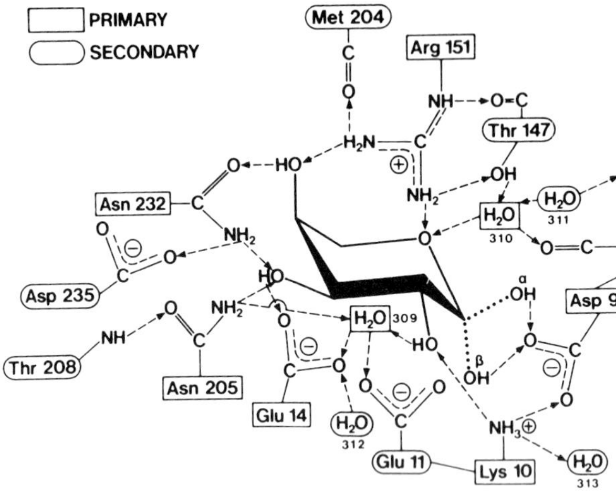

Fig. 3 Threedimensional network in an arabinose receptor from (10)

5. Previous work on lectins of tumor cells

It had been suggested previously and shown indirectly that certain malignant cells contain lectin-like carbohydrate binding moieties which can be characterized with known carbohydrates (8). It has also been proposed to use glycoconjugates for targeting drugs or other reagents (11). In all these tests the presence of lectins in tumor cells has been inferred indirectly by saccharide inhibition of hemaglutination in extracts, by saccharide inhibition of homotypic and heterotypic cell aggregation in vitro or by visualizing activities via labelled glycoconjugates or neoglycoproteins (8). In this respect primarily activities for ß-galactosides were detected on several human tumor cell lines (12).

6. Biochemical characterization of tumor specific lectin patterns

During the last two years we have set out to characterize the lectin pattern of tumor cells by isolation and description of tumor-specific and tumor associated lectins. The lectin pattern of a tumor in principle can be characteristic for

the type of tumor as compared to the nontransformed cell type,
the developmental stage of the particular tumor,
the tissue environment of the particular tumor.

In order to have a first orientation on the lectin pattern of malignant cells, we studied a number of rodent tumors (13). Three entirely different tumor types were investigated biochemically for the presence and characteristics of endogenous carbohydrate-binding proteins in an inbred Brown Norway rat, an outbred Sprague-Dawley rat, and an outbred Han:NMRI mouse. The patterns under investigation included specificities for a α- and ß-galactosyl, α-fucosyl moieties, respectively, and specificities for heparin, analyzed by affinity chromatography on resins with immobilized sugars or glycoproteins and polyacrylamide gel electrophoresis in the presence of sodium dodecyl sulfate. The patterns were divided into categories according to dependence of the binding activity on the presence of Ca^{2+} and dependence on extraction conditions. Rhabdomyosarcoma revealed only Ca^{2+}-independent activities, i.e., activities with specificity for ß-galactosides at a molecular weight of 12,000, with specificity for α-galactosides at molecular weights of 29,000, 43,000, and 45,000, with specificity for heparin molecular weights of 13,000 and 16,000, and with specificities for mannose and fucose at molecular weights ranging from 62,000 to 70,000. For the spontaneous mammary adenocarcinoma the pattern was entirely different and more diverse, including species with the Ca^{2+} requirement (Table 1). Extracts with the use of 0.2 M NaCl (salt) and 2 % Triton X-100 (detergent) from teratoma contained at least nine different carbohydrate-binding proteins (Table 2). The only similarities between the pattern of endogenous carbohydrate-binding proteins from teratoma and from mammary adenocarcinoma were ß-galactoside-binding proteins, one with a Ca^{2+} requirement and one without a Ca^{2+} requirement, and the heparin-binding proteins. These heparin-binding proteins were the only types of carbohydrate-binding proteins common to all three tumor types. The analysis indicates that certain bands represent newly identified proteins capable of binding to galactose-, mannose- or fucose-containing glycoconjugates, respectively. When assayed with rabbit erythrocytes, the different fractions showed agglutination acitivity. They can thus be termed "endogenous lectins".

Once the techniques and the principal characteristics of tumor lectins were established in the rodent system, we set out to extend these studies to a human tumor (14). Salt and detergent extracts of a malignant epithelial tumor, obtained by extraction of acetone powder, were fractionated on different sets of Sepharose columns covalently derivatized with lactose, asialofetuin, melibiose, mannan, fucose, and heparin. Successive elution by chelating reagent and specific sugar resulted in isolation

Table 1. Lectin pattern of mammary adenocarcinoma from affinity columns (13)

	Molecular weights						
	EDTA			Sugar			
Extract	La	In	Fu	La	La	Ma	Fu
Salt	64	35		52	52		62
	35		-	35	35	-	45
				22			42
				16			30
Triton	35		140	64		46	62
	32	-		16	-	44	13
	14						

Apparent molecular weight in thousand;
EDTA: lectins eluted by EDTA; sugar: lectins eluted by corresponding sugar.
La: Lactose; In: Invertase; Fu: Fucose;
Ma: Mannose; As: Asialofetuin.

Table 2. Lectin pattern of teratoma(13)

	Molecular weights					
	EDTA				Sugar	
Extract	La	As	Ma	Fu	La	Fu
Salt	24	29	31	-	64	100
Triton	32	32	32	32	34	-

Legend as in Table 1.

of different Ca^{2+}-dependent and Ca^{2+}-independent endogenous carbohydrate-binding proteins, as analyzed by gel electrophoresis. It appears from the analysis that certain bands represent newly identified proteins capable of binding to lactose (at M_r 64,000), melibiose (at M_r 28,000), and fucose (at M_r 62,000 and 70,000), (Table 3). Other carbohydrate-binding proteins isolated from this human tumor have been identified in normal, especially embryonic tissues of different nonhuman vertebrates. The carbohydrate-binding proteins are assayable as agglutinin with rabbit erythrocytes and show no detectable enzymatic activity. They can thus be defined as lectins. The presence of a complex pattern of endogenous lectins and their biochemical characteristics may contribute to an understanding of intercellular interaction during the complex process of metastatic spread and may furthermore allow a new tool for diagnosis and a lectin-based therapy.

Table 3. Lectin pattern of a human epithelial tumor (14)

	Molecular weights							
	EDTA					Sugar		
Extract	La	As	Me	Ma	Fu	La	Me	Fu
Salt	29	29	45 43 29 28	31	62	64 14	64	
Triton	70 35 29	29		31	70 62	64 35 16 14	64	62

Legend as in Table 1.

7. Differentiation between closely related tumors: Comparison of endogenous lectins in human embryonic carcinoma and yolk sac carcinoma as an example (15).

Carcinomas of the testis, expecially germ cell tumors, are one of the most common forms of cancer among young adult human males, and their incidence seems to be increasing (16, 17). This explains the considerable importance for understanding of the biology of this malignancy. In the histological classification system proposed by the World Health Organization embryonal carcinoma, teratoma, teratocarcinoma, choriocarcinoma, yolk sac tumor and seminoma (18, 19) are classified as testicular germ cell carcinomas. Although the classification and histogenesis is currently still controversial (20, 21), it is recognized that embryonal carcinoma (EC) cells are a highly proliferative and invasive population of stem cells that closely resemble normal uncommitted embryonic cells in their ability to differentiate in various morphological lines (22, 23). They can be serially transplanted into nude mice without histological changes (24, 25, 26).

We have compared the lectin pattern of yolk sac tumor and of two strains of human embryonic carcinomas as an example for closely related human tumors and indeed find characteristic differences which are shown in table 4.

The two types of human germ cell tumor, the embryonal carcinoma and the yolk sac carcinoma, represent different stages of differentiation from a common cellular origin (18, 19). Their endogenous lectin pattern reveals quantitative and, notably, qualitative differences. Whereas a Ca^{2+}-dependent carbohydrate-binding protein with an apparent molecular weight of 66,000 could only be isolated from salt extracts of the embryonic carcinoma, the salt extract of the yolk sac tumor contained a unique Ca^{2+}-independent fucose-binding protein of apparent molecular weight of 62,000. In relation to the embryonic carcinoma, the yolk sac tumor is more abundant in expression of a protein of apparent molecular weight of 56,000, in which the relative proportion of elution of EDTA and sugar indicated Ca^{2+}-independence, because partial elution by EDTA has been shown to occur without clear intrinsic requirement for EDTA (27). Also the relative expression of the Ca^{2+}-independent ß-galactoside binding proteins at apparent molecular weight of 14,000 and 32,000 differ between these two tumor types. EDTA-elution from the different resins yielded a protein of apparent molecular weight of 29,000 and 31,000, respectively, with broad specificity, not uncharacteristic for several endogenous lectins (28, 29, 30). Since this difference was consistently determined and the yolk sac tumor contained an additional band at apparent molecular weight of 31,000 after EDTA-elution from mannan-Sepharose that resembles a mannan-binding protein from human serum (31), it is possible that these two bands represent different stages of processing of the same protein, described for the mannan-binding protein from rat liver (32).

Table 4. Lectin pattern of three different human germ cell tumors (15)

	Molecular weights									
	EDTA					Sugar				
Tumor type	La	As	Me	Ma	Fu	La	As	Me	Ma	Fu
Yolk	56		56		56	56	56	56		62
sac	29	29	29	29	29	32	29	31		56
						29		29		29
						14				
	29	29	29	31		32				
				29	29	29				
						14				
EC (H23)	31	31	66	66	31	32				
			56	31		31				
			31			14				
						32				
						14				
H 12.1*	31	31	31	31	70	35				
					31	14				
	31	31	31	31	31	35				
						14			68	

*embryonal carcinoma with/without syncytiotrophoblastic giant cells and with/without immature teratoma

Legend as in Table 1; the upper part in each block refers to salt extract, the lower to triton extract.

It is remarkable that a protein at apparent molecular weight of 31,000 was also present in xenografts of a human teratocarcinoma, derived from the cell line H 12.1 (33). The histology of these xenografts had revealed embryonal carcinoma with/without syncytiotrophoblastic giant cells and with/without immature teratoma in contrast to the histology of pure embryonal carcinoma for xenografts of H 23. Differences in histology appear to be reflected in four qualitative differences. Whereas the carbohydrate-binding proteins at apparent molecular weight of 56,000 and 66,000, present in H 23, were undectable in H 12.1 (Table 4), these xenografts, however, contained an additional fucose-binding protein (Mr 70,000) and a unique mannan-binding protein (M_r 68,000) that had been implicated in cell-cell recognition in these tumor cells

(33). It therefore constitutes an example of a functional tumor marker.

8. Lectins and developmental stage

As shown in the previous chapter, two tumor types of common origin exhibit different lectin patterns once they have differentiated in different directions. We now would like to ask the question: Are there differences in same tumor type during its maturation or in different clones of that tumor with slightly different phenotpyic properties? To that end we have studied the lectins of three different clones of rat fibroblasts transformed with myoproliferative Sarcoma virus (34). These clones had different growth properties and different metastatic potential. The lectin patterns are shown in table 5.

All three different tumors contain various carbohydrate-binding proteins with specificities to α- and ß-galactosides, mannose and fucose. Since the presence of such proteins in malignant cells has not so far been investigated (for review, see (8)), comparison of their profile is only possible to normal adult and embryonic tissues. Carbohydrate-binding proteins at apparent molecular weight of 31 kDa, implicated in receptor-mediated endocytosis, have been isolated from different sources (35). Graduation of specificity, as known from the asialo-glycoprotein receptor (36), may explain the occurence of a similar protein in different fractions of the three clones. Some lectins, known from normal tissues, are virutally absent, e.g., the melibiose-specific lectin at apparent molecular weight of 45 kDa and 43 kDa (37).

The differences in the pattern between the clones suggests a differential expression of individual lectins, although at present little is known about their physiological role. In relation to sarcomas of different origin, analysis of rodent osteosarcoma revealed the Ca^{2+}-dependent mannan-binding activity at 31 kDa and lactose-specific binding proteins at 64 kDa and 14 kDa (unpublished observation), and analysis of rodent rhabdomyosarcoma showed Ca^{2+}-independent binding proteins with specificity to ß-galactosides at 12 kDa, with specificity to α-galactosides at 29 kDa, 43 kDa and 45 kDa and with specificities to mannose and fucose, respectively, in the range of 62 - 70 kDa (13). Thus, different types of sarcoma cells show biochemically distinguishable differences in the pattern of carbohydrate-binding properties. This may in the future prove useful for diagnosis.

Furthermore, affinity chromatography on immobilized lectins revealed differences in the composition of membrane glycoproteins with affinity to pea nut agglutinin and Ulex europaeus agglutinin, respectively. These alterations are of interest, because glycoproteins, together with specific carbohydrate-binding proteins in a carbohydrate-protein recognition system, are supposed to play a crucial role in mediation of recognitive processes in differentiation and metastasis (38, 39, 40). Changes in glycoprotein profile have been identified histochemically and biochemically in another model system, the Eb/ESb tumor system from

Table 5. Carbohydrate-binding proteins of the different clones of transformed fibroblasts (34)

Tumor type	Molecular weights							
	EDTA					Sugar		
	La	As	Me	Ma	Fu	La	Ma	Fu
5-8#1	31 14	31	31 29	31	68	34 31 14		62 20
	42 31 14			31		34 31 14	31 14	
5-20#20	31		64 31	31 14	31	34 31 14		
	42 31 14	31	31	31	31	34 31 14		
6-6#3+F	31 14	31	31	31	31	34 31 14		16
	14	31	31	31	31	34 31 14		

Legend as in Table 1; the upper part in each block refers to salt extract, the lower to triton extract.

DBA/2 mice, indicating a correlation to the different metastatic potential (41, 42). Changes in the lectin pattern in comparison to normal tissues have been identified biochemically for solid tumors of rodents (13). In this respect, the differences in pattern of carbohydrate-binding proteins for the three MPV-transformed fibroblast clones enable the proposal that monitoring the pattern of these proteins may be potentially valuable as indicator of cell differentiation, as marker for tumor diagnosis and for the understanding of cellular interactions in tumor progression and metastasis.

9. Heterotypic and homotypic cell aggregation

We could recently show that human teratocarcinoma cells express a new mannan specific endogenous lectin which is located on the membrane (33). In addition there is a ß-galactoside specific lectin which also occurs elsewhere. No further carbohydrate-binding protein is isolatable on columns derivatized with asialofetuin, melibiose and L-fucose. Both protein species agglutinate trypsinized, glutaraldehyde-fixed rabbit erythrocytes in the absence of Ca^{2+} and can thus be defined as endogenous human teratocarcinoma lectins. Inhibition of heterotypic and homotypic aggregation of human teratocarcinoma cells by D-mannose, D-galactose and glycoproteins rich in one of these sugars is consistent with a functional role of these Ca^{2+}-independent lectins in cell aggregation. The inhibition of rosette formation is shown in table 6,

Table 6. Inhibition of rosette formation (33)

Inhibitor	% Inhibition of rosette formation
NAc-D-galactosamine	0
L-fucose	2
D-galactose	4
NAc-D-glucosamine	0
D-mannose	10
Fetuin	0
Asialofetuin	7
Asialo-agaloctofetuin	0
Mannan	21
Invertase	34
Invertase (periodated)	4
Lactose-BSA	7
Mannose-BSA	14
Mannan+D-galactose	32

Glutaraldehyde-fixed, trypsinized rabbit erythrocytes were used. Sugars were added at a concentration of 0.2 M, all glycoproteins at a concentration of 1 mg/ml. The results are averages from 8 - 10 independent experiments.

the inhibition of the homotypic aggregation is given in Fig. 4. This experiment suggests the proposal that homotypic as well as heterotypic aggregation of tumor cells occurs via sugar-lectin interaction. This is depicted in Fig. 5.

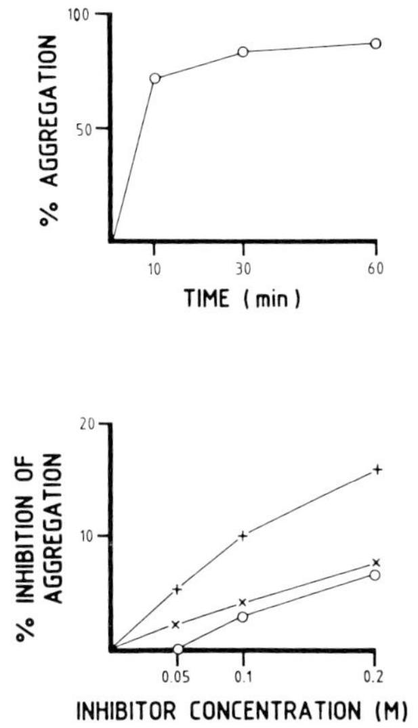

FIG. 4 INHIBITION OF HOMOTYPIC EC-CELL AGGREGATION BY SUGARS (33)
TOP: SPONTANEOUS REAGGREGATION OF EC-CELLS
BOTTOM: INHIBITION OF REAGGREGATION AFTER 15 MIN IN PRESENCE OF INHIBITOR
O—O L-FUCOSE, X—X D-GALACTOSE AND T—T D-MANNOSE

10. Histochemical Staining via Lectins (34)

If our new tumor markers are indeed situated on the cell surface, their sugar binding sites might be used for specific histological staining. For this purpose the corresponding sugars were derivatised with fluorescent bovine serum albumine (43). In this way human teratocarcinoma cells could be specifically visiualized by mannosylated and lactosylated fluorescent markers (33). In a similar way virus transformed rat fibroblasts could be differentiated (34). With the progressing knowledge of specific lectin patterns of tumor cells one might be able to develop an entire lectin based histology for tumors.

11. Future Prospects

Which stages of tumor growth and spread could be geared by endogenous lectins? A carbohydrate-lectin interaction can play a role in tumor growth at the primary site, because it has been shown that certain

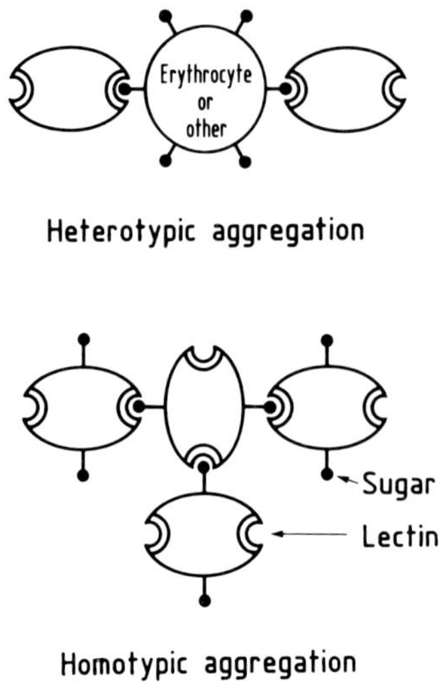

Fig. 5 Model for heterotypic and homotypic cell aggregation via the glycoconjugate-lectin system

vertebrate lectins exhibit mitogenic properties (44, 45). This provides an experimental basis for lectin impact on growth control and proliferation. If neoplastic cells behave like hemopoetic stem cells and if they expose both the lectin and the corresponding sugar (which seems likely from our aggregation experiments) then they could stimulate themselves autocatalytically to exponential growth. And this is what one indeed observes with tumor colonies in vitro and in vivo.

In relation to qualitative changes of glycoconjugate structures in the tumor, lectins can influence the release of certain tumor cells from the primary tumor due to altered homotypic adhesion. During transportation they may direct embolization by controlling heterotypic cell interactions. Furthermore, they can contribute to the selectivity in adhesive contacts between invading cells and parenchymal cells that establishes the organotropy of metastasis. These different steps may also include possible interactions between endogenous lectins of normal cells and glycoconjugates of tumor cells (46, 47). In order to provide a valuable tool for clinical application, the spectrum for lectins of tumor cell lines should be demonstrated to undergo characteristic changes. Our results, employing sugar inhibition of homotypic aggregation, emphasize this variation. While for example the aggregation of a human teratocarcinoma cell line (H 12.1) was strongly affected by D-mannose, the homotypic aggregation of human Wilms tumor and tongue carcinoma (TR 126) was strongly inhibited by melibiose. Only the line TR 126 showed additional strong sensitivity to N-acetyl galactosamine and L-fucose (47). This provides phenomenological evidence for at least quantitative differences in the lectin pattern present in these different cell lines and can guide selection of affinity resins for biochemical analysis. Besides a differential expression of lectins by different tumors, indicated by sugar inhibition of aggregation, lectin levels can also appear quantatively changed in malignant cells. In relation to their normal counterparts, we measured a decrease of the overall lectin activity for ß-galactosides of rodent lung and uterus carcinoma. The amount of lectin, however, detectable on the cell surface, increases for the tumor cells (48).

Besides the diagnostic prospect for specific histological staining via lectins, a lectin directed radio- or chemotherapy adoptive immunotherapy seems to be not unreasonable once one gains a more detailed knowledge on the lectin patterns of tumors. One also might speculate to be able to interfer therapeutically with the metastasation process via lectin inhibition. In contrast to conventional tumor markers the endogenous lectins are funtional tumor markers, themselves participating in processes of tumor growth and spread.

Also we are well aware that our findings fit beautifully with the recent demonstration on the role of carbohydrate structures as onco-developmental antigens (49). All what has been said there (49) would also apply to the corresponding lectins. In fact, only the two together, the glycoconjugate and its lectin make sense in biological function.

We are just beginning to study the intercellular language between complex glycoconjugates and lectins. The carbohydrate structures known so far at cell surfaces are probably only the first few letters of a script of perhaps several hundred characters, and the same would apply to the corresponding receptors, the lectins.

Acknowledgement:

We are gratefully indepted to the following scientists, who have cooperated experimentally or by supplying biological material or by stimulating discussion:

S.H. Barondes, Dept. of Psychiatry, Univ. Calif., San Diego
R. Bäthge, Med. Univ.-Klinik, Göttingen
J. Caspar, Med. Hochschule, Hannover
B. Damerau, Max-Planck-Inst.f.exp.Medizin, Göttingen
F. Deerberg, Zentralinstitut f. Versuchstierzucht, Hannover
F.R. Engelhardt, Max-Planck-Inst.f.exp.Medizin, Göttingen
G. Graupner, Inst.f.Physiolog.Chem., Univ. Erlangen

G.A. Nagel, Med. Univ.-Klinik, Göttingen
S. Rehm, Zentralinst.f.Versuchstierzucht, Hannover
D. Reile, Med. Hochschule, Hannover
D. Sartoris, Dept. of Radiology, Univ. Calif., San Diego
H.J. Schmoll, Med. Hochschule, Hannover
S. Schuhmacher, Med. Hochschule, Hannover
K. Vehmeyer, Med. Univ.-Klinik, Göttingen

REFERENCES

1) J.W. Becker, G.N. Reeke, Jr., B.A. Cunningham, and G.M. Edelman: New evidence on the location of the saccharide binding site of concavalin A. Nature 259, 406-409 (1976); Review Article: H. Lis and N. Sharon. Ann. Rev. Biochem. 42, 541-574 (1973). I.J. Goldstein and C.E. Hayes: The lectins: Carbohydrate-binding proteins of plants and animals. Adv. in Carbohydrate Chem. and Biochem. 35, 127-339 (1978).
2) S.H. Barondes: Lectins: Their multiple endogenous cellular functions. Ann. Rev. Biochem. 50, 207-231 (1981).
3) T.P. Nowak, P.L. Haywood, and S.H. Barondes: Developmentally regulated lectin in embryonic chick muscle and a myogenic cell line. Biochem. Biophys. Res. Comm. 68, 650-657 (1976).
4) N. Sharon: Surface carbohydrates and surface lectins are recognition determinants in phagocytosis. Immunology Today 5, 143-147 (1984) and N. Sharon: Carbohydrates as recognition determinants in phagocytosis and in lectin-mediated killing of target cells. Biol. Cell. 51, 239-246 (1984).
5) T.T.F. Huang, jr., E. Ohzu, and R. Yamagimachi: Evidence suggesting that L-fucose is part of a recognition signal for sperm-zona Pellucida attachment in mammals. Gamete Res. 5, 355-361 (1982).
6) C. Glabe, L. Grabel, V. Vacquier and S. Rosen: Carbohydrate specificity of sea urchin sperm binding. J. Cell Biol. 94, 123-128 (1982).
7) R.L. Hudgin, W.E. Pricer, G. Ashwell, R.J. Stockert, and A.G. Morell: The isolation and properties of a rabbit liver binding protein specific for asialoglycoproteins. J. Biol. Chem. 249, 5536-5543 (1974).
8) M. Monsigny, C. Kieda, and A. Roche: Membrane glycoproteins, glycolipids and membrane lectins as recognition signals in normal and malignant cells. Biol. Cell 47, 95-110 (1983).
9) H.J. Gabius, R. Engelhardt and F. Cramer: Endogenous lectins of bovine pancreas. Hoppe Seyers Zeitschrift Physiol. Chem. 365, 633-638 (1984).
10) F.A. Quiocho and N.K. Fyas: Novel stereospecificity of L-arabinose-binding protein. Nature 310, 381-386 (1984).
11) A.C. Roche, C. Marteau, P. Midoux, F. Delmotte, and M. Monsigny: Uptake of glucosylated serumalbumin-bound daunorubicin by Lewis lung carcinoma cells (in preparation) and A.C. Roche, M. Barzilay, P. Midoux, S. Junqua, N. Sharon, and M. Monsigny: Sugar-specific endocytosis of glycoproteins by Lewis lung carcinoma cells. J. Cell. Biochem. 22, 131-140 (1983).

12) A. Raz and R. Lotan: Lectin-like activities with human and murine neoplastic cells. Cancer Res. 41, 3642-3647 (1981).
13) H.-J. Gabius, R. Engelhardt, S. Rehm and F. Cramer: Biochemical characterization of endogenous carbohydrate binding proteins from spontaneous murine Rhabdomyosarcoma mammary adenocarcinoma and ovarian teratoma. J. Nat. Canc. Inst. 73, 1349-1357 (1984).
14) H.-J. Gabius, R. Engelhardt, F. Cramer, R. Bätge and G.A. Nagel: Pattern of endogenous lectins in a human epothelial tumor. Cancer Res. 45, 253-257 (1985).
15) H.-J. Gabius, R. Engelhardt, J. Casper, H.J. Schmoll, G.A. Nagel and F. Cramer: Endogenous lectins of human embryonic carcinoma and yolk sac carcinoma. Cell and Tissue Res., in preparation.
16) J. Clemmensen: Testis cancer incidence - suggestion for a world pattern. In: Early detection of testicular cancer, N.E. Skakkebaeck, J.G. Berthelsen, K.M. Grigor and J. Visfeldt (eds.), pp. 111-112, Copenhagen: Scriptor (1981).
17) A.B.W. Nethersell, L.K. Drake and K. Sikora: The increasing incidence of testicular cancer in East Anglia. Brit. J. Cancer 50, 377-380 (1984).
18) F.K. Mostofi: Pathology of germ cell tumors of testis. Cancer 45, 1735-1754 (1980).
19) B. Norgaard-Pedersen and D. Rayhavan: Germ cell tumors: A collaborative review. Oncodev. Biol. Med. 1, 327-358 (1980).
20) F.K. Mostofi: Comparison of various clinical and pathological classification of tumors of testes. Sem. Oncol. 6, 26-30 (1979).
21) R.A. Risdon: Germ cell tumors of testis. J. Pathology 141, 355-361 (1983).
22) L.E. Nochomovitz, F.E. Dela Torre and J. Rosai: Pathology of germ cell tumors of the testis. Urol. Clin. N. Am. 4, 359-378 (1977).
23) D. Solter and I. Damjanov: Teratocarcinoma and the expression of oncodevelopmental genes. Meth. Cancer Res. 18, 227-332 (1979).
24) P.W. Andrews, I. Damjanov, D. Simon, G.S. Banting, C. Carlin, N.C. Dracopoli and J. Fogh: Pluripolent embryonal carcinoma clones derived from the human teratocarcinoma cell line Fera-2. Lab. Invest. 50, 147-162 (1984).
25) D.L. Bronson, P.W. Andrews, D. Solter, J. Cervenka, P.H. Lange, and E.C. Fraley: Cell line derived from a metastasis of a human testicular germ cell tumor. Cancer Res. 40, 2500-2506 (1980).
26) P. Monaghan, D. Raghavan, and A.M. Neville: Ultrastructural studies of xenografted human germ cell tumors. Cancer 49, 683-697 (1982).
27) C.F. Roff and J.L. Wang: Endogenous lectins from cultured cell. J. Biol. Chem. 258, 10657-10663 (1983).
28) T. Imamura, S. Toyoshima and T. Osawa: Lectin-like molecules on the murine macrophage cell surface. Biochim. Biophys. Acta 805, 235-244 (1984).
29) T.B. Kuhlenschmidt and Y.C. Lee: Specificity of chicken liver carbohydrate-binding protein. Biochemistry 23, 3569-3575 (1984).
30) C.P. Stowell and Y.C. Lee: The binding of D-glucosyl-neoglycoproteins to the hepatic asialoglycoprotein receptor. J. Biol. Chem. 253, 6107-6110 (1978).

31) N. Kawasaki, T. Kawasaki, and I. Yamashina: Isolation and characterization of a mannan-binding protein from human serum. J. Biochem. 94, 937-947 (1983).
32) M.D. Brownell, K.J. Colley, and J.U. Baenzinger: Synthesis, processing and secretion of the core-specific lectin by rat hepatocytes and hepatoma cells. J.Biol. Chem. 259, 3925, 3932 (1984).
33) H.-J. Gabius, R. Engelhardt, J. Casper, D. Reile, S. Schumacher, H.J. Schmoll, G. Graupner, and F. Cramer: Cell surface lectins of transplantable human teratocarcinoma cells: Purification of a new mannan-specific endogenous lectin. Tumour Biol., in press.
34) H.-J. Gabius, K. Vehmeyer, R. Engelhardt, G.A. Nagel, and F. Cramer: Carbohydrate-binding proteins of tumor lines with different growth properties. I. Differences in their pattern for three clones of rat fibroblasts transformed with a myeloproliferative sarcoma virus. Cell and Tissue Res., in press.
35) G. Ashwell, J. Harford: Carbohydrate-specific receptors of liver. Ann. Rev. Biochem. 51, 531-554(1982).
36) C.P. Stowell and Y.C. Lee: The binding of D-glucosyl-neoglycoproteins to the hepatic asialoglycoprotein receptor. J. Biol. Chem. 253, 6107-6110 (1978).
37) M.M. Roverson and S.H. Barondes: Lectin of embryos and oocytes of Xenopus laevis. J. Biol. Chem. 257, 7520 (1982).
38) F.L. Harrison and C.J. Chesterton: Factors mediatin cell-cell recognition and adhesion. FEBS Letters 122, 157-165 (1980).
39) G. Uhlenbruck: The Thomsen-Friedenrich (TF) receptor. An old history with new mystery. Immunol. Commun. 10, 251-254 (1981).
40) A. Raz and R. Lotan: On the possible role of tumor associated lectins in metastasis. In: T. Galeotti, A. Cittadini, G. Neri and S. Papa (eds.): Membranes in tumor growth, pp. 213-221, Elsevier /North Holland Biomedical Press, Amsterdam (1982).
41) P. Altevogt, M. Fogel, R. Cheingsong-Popov, J. Dennis, P. Robinson, and V. Schirrmacher: Different patterns of lectin binding and cell surface sialylation detected on related high and low metastatic tumor lines. Cancer Res. 43, 5138-5144 (1983).
42) R. Schwartz, V. Schmirrmacher, and P.F. Mühlradt: Glycoconjugates of murine tumor lines with different metastatic patterns. Int. J. Cancer 33, 503-509 (1984).
43) C. Kieda, A.C. Roche, F. Delmotte, and M. Monsign: Lymphocyte membrance lectins. Direct visualization by the use of fluoresceinyl-glycosylated cytochemical markers. FEBS Letters 99, 329-332 (1979).
44) A. Novogrodski and A. Ashwell: Lymphocyte mitogenesis induced by a mammalian liver protein that specifically binds desialylated glycoproteins. Proc. Natl. Acad. Sci. USA 74, 676 (1977).
45) J.S. Lipsick, E.C. Beyer, S.H. Barondes, and N.O. Kaplan: Lectins from chicken tissues are mitogenic for Thy-1 negative murine spleen cells. Biochem. Biophys. Res. Commun. 97, 56 (1980).

46) V. Schirrmacher, P. Altevogt, M. Fogel, J. Dennis, C.A. Waller, D. Barz, R. Schwartz, R. Cheingsong-Popov, G. Springer, P.J. Robinson, T. Nebe, W. Brossmer, I. Vlodavsky, N. Paweletz, H.P. Zimmermann, and G. Uhlenbruck: Importance of cell surface carbohydrates in cancer cell adhesion, invasion and metastasis. Invasion Metastasis 2, 313 (1982).
47) H.-J.Gabius, unpublished observation.
48) T. Feizi: Demonstration by monoclonal antibodies that carbohydrate structures of glycoproteins and glycolipids are onco-developmental antigens. Nature 314, 53-57 (1985).

CONTROL OF GENE EXPRESSION BY OLIGONUCLEOSIDE METHYLPHOSPHONATES

Paul S. Miller, Cheryl H. Agris, Laure Aurelian*, Kathleen R. Blake, Shwu-Bin Lin, Akira Murakami, M. Parameswara Reddy, Cynthia Smith*, and Paul O.P. Ts'o
Division of Biophysics, Johns Hopkins University, Baltimore, Maryland, and *Department of Pharmacology, University of Maryland, Baltimore, Maryland

A major goal in understanding the processes of aging, cancer and differentiation is to understand gene expression and thus the function of various proteins in the overall biochemical processes of the cell. One of the classical ways to study gene expression is through the use of temperature sensitive mutants. Although this approach has been particularly effective in studying gene expression in bacteria and viruses, it is technically more difficult in eukaryotes, particularly mammalian cells. It would be desirable to have an alternative approach which would allow selective inhibition of gene expression either at the level of transcription or at the mRNA level. Recent studies have shown that such regulation may be achieved through the use of complementary DNAs (cDNAs) or anti-sense RNAs. The expression of mRNA can be regulated both in the test tube and in cells by cDNAs which selectively hybridize to a target mRNA. Control of cell-free mRNA translation in this manner is termed hybridization arrest (1). This procedure has been used to study the location and arrangement of adenovirus 2 genes within the viral genome and to analyze mRNA populations in mouse liver (2). Hybridization arrest has also been used to study the function of the 3'-non-coding region of globin mRNA (3). Recent studies in rabbit reticulocyte lysates suggest a helix destabilizing activity can disrupt cDNA hybrids with the coding region of mRNA and that effective inhibition of translation occurs only when the cDNA hybridizes to nucleotides including the AUG initiation codon or the 5'-terminal nucleotides of the mRNA (4).

Anti-sense RNAs can specifically block globin mRNA translation when injected into frog oocytes (5). The anti-sense RNA was synthesized in vitro by transcription of an inverted globin cDNA clone. As in the case in reticulocyte lysates, these studies showed that the 5'-region including the AUG initiation codon must be covered in order to inhibit translation.

Anti-sense RNAs transcribed from inverted cDNA clones, carried by plasmids can be used to specifically block targeted mRNA function in

B. Pullman et al. (eds.), Interrelationship Among Aging, Cancer and Differentiation, 207–219.

bacterial and mammalian cells which have been transfected with the plasmid. For example, Coleman et al. (6) find reductions in the amount of lpp protein or OmpC protein in E.coli cells transfected with plasmids that code for anti-sense lpp mRNA or anti-sense OmpC mRNA. Production of Herpes simplex virus type 1 (HSV-1) thymidine kinase (TK) is reduced dramatically in HSV-1 infected TK$^-$ mouse L-cells which have been transformed by a plasmid which encodes anti-sense HSV-1 TK mRNA (7). It is interesting to note that arrest of translation by anti-sense RNA also occurs naturally during osmoregulation of the Omp F protein of E.coli (8). In this case RNA complementary to the 5' end of Omp F mRNA is produced, and this anti-sense RNA inhibits translation of OmpF mRNA.

A somewhat different approach to the control of gene expression which is receiving increasing attention involves the use of synthetic, sequence-specific oligodeoxyribonucleotides. Zamecnik and Stephenson were the first to demonstrate that oligodeoxyribonucleotides (13 mers) complementary to the 5' and 3' terminal redundant sequences of Rous sarcoma virus (RSV) RNA could inhibit RSV protein synthesis in a cell-free translating system (9). Additional experiments showed that virus protein synthesis and replication are inhibited by the oligomer when it is applied directly to RSV-infected chicken fibroblasts (10).

A potential problem with this approach concerns the stability of the oligomer. Oligodeoxyribonucleotides are readily hydrolyzed by nucleases found in cells and the serum used in cell culture medium. Degradation of the oligomer would of course result in loss of selectivity and/or inhibitory activity. In addition to this problem, it is not known how well the negatively charged oligomers can pass through the phospholipid plasma cell membrane. It is conceivable that some restriction to uptake may occur as the length and net charge of the oligomer increases.

Research in our laboratory has focused on oligonucleotide analogs which can be taken up intact by cells in culture. As shown in Figure 1, these analogs contain a 3'-5' linked methylphosphonate internucleotide bond which replaces the phosphodiester bond found in naturally occurring nucleic acids (11,12). The methylphosphonate group is non-charged. As a result, the oligomer is quite lipophilic and is able to penetrate the plasma membrane of cells. The methylphosphonate linkage is also resistant to nuclease hydrolysis and therefore the oligomers have very long halflives in cell culture medium and inside cells (13). Previous studies with analogs ranging from three to seven nucleoside units in length have shown that they can inhibit tRNA aminoacylation and mRNA translation in cell-free systems, and that they can selectively inhibit protein synthesis in permeable E.coli cells (13,14).

We have recently developed solid phase synthetic procedures which allow the synthesis of oligomers up to 15 nucleoside units in length (15,16,17). We have also developed methods to characterize the

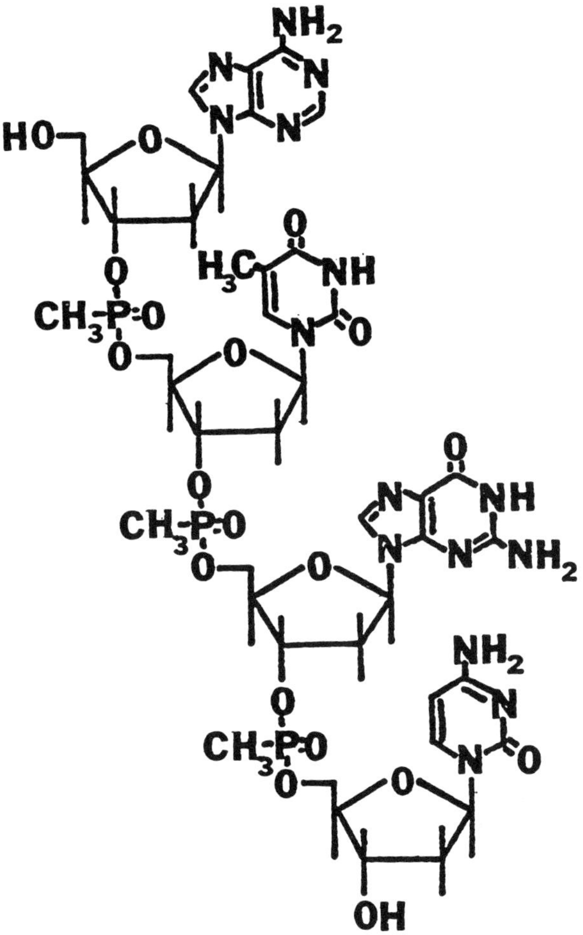

d-A$\underline{p}$T$\underline{p}$G$\underline{p}$C

Figure 1.
Structure of an oligonucleoside methylphosphonate. The symbol $\underline{p}$ represents (3'-5') methylphosphonate internucleoside bond.

chainlength and sequence of the methylphosphonate oligomers and to study their interactions with mRNA (18). Using these methods we have prepared a number of sequence specific oligodeoxyribonucleoside methylphosphonates complementary to functional regions of virus mRNAs. The effects of these oligomers on mRNA translation and processing has been studied both in the test tube and in virus-infected cells growing in culture.

Inhibition of VSV mRNA Translation by Oligodeoxyribonucleoside Methylphosphonates

Methylphosphonate oligomers complementary to the initiation codon regions of Vesicular stomatitis virus (VSV) N, NS and G protein mRNAs were synthesized (see Figure 2). The effects of these oligomers on cell-free translation of VSV mRNA in a rabbit reticulocyte lysate are currently being studied. The results of our studies with d-ApACAGACAT which is complementary to the N protein mRNA are shown in Table 1. At low concentrations (50 to 100 µM), the oligomer selectively inhibits N protein synthesis when N, NS and M protein mRNAs are translated simultaneously in a reticulocyte lysate.

The effects of these oligomers on the synthesis of the five VSV proteins in VSV-infected mouse L-cells were determined. Figure 3 shows the results obtained with d-ACAGACAT which is complementary to N protein mRNA. In contrast to the results obtained *in vitro*, the oligomer inhibits synthesis of all five virus proteins to approximately the same extent. At these concentrations, the oligomer has no detectable inhibitory effects on cellular protein synthesis by mouse L-cells and is not cytotoxic to the mouse L-cells as determined from mass culture growth curves. Thus the methylphosphonate oligomer appears to specifically inhibit virus function. The observation that synthesis of all five virus proteins is inhibited may be due to the known requirement of N protein for synthesis of VSV proteins in infected cells.

Table 1.

Effect of dApACAGACAT on Translation of Vesicular Stomatitis Virus mRNA at 30°C in a Rabbit Reticulocyte Lysate

Oligomer Concentration	% Inhibition(a)		
µM	N	NS	M
50	23	-4	-16
100	33	-1	-19
150	77	38	43

(a) Negative sign indicates stimulation of synthesis.

```
N  5'----------------------A-A-A-A-U-G-U-C-U-G-U-U-A-C-A--------------'3
                             3'-TpApCpApGpApCpA-'5  ANTI N

NS 5'----------------------A-U-C-A-U-G-G-A-U-A-A-U-C-U-C--------------'3
                             3'-TpApCpCpTpApTpTpA-'5 ANTI NS

G  5'----------------------A-C-U-A-U-G-A-A-G-U-C-G-C-U-U--------------'3
                             3'-TpApCpTpTpCpApCpG-'5 ANTIG
```

Figure 2.

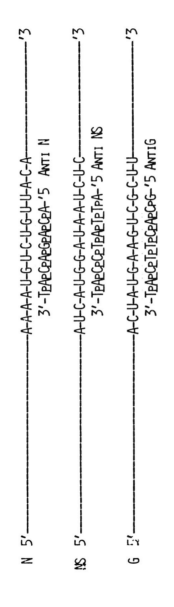

Partial nucleotide sequence of the initiation codon regions of Vesicular stomatitis virus N, NS, and G protein mRNAs. The sequence of the complementary oligonucleoside methylphosphonate is shown below each mRNA sequence.

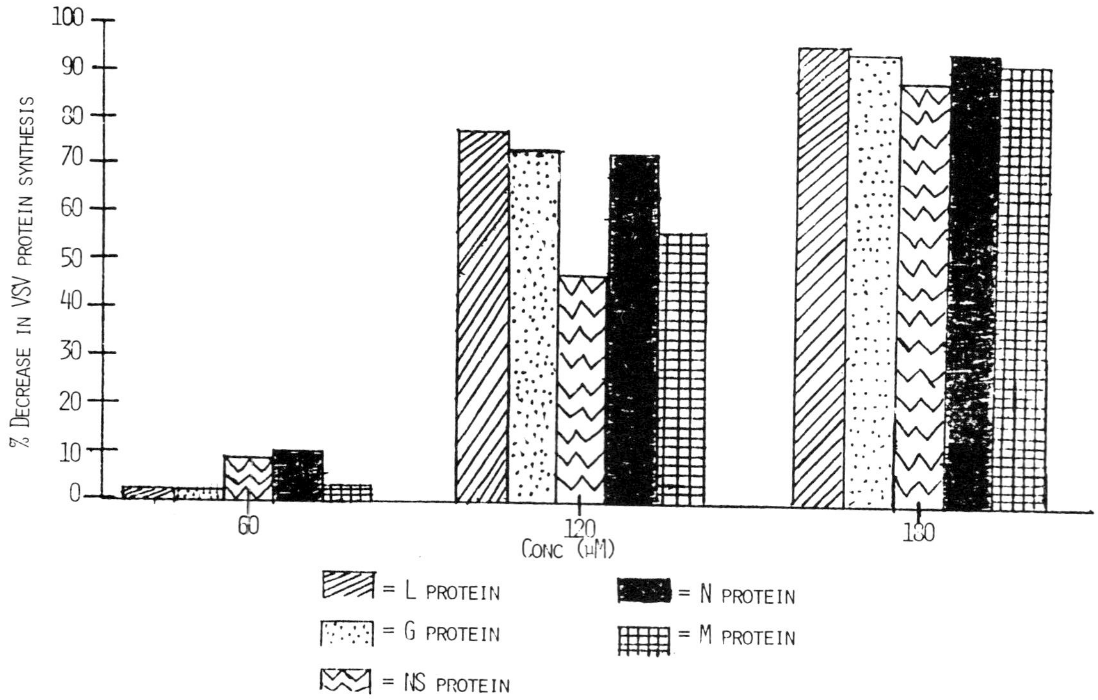

Figure 3.

Effect of d-ACAGACAT on Vesicular stomatitis virus protein synthesis in VSV-infected mouse L-cells.

Similar selective inhibition of VSV versus L-cell protein synthesis was also obtained with the methylphosphonate oligomers complementary to the NS and G protein mRNAs. Again inhibition of all five virus proteins was observed. In addition to their effects on VSV protein synthesis, each of the oligomers reduces virus titer approximately one log unit at an oligomer concentration of 150 μM.

The results of these experiments suggest that it may be possible to selectively control mRNA translation in virus infected cells. Further experiments are currently in progress to characterize the observed inhibition.

Inhibitory Effects of Oligodeoxyribonucleoside Methylphosphonates Complementary to the Splice Junction of Virus pre-mRNA

To explore the possibility of controlling gene expression at the level of mRNA processing, we have synthesized oligonucleoside methylphosphonates which are designed to interact with the donor and acceptor splice junction sequences of SV40 large T-antigen precursor mRNA (pre-mRNA). We have tested the methylphosphonate oligomers shown in Figure 4 which are complementary to the donor splice junction of SV40 large T-antigen and to the 5'-terminal sequences of U_1 RNA. U_1 RNA has been implicated as an agent involved in the splicing of mammalian and viral pre-mRNAs (19). The effects of the compounds on large T-antigen synthesis in SV40-infected African green monkey kidney cells (BSC40) were determined. The level of large T-antigen was measured by an immunoprecipitation technique using a monoclonal antibody directed against large T-antigen. As shown in the table accompanying Figure 4, the splice junction-complementary oligomer, d-AATACCTCA, and the two U_1 RNA-complementary oligomers each reduce the level of large T-antigen in the cells, while the non-specific oligomer, d-TTTTTT, had no effect. None of the oligomers had any effect on overall protein synthesis by the BSC40 cells.

We have also prepared a methylphosphonate oligomer, d-TpCCTCCTG, which is complementary to the acceptor splice junctions of Herpes simplex virus type 1 immediate early mRNA 4 and 5. The immediate early proteins are believed to be necessary for regulating the early and late genes of HSV-1 during infection. The oligomer was tested for its effects on HSV-1 in virus infected Vero and human cells. Virus growth was inhibited two log units by 300 μM oligomer. The oligomer selectively inhibits HSV-1 protein synthesis and DNA synthesis in the infected cells. It has no appreciable effect on cellular protein or DNA synthesis. Of particular interest is the effect of d-TpCCTCCTG on the virus proteins. Immediate early mRNA 4 codes for a 68 K protein. This protein is virtually absent in virus infected cells treated with the oligomer. Additional studies are in progress to further characterize the inhibitory effects of this oligomer.

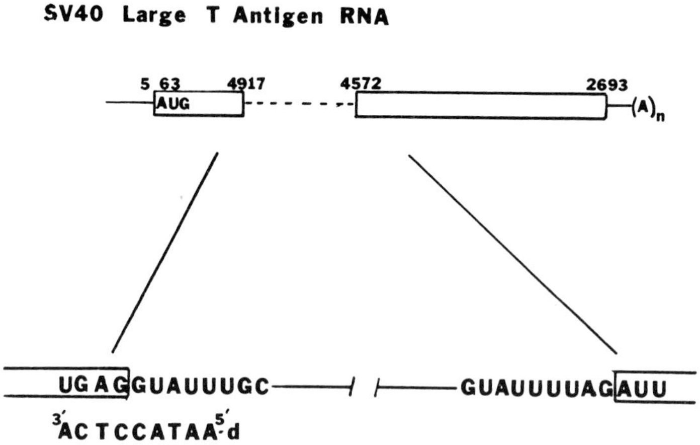

OLIGOMER (100 μM)	% INHIBITION OF T-ANTIGEN IN SV40 INFECTED CELLS
d-ApApTpApCpCpTpCpA	19%
d-CpCpApGpGpT (U1 RNA)	52%
d-GpGpTpApApG (U1 RNA)	29%
d-TpTpTpTpTpT	0%

Figure 4.

Partial nucleotide sequence of the splice junctions of SV40 large T-antigen pre-mRNA. The sequence of the methylphosphonate oligomer complementary to the donor splice junction is shown below the premRNA sequence. The table shows the effects of this oligomer and oligomers complementary to U_1 RNA on large T-antigen synthesis in SV40-infected BSC40 cells.

Oligodeoxyribonucleoside Methylphosphonates Which Can Crosslink with mRNA

In the experiments described above, rather high concentrations, 100 to 300 μM, of oligomers are required to obtain significant inhibitory effects. This is due to the equilibrium nature of the interaction between the oligomer and its target mRNA. It appears more effective inhibition could be obtained if the oligomer could crosslink with the mRNA. To test this idea we have prepared oligonucleoside methylphosphonates derivatized at their 5'-end with aminomethyltrimethylpsoralen (AMT), a photoactivatable crosslinking reagent. The structure of the oligomer is shown in Figure 5. We have prepared d-AMT-pTpGCACCAT which is is complementary to the initiation codon region of rabbit β globin mRNA and partially complementary to the initiation codon region of α globin mRNA. The AMT group is opposite a U residue when the oligomer is complexed with β globin mRNA, whereas it lies opposite a G residue when complexed with α globin mRNA. Since the 3,4 double bond of psoralen can form cyclobutane type adducts with the 5,6 double bonds of U and C residues upon irradiation with 365 nm light, the d-AMT-pTpGCACCAT should be able to form a photoadduct with β globin mRNA.

When a mixture of [^{32}P]-labeled d-AMT-p-TpGCACCAT and rabbit globin mRNA are photoirradiated and then subjected to agarose gel electrophoresis, radioactivity is observed to migrate with the mRNA (see Figure 6). No radioactivity is observed in the position of mRNA in the absence of irradiation or in the absence of mRNA. No degradation of the mRNA as a result of irradiation was detected in these experiments. These results are consistent with covalent bond formation between the mRNA and the d-AMT-pTpGCACCAT.

The effect of d-AMT-pTpGCACCAT on mRNA translation was also tested. After photoirradiation, translation was inhibited 75% (β globin) and 56% (α globin) by 25 μM AMT-pTpGCACCAT. In contrast, 25 μM d-TpGCACCAT had no inhibitory effect. Comparable inhibition by d-TpGCACCAT, 70% β globin, 73% α globin, is observed at 200 μM concentration. The nonselective inhibition by d-AMT-p-TpCGACCAT was expected since translation of α and β globin mRNA is coordinated in the reticulocyte system (20,21). The results of these experiments suggest that crosslinking will be an effective way to increase the efficiency and possibly the selectivity of inhibition by methylphosphonate oligomers. In addition, covalent bond formation between the oligomer and target mRNA should provide a way for studying the mechanism of inhibition in a definitive manner.

The results of our experiments suggest that oligodeoxyribonucleoside methylphosphonates may be used to specifically control gene expression in living cells by interfering with translation or splicing of mRNA. Because the specificity of these compounds resides in their ability to bind to complementary nucleic acid sequences, it should be possible to use nucleic acid sequence information to design novel derivatives of methylphosphonate oligomers which can be used to probe the function of specific proteins in normal, transformed or virus-infected cells. These oligonucleotide analogs may eventually find use as anti-viral or chemotherapeutic agents.

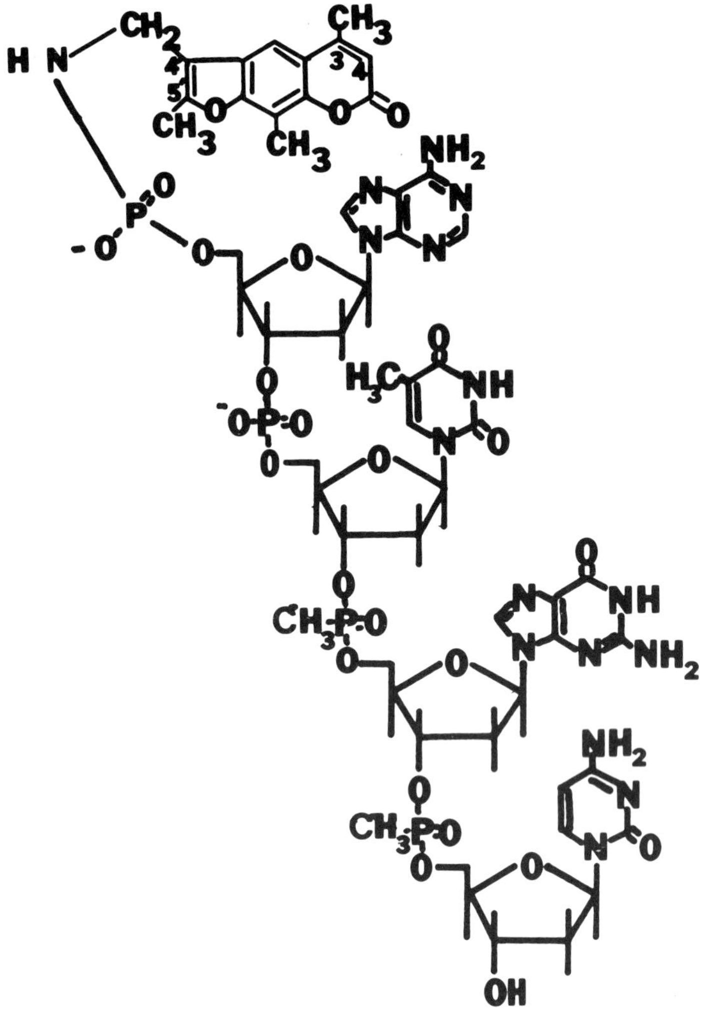

Figure 5.

Structure of an oligodeoxyribonucleoside methylphosphonate derivatized with trimethylaminomethylpsoralen: d-AMT-pApTGC.

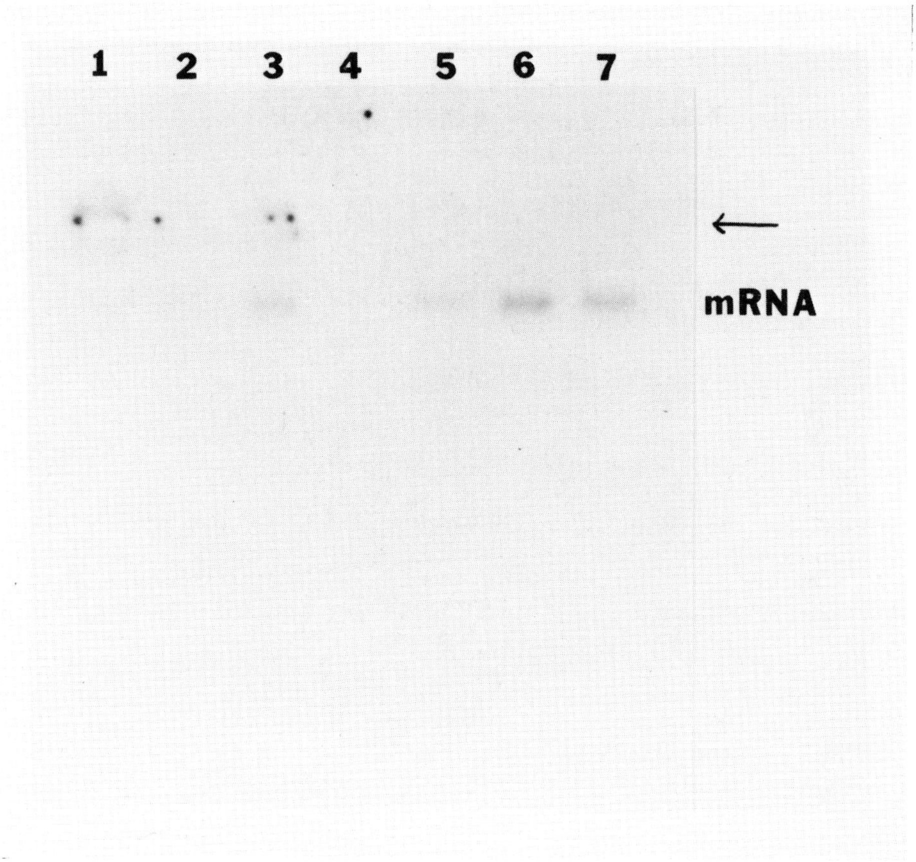

Figure 6.

Photocrosslinking of [^{32}P]-labeled d-AMT-pTpGCACCAT (0.056 μM) with rabbit globin mRNA (0.150 μM) after irradiation at 0°C. The reaction mixtures were electrophoresed on a 2.5% agarose gel. Lanes 1-3: Oligomer and mRNA irradiated for 40, 80 and 120 min. respectively. Lanes 4-7: Oligomer and mRNA were preannealed for 30 sec. at 100° and then irradiated for 0, 40, 80 and 120 min. respectively. The positions of the top of the gel (←--) and the mRNA are indicated.

Acknowledgement

The authors wish to thank Dr. Thomas Kelly and Mr. Ronald Wides for their help in working with SV40. This research was supported in part by a grant from the National Institutes of Health (GM 31927) and by the Albert Szent-Gyorgyi Foundation.

References

1. Paterson, B.M., Roberts, B.E. and Kuff, E.L.:1977, Proc. Natl. Acad. Sci. 74, pp. 4370-4374.

2. Hastie, N.D. and Held, W.A.:1978, Proc. Natl. Acad. Sci. 75, pp. 1217-1221.

3. Kronenberg, H.M., Roberts, B.E. and Efstratiadis, A.:1979, Nucleic Acids Res. 6, pp. 153-166.

4. Liebhaber, S.A., Cash, F.E. and Shakin, S.H.:1984, J. Biol. Chem. 259, pp. 15597-15602.

5. Melton, D.A.:1985, Proc. Natl. Acad. Sci. 82, pp. 144-148.

6. Coleman, J., Green, P.J. and Inouye, M.:1984, Cell 37, pp. 429-436.

7. Izant, J.G. and Weintraub, H.:1984, Cell 36, pp. 1007-1015.

8. Mizuno, T., Chou, M.-Y., and Inouye, M.:1984, Proc. Natl. Acad. Sci. 81, pp. 1966-1970.

9. Zamecnik, P.C. and Stephenson, M.L.:1978, Proc. Natl. Acad. Sci. 75, pp. 280-284.

10. Stephenson, M.L. and Zamecnik, P.C.:1978, Proc. Natl. Acad. Sci. 75, pp. 285-288.

11. Miller, P.S., Yano, J., Yano, E., Carroll, C., Jayaraman, K. and Ts'o, P.O.P.: 1979, Biochemistry 18, pp. 5134-5143.

12. Kan, L.-S., Cheng, D.M., Miller, P.S., Yano, J. and Ts'o, P.O.P.: 1980, Biochemistry 19, pp. 2122-2132.

13. Miller, P.S., McParland, K.B., Jayaraman, K., and Ts'o, P.O.P.: 1981, Biochemistry 20, pp. 1874-1880.

14. Jayaraman, K., McParland, K., Miller, P. and Ts'o, P.O.P.:1981, Proc. Natl. Acad. Sci. 78, pp. 1537-1541.

15. Miller, P.S., Agris, C.H., Blandin, M., Murakami, A., Reddy, M.P., Spitz, S.A., and Ts'o, P.O.P.:1983, Nucleic Acids Res. 11, pp 5189-5204.

16. Miller, P.S., Agris, C.H., Murakami, A., Reddy, M.P., Spitz, S.A., and Ts'o, P.O.P.:1983, Nucleic Acids Res. 11, pp. 6225-6242.

17. Miller, P.S., Reddy, M.P., Murakami, A., Blake, K.R., Lin, S.B., and Agris, C.H.:1985, Biochemistry, submitted for publication.

18. Murakami, A., Blake, K.R., and Miller, P.S.:1985, Biochemistry 24, in press.

19. Lerner, M.R., Boyle, J.A., Mount, S.M., Wolin, S.L., and Steitz, J.A.:1980, Nature 283, pp. 220-223.

20. Blake, K.R., Murakami, A. and Miller, P.S.:1985, Biochemistry, submitted for publication.

21. Blake, K.R., Murakami, A., Spitz, S.A., Glave, S.A., Reddy, M.P., Ts'o, P.O.P., and Miller, P.S.:1985, Biochemistry, submitted for publication.

GENETIC MECHANISMS IN TUMOR PROGRESSION, HETEROGENEITY, AND METASTASIS

Isaiah J. Fidler
Department of Cell Biology
The University of Texas System Cancer Center
M. D. Anderson Hospital and Tumor Institute at Houston
6723 Bertner Avenue (173)
Houston, Texas 77030, U.S.A.

ABSTRACT. By the time of diagnosis, most malignant neoplasms are heterogeneous and contain numerous subpopulations of cells with diverse biologic characteristics, which include the ability to invade and produce metastasis. The metastatic process is not random. Rather, metastases result from the survival and proliferation of a few specialized metastatic cells that preexist within parent neoplasms. Diversity for the metastatic phenotype may be a consequence of the multicellular origin of a neoplasm, or it may be the result of continuous evolution and progression of tumors either unicellular or multicellular in origin. Metastatic variants, in general, are less phenotypically stable than benign or nonmetastatic clones, and they also exhibit higher rates of spontaneous mutations. Although metastases may have a clonal-unicellular origin, the genetic instability of these cells coupled with host selection pressure can rapidly generate biologic diversity among and within metastases. These data support the concept that tumor evolution and progression toward increased malignancy could be regulated by genetic mechanisms.

INTRODUCTION

The spread of cancer cells from the primary neoplasm to distant organ sites remains the most devastating aspect of neoplasia. Therefore, the major challenge facing the oncologist in treating cancer is how to eradicate metastases. Despite major advances in surgical treatment of primary neoplasms and in development of aggressive adjuvant chemotherapy, the majority of deaths from malignant melanoma are due to the uncontrolled proliferation of metastases that are resistant to conventional therapies (1). There are several reasons for the present failure in treating metastases. First, by the time of diagnosis, metastatic lesions may exist in many organs to which selective therapeutic agents cannot be delivered. Second, metastases may be large and contain at least 10^8 cells. Thus, the destruction of even 99% of their population would leave 10^6 cells to once again proliferate (2). Third, most malignant neoplasms are heterogeneous

B. Pullman et al. (eds.), Interrelationship Among Aging, Cancer and Differentiation, 221–231.

and consist of multiple subpopulations of cells with different biologic properties. Cells obtained from individual tumors can differ with respect to cell surface properties, antigenicity, immunogenicity, growth rate, karyotype, sensitivity to various cytotoxic drugs, and the ability to invade and metastasize (review 1-5). Biologic heterogeneity is not confined to cells in primary tumors and is equally prominent among cells that populate metastases (1,5). Indeed, many clinical observations suggest that multiple metastases proliferating in different organs or even within the same organ are diverse in many characteristics such as hormone receptors, antigenicity and immunogenicity, and sensitivity to various chemotherapeutic drugs (1,5).

We have been investigating the mechanisms that regulate tumor progression, the development of biologic heterogeneity in neoplasms, and the pathogenesis of cancer metastasis. Some of the progress gained in these investigations is summarized below.

The Pathogenesis of Metastasis

Metastasis may be defined as the formation of a lesion that lacks continuity with the parent tumor. This capacity to metastasize is the sine qua non of truly malignant tumor cells. To produce a clinically relevant metastasis, malignant tumor cells must complete a sequence of potentially lethal interactions with host homeostatic mechanisms. Although the process of cancer metastasis is a dynamic one that passes from one phase to another without interruption, it can be divided into a series of sequential steps (6). Metastasis begins with the local invasion of host stroma by several different mechanisms. Rapidly proliferating tumors may create mechanical forces that propel fingerlike cords of cells along fascial planes or other areas of low resistance (7). In contrast, tumors that grow within the major body cavities can shed cells that seed the mucosal and serosal surfaces of other organs, thus establishing secondary growths. Thin-walled venules, like lymphatic channels, offer very little resistance to penetration by tumor cells and provide the most common pathway for access into the circulation. Although clinical observations have suggested that carcinomas frequently metastasize and grow through the lymphatic system, malignant tumors of mesenchymal origin more frequently spread by the hematogenous route. However, the presence of numerous venolymphatic anastomoses makes this conceptualization questionable (8). Detachment and embolization of small tumor cell aggregates next occurs. Most tumor cells in the circulatory system are rapidly destroyed (9). Thus, the presence of tumor cells in the blood seems to be of little prognostic value. Once the tumor cells have survived the hostile environment of the circulation, they come to rest in the capillary beds of organs, either by adhering to capillary endothelial cells or by adhering to subendothelial basement membrane that may be exposed (10). Extravasation next occurs, probably by the same mechanisms that influence initial invasion. Newly established metastases may give rise to other metastases, the so-called metastasis of metastases (1,6,11).

Failure to complete any step in metastasis leads to the elimination of the disseminating tumor cell. For example, the presence of tumor cells in the circulation does not predict that metastasis will occur, because most tumor cells that enter the bloodstream are rapidly eliminated. Using radiolabeled tumor cells, we found that 24 hr after entry into the circulation, <1% of the cells were still viable and <0.1% of tumor cells placed into the circulation survived to produce metastases (9). Observations such as this prompted us to question whether the 0.1% of the circulating tumor cells responsible for the development of metastases survived at random, or whether the metastases resulted from the selective survival and growth of preexistent subpopulations of cells endowed with special properties (12). Data generated by us and many others support the latter possibility.

The Heterogeneous Nature of Malignant Neoplasms

During the last decade, the concept that neoplasms are heterogeneous and contain multiple subpopulations of cells with different biologic properties has gained wide acceptance (review 1-5). This concept, however, is not new. Almost a century ago, Paget (13) analyzed autopsies of more than 700 women with breast cancer and concluded that the organ distribution of metastasis was not random and hence not owed to chance. Paget postulated that a few tumor cells ("seed") traveling by vascular routes had affinity for growth in the environment provided by certain organs ("soil"). Only when the "seed" and "soil" were matched did metastases develop. Recently, this hypothesis has received considerable support (14). Site-specific metastasis occurs with many transplantable experimental tumors (1,4-7,10-15) and has been reported recently in autochthonous human tumors in patients with peritoneovenous shunts (16).

The first direct evidence for metastatic heterogeneity of neoplasms was provided by Margaret Kripke and I in 1977 working with the murine B16 melanoma (12). Based on the modified fluctuation assay of Luria and Delbruck (17), research showed that different tumor cell clones, each derived from individual cells isolated from the parent tumor, varied dramatically in their potential to produce lung metastases following intravenous inoculation into syngeneic mice. Control subcloning procedures demonstrated that the observed metastatic diversity was not a consequence of the in vitro cloning procedure (12).

The validity of these initial observations has been substantiated by numerous subsequent experiments. Exactly comparable data have been reported with another murine melanoma of more recent origin than that of the B16 melanoma (18). The K-1735 melanoma was isolated by Kripke in a C3H/HeN mouse that had been exposed to UV radiation (19). The primary K-1735 tumor was established in culture after one passage in an immunodeficient mouse. Clones were produced from parent cells after five in vitro passages. The clones differed markedly from each other and from the parent line in their propensity for metastasis and the preferential localization of the metastases in the lungs, lymph nodes, and other organs, such as the brain (18). The emergence of heterogeneity with regard to metastatic potential in neoplasms does not require a long latent interval. This conclusion is based on the results of experiments showing that fibroblasts transformed in vitro by an oncogenic mouse sarcoma virus develop a similar spectrum of metastatic variability within a month of initial transformation (20). The finding that preexisting tumor cell subpopulations growing in the same tumor exhibit heterogeneous metastatic potential has since been confirmed in numerous laboratories using a wide range of experimental animal tumors of different origins and histological classification (review 1,3,7). Moreover, investigators using young nude mice as models for metastasis of human neoplasms have recently demonstrated that several human tumor lines also contain subpopulations of cells with different metastatic potentials (21,22).

Additional recent studies from my laboratory illustrate that the selection of tumor cells with increased metastatic potential can occur during the process of metastasis. The metastatic capability of cells harvested from spontaneous metastases was compared with that of cells harvested from the parent tumor. Using four different tumor systems and three variant lines from these tumors, my coworkers and I consistently demonstrated that cells populating spontaneous metastases were more metastatic than their respective heterogeneous parent lines (23). These data reinforce those of earlier studies (24,25) and confirm that the process of metastasis selects for tumor cells with increased metastatic potential (26).

Origin of Biologic Heterogeneity in Primary Neoplasms

There are several possible mechanisms that could explain the origin of tumor cell heterogeneity. Heterogeneity could certainly result from the multicellular origin of the tumor (27). However, most human cancers are thought to result from the clonal expansion of single transformed cells (28). For example, studies using specific allotypic immunoglobulins or glucose-6-phosphate dehydrogenase polymorphism have shown that chronic myelogenous leukemia, Burkitt's lymphoma, and multiple myeloma probably arise from single cells, whereas colonic carcinomas are thought to have a muticellular origin (29). For tumors of unicellular origin, heterogeneity could result from the process of tumor progression (30).

Clinical and histological observations of neoplasms have suggested that tumors undergo a series of changes during the course of disease. For example, a tumor initially diagnosed as benign can, over a long period, evolve into a malignant tumor. To explain the process of tumor evolution and progression originally proposed by Foulds (30), Nowell (28) suggested that acquired genetic variability within developing clones of tumors, coupled with host selection pressures, can bring about the emergence of new tumor variants with increased growth autonomy, that is, malignancy.

The hypothesis forwarded by Nowell (28) predicts that tumor cells progressing from the benign to the malignant state will exhibit increased genetic instability. To test this hypothesis, we examined the metastatic stability and the rates of mutation to ouabain resistance or 6-thioguanine resistance of paired metastatic and nonmetastatic cloned lines isolated from four different mouse neoplasms (31). In all the tumors, which incidentally were of different histological classifications, highly metastatic cells were phenotypically less stable than their nonmetastatic counterparts isolated from the same neoplasm. Moreover, in highly metasatic clones, the rate of spontaneous mutation was severalfold higher than it was in weakly metastatic clones (31). These results are in accord with the hypothesis that tumor progression occurs as a result of acquired genetic alterations. Similar data have now been published by other investigators (32-34). Additional evidence that genetic mechanisms can be responsible for tumor progression comes from mutagenesis experiments using nitrosoguanidine (35) or ultraviolet radiation (36). In these experiments, the treatment of tumor cell populations with mutagens resulted in the emergence of new tumor cell variants, some of which exhibited increased tumorigenicity and metastatic capacities. Others, however, had decreased tumorigenic potential (34). Collectively these types of data suggest that the more metastatic a tumor cell population is, the greater the likelihood that its constituent cells will undergo spontaneous mutations and thus biologic diversification.

Tumors of unicellular origin may exhibit metastatic heterogeneity at early stages in their development. We base this conclusion on data generated by our studies of the in vivo behavior of murine embryo fibroblasts transformed by an oncogenic virus (20). Six colonies of BALB/c embryo fibroblasts, each derived from a single infected cell, were grown as individual cell lines. Injection of cells from each clone resulted in marked differences with regard to the production of lung nodules. Similarly, when the clones from two colonies (one of high and one of low experimental metastatic capacity) were subcloned and evaluated in the same manner, both clones exhibited a pattern of metastatic heterogeneity. Interestingly, the clone with higher metastatic capacity exhibited a greater degree of variability than the clone with lower metastatic capacity. Thus, despite originating from a single cell, by the time of the first subcloning 6 weeks after initial transformation, the clone already contained subpopulations of cells with different metastatic properties (20).

Origin and Development of Heterogeneity in Metastases

Biologic heterogeneity is not confined to primary tumors. Quite the contrary, biologic heterogeneity is most prominent in and among different metastases. Multiple metastases proliferating in the same host, even in the same organ, often exhibit diversity in many characteristics, such as hormone receptors, antigenicity or immunogenicity, and response to various chemotherapeutic agents (1-4). This diversity may be due to the nature of the pathogenesis of metastasis, to the process of tumor evolution and progression, or to both (1,37).

Pathologists have long been aware that primary neoplasms exhibit different morphological appearances in different areas of the tumor (38). For this reason, the malignant or benign nature of a tumor cannot be determined without examining of multiple sections. The zonal differences in tumors are not restricted to morphology alone, but include biologic characteristics such as growth rate, sensitivity to cytotoxic drugs, antigenicity, and pigmentation (39). Thus, tumor cell aggregates that enter the circulation from one zone of the tumor may be qualitatively different from cell aggregates that enter from another zone. Moreover, if an embolic aggregate originates from a relatively homogeneous zone of a primary tumor, it could exhibit a degree of uniformity for a specific characteristic. Under these conditions, irrespective of whether only one tumor cell or several tumor cells survive to proliferate in distant organs, the resulting metastasis is analogous to a primary tumor of unicellular origin with regard to the subsequent development of heterogeneity. If a mixed tumor cell embolus derived from an area of zonal junctions enters the circulation, the unicellular or multicellular origin of the metastasis would depend on whether a single cell or multiple cells survived to give rise to the metastasis.

To determine whether individual metastases are clonal in origin and whether different metastases can be produced by different progenitor cells, we carried out a series of experiments based on the fact that X-irradiating tumor cells induces random chromosome breaks and rearrangements (37). We examined the metastases that arose from subcutaneous tumors produced by K-1735 mouse melanoma cells that had been X-irradiated to induce chromosomal breaks and rearrangements. We reasoned that if a metastasis were derived from a single cell, all the chromosome spreads examined within an individual metastasis would exhibit the same karyotype. In contrast, if a metastasis had been formed from more than one progenitor cell, its constituent cells would exhibit different chromosomal arrangements, assuming that the different cells involved carried distinguishable karyotypic markers. We analyzed the cellular composition of 21 individual metastases after establishment of cell lines from individual lesions. In 11 lesions we found unique karyotypic patterns of abnormal, marker chromosomes, suggesting that each metastasis originated from a single progenitor cell. Moreover, the finding that different metastases are populated exclusively by cells with different chromosome markers (37) indicates that different metastases originate from different progenitor cells.

In the next set of experiments, we examined whether melanoma metastases arose as a consequence of individual cells surviving in the bloodstream or whether metastases resulted from homogeneous clumps (i.e., a multicellular embolus of cells with the same chromosome marker) surviving in the circulation (40). Mice were injected with mixed aggregates of cells from two highly metastatic populations of K-1735 melanoma cells with unique karyotypes. Several weeks later, lung metastases were harvested, established as individual cell lines in culture, and chromosome analysis for at least 60 spreads of each individual line was performed. In all cases, a solitary metastasis consisted of cells with the same karyotype (40).

Analysis of the distribution and fate of circulating tumor emboli has demonstrated that multicellular aggregates are more likely to give rise to a metastasis than a unicellular tumor embolus (41). The aggregates introduced into the circulation of the mice were large and contained greater than 20 cells of both cell lines. The results therefore indicate that the metastases resulted from the proliferation of a single viable cell within an embolus. Thus, whether an embolus is homogeneous or heterogeneous, metastases can have a unicellular origin (37,40). Similar results have been obtained in experiments using B16 melanoma cell clones bearing identifiable biochemical markers (42,43). This study not only revealed that the majority of metastases are of clonal origin, but also that variant clones with diverse phenotypes are formed rapidly to generate significant cellular diversity within individual metastases (42).

These observations indicate that different metastases can arise from different progenitor cells (i.e., interlesional heterogeneity), and they account for the well-documented differences in the behavior of individual metastases in the same patient, including differences in response to therapy. However, even within individal metastases of proven clonal origins, heterogeneity can develop rapidly to create significant intralesional heterogeneity.

Generation of Biologic Diversity in a Metastasis

The findings that metastatic cells exhibit higher rates of spontaneous mutations than do nonmetastatic cells (31-34) and that heterogeneity develops more rapidly in tumors containing few subpopulations of cells (44,45) suggest that tumor evolution and progression will accelerate in solitary metastases, especially when such lesions are of clonal origin (37). The results of two major experiments reveal that this indeed is the case.

Isolation of multiple clones from individual lung metastases produced by the B16 melanoma revealed that small metastases, 6-25 days old exhibit intralesional clonal homogeneity (42). At the same time, the metastatic phenotypes of clones isolated from different metastases in the same animal differed significantly (i.e., interlesional clonal heterogeneity). In contrast, large metastases, 40-45 days old, were heterogeneous. These data indicate that in the B16 melanoma individual lung metastases are populated initially by cells with uniform metastatic phenotypes (42,43). This is in agreement with the data discussed earlier showing that the majority of experimental metastases result from the arrest and proliferation of a single tumor cell (37,40).

Similar findings were obtained with a different murine tumor (46). A highly metastatic cloned line of the K-1735 melanoma with a stable and unique marker chromosome was used. The line was cloned immediately after establishment in vitro. The cells were also injected subcutaneously, and the skin tumor and spontaneous lung metastases were recovered 60-90 days later. The experimental metastatic ability was determined for each of these lines produced in vitro or selected in vivo. Six of the ten clones derived in vitro and six of seven lines established from spontaneous metastases differed significantly from the parental cell line in their metastatic ability. The relative sensitivities to amsacrine, vincristine, bleomycin, and doxorubicin also varied greatly among and within different metastases, in clones isolated from these lesions and grown in vitro and in the original parent tumor cell line. These findings clearly indicate that even with metastases of clonal origin, heterogeneity for metastasis and sensitivity to chemotherapy can develop rapidly (46).

Conclusions

Neoplasm heterogeneity, however it originates, has important implications for studying metastasis and for developing new treatment modalities. The most formidable obstacle to successfully treating of metastases may well be the biologic heterogeneity of the cells of metastases. This phenotypic diversity, which allows selected variants to develop, means that different metastases can exhibit different levels of sensitivities to therapeutic agents. That different metastases can originate from different progenitor cells accounts for the biologic diversity that exists among various metastases. However, even within a solitary metastasis of proven clonal origin, heterogeneity for various characteristics can develop rapidly, possibly resulting from phenotypic instability associated with clonal populations or from a high rate of spontaneous mutation exhibited by cells populating metastases, or both.

The pathogenesis of metastasis is a complex and selective process, and the growth of metastases represents the end point of many destructive events that few tumor cells survive. The data demonstrating metastatic heterogeneity in neoplasms and those showing that the outcome of metastasis is also dependent on host factors support the concept that metastasis is not a random process. Notwithstanding its implications for the value of current therapeutic strategies, this finding is cause for optimism. Metastasis is a selective biologic process, governed by mechanisms that can be studied and ultimately understood in sufficient detail to allow rational therapeutic intervention.

REFERENCES

1. Fidler, I. J. 'The evolution of biological heterogeneity in metastatic neoplasms.' In: Nicolson, G. L. and Milas, L. (eds.), Cancer Invasion and Metastasis: Biologic and Therapeutic Aspects. New York, Raven Press, 1984, p. 5.
2. Fidler, I. J. and Hart, I. R. 'Biological diversity in metastatic neoplasms: Origins and implications.' Science **217**:998, 1982.
3. Heppner, G. 'Tumor heterogeneity.' Cancer Res **214**:2259, 1984.
4. Fidler, I. J., and Poste, G. 'The heterogeneity of metastatic properties in malignant tumor cells and regulation of the metastatic phenotype.' In: Owen, A., Coffey, D. S., and Baylin, S. B. (eds.), Tumor Cell Heterogeneity. New York, Academic Press, 1982, p. 127.
5. Hart, I. R., and Fidler, I. J. 'The implications of tumor heterogeneity for studies on the biology and therapy of cancer metastasis.' Biochim. Biophys. Acta **651**:37, 1981.
6. Poste, G., and Fidler, I. J. 'The pathogenesis of cancer metastasis.' Nature **283**:139, 1980.
7. Poste, G. 'Experimental systems for analysis of the malignant phenotype.' Cancer Metastasis Rev **1**:141, 1982.
8. Fisher, B., and Fisher, E. R. 'The interrelationship of hematogenous and lymphatic tumor cell dissemination.' Surg. Gynecol. Obstet. **122**:791, 1966.
9. Fidler, I. J. 'Metastasis: Quantitative analysis of distribution and fate of tumor emboli labeled with ^{125}I-5-iodo-2'-deoxyuridine.' JNCI **45**:773, 1970.
10. Nicolson, G. L. 'Organ colonization and the cell surface properties of malignant cells.' Biochim. Biophys. Acta **695**:113, 1982.
11. Fidler, I. J., Gersten, D. M., and Hart, I. R. 'The biology of cancer invasion and metastasis.' Adv. Cancer Res **28**:149, 1978.
12. Fidler, I. J., and Kripke, M. L. 'Metastasis results from preexisting variant cells within a malignant tumor.' Science **197**:893, 1977.
13. Paget, S. 'The distribution of secondary growths in cancer of the breast.' Lancet **1**:571, 1889.
14. Hart, I. R. '"Seed and soil" revisited: Mechanisms of site-specific metastasis.' Cancer Metastasis Rev 1:5, 1982.
15. Price, J. E., Carr, D., Jones, L. L., Messer, P., and Tarin, D. 'Experimental analysis of factors affecting metastatic spread using naturally occurring tumors.' Invasion and Metastasis **2**:77, 1982.
16. Tarin, D., Price, J. E., Kettlewell, M. G. W., Souter, R. G., Vass, A. C. R., and Crossley, B. 'Mechanisms of human tumor metastasis studied in patients with peritoneovenous shunts.' Cancer Res. **44**:3584, 1984.
17. Luria, S. E., and Delbruck, M. 'Mutations of bacteria from virus sensitivity to virus resistance.' Genetics **28**:491, 1943.

18. Fidler, I. J., Gruys, E., Cifone, M. A., Barnes, Z., and Bucana, C. 'Demonstration of multiple phenotypic diversity in a murine melanoma of recent origin.' JNCI **67**:947, 1981.
19. Kripke, M. L. 'Speculations on the role of ultraviolet radiation in the development of malignant melanoma.' JNCI **63**:541, 1979.
20. Fidler, I. J., and Hart, I. R. 'The origin of metastatic heterogeneity in tumors.' Eur. J. Cancer **17**:487, 1981.
21. Kozlowski, J. M., Hart, I. R., Fidler, I. J., and Hanna, N. 'A human melanoma line heterogeneous with respect to metastatic capacity in athymic nude mice.' JNCI **72**:913, 1984.
22. Kozlowski, J. M., Fidler, I. J., Campbell, D., Xu, Z., Kaighn, M. E., and Hart, I. R. 'Metastatic behavior of human tumor cell lines grown in the nude mouse.' Cancer Res. **44**:3522, 1984.
23. Talmadge, J. E., and Fidler, I. J. 'Enhanced metastatic potential of tumor cells harvested from spontaneous metastases of heterogeneous murine tumors.' JNCI **69**:975, 1982.
24. Fidler, I. J. 'Selection of successive tumor lines for metastases.' Nature **242**:148, 1973.
25. Nicolson, G. L., and Custead, S. E. 'Tumor metastasis is not due to adaptation of cells to a new organ environment.' Science **215**:176, 1982.
26. Talmadge, J. E., and Fidler, I. J. 'Cancer metastasis is selective or random depending on the parent tumor population.' Nature **27**:593, 1982.
27. Reddy, A. L., and Fialkow, P. J. 'Multicellular origin of fibrosarcomas in mice induced by the chemical carcinogen 3-methylcholanthrene.' J. Exp. Med. **150**:878, 1980.
28. Nowell, P. C. 'The clonal evolution of tumor cell populations.' Science **194**:23, 1976.
29. Fialkow, P. J. 'Clonal origin of human tumors.' Biochim. Biophys. Acta. **458**:283, 1976.
30. Foulds, L. 'The experimental study of tumor progression: A review.' Cancer Res. **14**:327, 1954.
31. Cifone, M. A., and Fidler, I. J. 'Increasing metastatic potential is associated with increasing genetic instability of clones isolated from murine neoplasms.' Proc. Natl. Acad. Sci. USA **78**:6949, 1982.
32. Bosslet, K., and Schirrmacher, V. 'High frequency generation of new immunoresistant tumor variants during metastasis of a cloned murine tumor line (ESb).' Int. J. Cancer **29**:195, 1982.
33. Harris, J. F., Chambers, A. F., Hill, R. P., and Ling, V. 'Metastatic variants are generated spontaneously at a high rate in mouse KHT tumor.' Proc. Natl. Acad. Sci. USA **79**:5547, 1982.
34. Hill, R. P., Chambers, A. F., and Ling, V. 'Dynamic heterogeneity: Rapid generation of metastatic variants in mouse B16 melanoma cells.' Science **224**:998, 1984.
35. Boon, T., Snick, J. V., and Pel, A. V. 'Immunogenic variants obtained by mutagenesis of mouse mastocytoma P815. II. T lymphocyte mediated cytolysis.' J. Exp. Med. **152**:1184, 1980.

35. Fisher, M. S., and Cifone, M. A. 'Enhanced metastatic potential of murine fibrosarcomas treated in vitro with ultraviolet radiation.' Cancer Res. **41**:3018, 1981.
37. Talmadge, J. E., Wolman, S. R., and Fidler, I. J. 'Evidence for the clonal origin of spontaneous metastases.' Science **217**:361, 1982.
38. Clark, W. H., Elder, D. E., Guerry, D., Epstein, M. N., Greene, M. H., and van Horn, M. 'A study of tumor progression: The precursor lesions of superficial spreading and nodular melanoma.' Hum. Pathol. **15**:1147, 1984.
39. Fidler, I. J., and Hart I. R. 'Biological and experimental consequences of the zonal composition of solid tumors.' Cancer Res. **41**:3266, 1981.
40. Fidler, I. J., and Talmadge, J. E. 'The origin and progression of cancer metatases.' In: Bishop, J.M., Rowley, J.D., and Greaves, M. (eds.) Genes and Cancer. New York, Alan R. Liss, 1984, p. 239.
41. Fidler, I. J. 'The relationship of embolic homogeneity, number, size, and viability to the incidence of experimental metastasis.' Eur. J. Cancer **9**:223, 1973.
42. Poste, G., Tzeng, J., Doll, J., Greig, R., Rieman, D., and Zeidman, I. 'Evolution of tumor cell heterogeneity during progressive growth of individual lung metastases.' Proc. Natl. Acad. Sci. USA **79**:6574, 1982.
43. Poste, G., and Greig, R. 'On the genesis and regulation of cellular heterogeneity in malignant tumors.' Invasion and Metastasis **2**:137, 1982.
44. Poste, G., Doll, J., and Fidler, I. J. 'Interactions between clonal subpopulations affect the stability of the metastatic phenotype in polyclonal populations of the B16 melanoma cells.' Proc. Natl. Acad. Sci. USA **78**:6626, 1981.
45. Poste, G., Greig, R., Tzeng, J., Koestler, T., and Corwin, S. 'Interactions between tumor cell subpopulations in malignant tumors.' In: Nicolson, G.L., and Milas, L., (eds.), Cancer Invasion and Metastasis: Biologic and Therapeutic Aspects. New York, Raven Press, 1984, p. 223.
46. Talmadge, J. E., Benedict, K., Madsen, J., and Fidler, I. J. 'The development of biological diversity and susceptibility to chemotherapy in cancer metastases.' Cancer Res. **44**:3801, 1984.

GENES, FRAGILE SITES, CHROMOSOMAL TRANSLOCATIONS, AND CANCER IN AGING

J. Whang-Peng, E. Lee, C.S. Kao-Shan, R. Boccia, and T. Knutsen
Cytogenetic Oncology Section, Medicine Branch, NCI, NIH,
Bethesda, MD 20205

ABSTRACT

Use of cell synchronization techniques has allowed high resolution analysis of chromosomes and made it possible to show that the majority of tumors have cytogenetic abnormalities. Many non-random chromosomal abnormalities, both numerical and structural, have been correlated with specific neoplasms. Recently a remarkable concordance between the chromosomal location of human cellular oncogenes and breakpoints involved in various forms of cancer has been demonstrated. These chromosomal abnormalities may arise as a result of the action of specific agents on heritable or constitutional fragile sites. In a study of chromosome aberrations and fragile sites in different age groups, we have shown that there are significant increases in the degree of hypodiploidy and both minor and major aberrations with increasing age. In our study of fragile sites, we found increasing expression of breakpoints with increasing age, although the frequency of breaks at oncogene, c-fra, h-fra sites, and on the X chromosome did not change. The most common site of chromosome breakage occurred at 3p14.2, followed by breaks at 1q21.3, 7q32.3, and 11q13.3. These four sites are at or near known cancer break sites. The consequence of chromosome rearrangement is unknown, but may cause a change in a constitutional gene, cause a translocation of an oncogene, an alteration in gene dosage, or even activation of a normally quiescent oncogene which may lead to malignant transformation.

INTRODUCTION

Through the examination of elongated chromosomes (more than 500 bands per haploid set) using high resolution techniques, it has been possible to demonstrate that nearly all tumor cells contain cytogenetic abnormalities. These abnormalities may be numerical and/or structural. The structural abnormalities can be either simple or complex rearrangements, and result from chromosomal deletion, translocation, or inversion. To date, many non-random chromosomal abnormalities have been correlated with specific neoplasms, and a remarkable agreement between the chromosomal location of human cellular oncogenes and breakpoints

B. Pullman et al. (eds.), Interrelationship Among Aging, Cancer and Differentiation, 233–243.

described for various forms of cancer has been established (1). Individuals with constitutional chromosomal abnormalities of either the sex chromosomes or of the autosomes are known to be at high risk for leukemia. In addition, there is a predisposition to cancer for individuals with inherited chromosome instability syndromes. The data obtained from our previous study and reports by other groups (2,3) show an increased incidence of breaks at points identical to those involved in chromosomal translocations associated with various tumors. These data suggest that the break points or fragile sites may precede the chromosomal rearrangements found in human neoplasms. The mechanism(s) triggering chromosome translocation in cancer is not known. It is conceivable that such rearrangements arise as a result of exposure to specific agents (carcinogens, ionizing radiation, virus, etc.) (4,5) and are facilitated by the presence of the heritable or constitutional fragile sites. In this report, we wish to examine the influence of aging on the frequency of the expression of fragile sites and to then determine whether or not these these fragile sites and other important sites can be correlated with chromosomal translocations and induction of malignancy.

MATERIALS AND METHODS

Chromosomal Aberrations

For this part of the study, normal volunteers (174 females, 172 males) ranging in age from 1 day to 94 years were used. Phytohemagglutinin (PHA)-stimulated peripheral blood lymphocytes were harvested for chromosome analysis after 72 hours incubation at 37°C. Air-dried slides were made and the slides stained with Giemsa stain. Metaphases were then scored for chromosome number, and for minor aberrations (chromosomal breaks and fragments) and major aberrations (dicentric chromosomes, ring chromosomes, extensive fragmentation, and chromatid exchanges). Those metaphases with fewer than 46 chromosomes were classified as hypodiploid, while those with more than 46 were described as hyperdiploid. The results were divided into three groups, according to the age of the volunteers: 1 day to 19 years, 20-49 years, and 50-94 years, and the variations with age examined.

Fragile Sites

A total of 44 normal volunteers, ranging in age from 21 to 61 years were used to study the frequency and location of fragile sites. Whole blood cultures for the fragile sites were prepared as described by Yunis and Soreng (1). Heparinized peripheral blood (0.2 ml) was added to 5 ml MEM medium (GIBCO) supplemented with penicillin, streptomycin, glutamine, and 10% fetal calf serum; 0.1 ml PHA (Burroughs-Wellcome) was added to each culture. After 72 hours of incubation at 37°C, fluorodeoxyuridine (Sigma Chemical Co) was added to a final concentration of 1 X 10^{-7} M. The cultures were incubated an additional 24 hours at 37°C with the addition of caffeine (Sigma Chemical Co., final concentration: 2.2 mM) for the last six hours. Colcemid (GIBCO) was added for the final 20 minutes of incubation (final concentration: 0.005 μg/ml). Air-dried slides were prepared after the usual exposure to hypotonic solution (0.075M KCl) and fixative (3:1 methanol:glacial acetic acid). The metaphases were then scored for the total number of breaks as well as for the chromosomal location of the breaks. For those metaphases with extensive fragmentation the number of breaks was scored as >60. The results were then divided into three groups according to the age of the volunteers (21-25, 26-50, and 51-61 years) and examined for frequency of fragile sites, chromosomal location of the 10 most common breakpoints, and for breakpoints corresponding to the location of oncogenes, cancer sites, constitutional fragile sites, and heritable fragile sites.

RESULTS

Chromosomal aberrations

The results of the studies of ploidy and chromosomal aberrations in the peripheral blood lymphocytes of normal volunteers are shown in Table I.

Table I

Ploidy and Chromosomal Aberrations in Different Age Groups (174 Females, 172 Males)

Age Years	# Individuals (# Cells)	% Hypo(Range)	%Hyper(Range)	% Minor Ab. (Range)	% Major Ab. (Range)
0-19	151 (8462)	3.55(0-25)	0.14(0-2.0)	1.42(0-14.2)	0.08(0-4.0
20-49	132 (6344)	4.45(0-33.3)	0.43(0-19.0)	2.66(0-42.9)	0.32(0-6.7)
50-89	63 (2883)	8.08(0-26.5)	0.41(0-4.5)	2.29(0-14.0)	0.45(0-4.5)

There is a significant increase in the degree of hypodiploidy and in the frequency of both major and minor aberrations in both males and females with increasing age; this is especially true if females are considered separately (Figure 1). There was no significant increase, however, in the degree of hyperdiploidy.

Fragile sites

The number of fragile sites found per metaphase for each of the three age groups is shown in Table II. The median number of fragile sites per cell increases with age although there is no difference in the median number of cells with extensive breaks.

Figure 1. Ploidy and Chromosomal Aberrations found in different age groups in females.

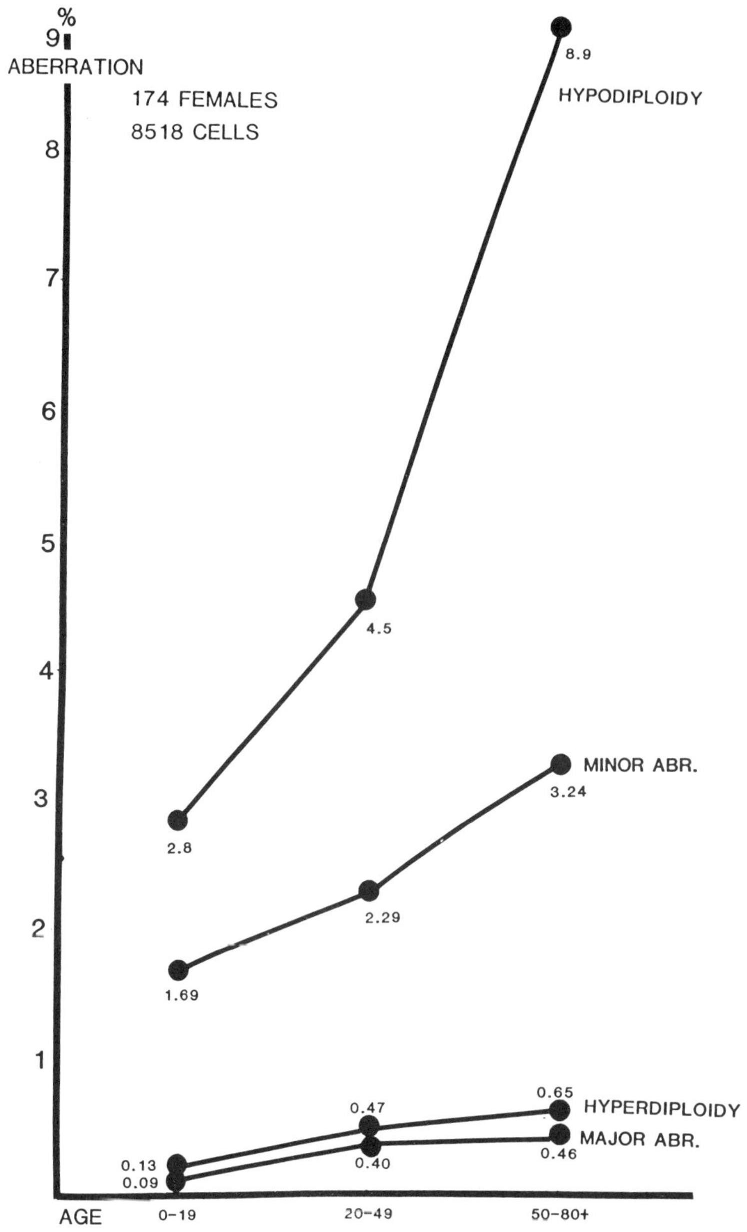

Table II

Number of Fragile Sites in Each Metaphase

Age Years	No. Sites/Metaphase Median (Range)	Extensive Breaks (>60)/Individual Median (Range)
21-25	7.7 (1.697-22.32)	2 (0-8)
26-50	8.59 (3.297-17.05)	1 (0-6)
51-62	9.62 (2.98-16.9)	2 (0-7)

Figures 2, 3, and 4 show representative metaphases with different numbers of fragile sites; arrows indicate location of break sites.

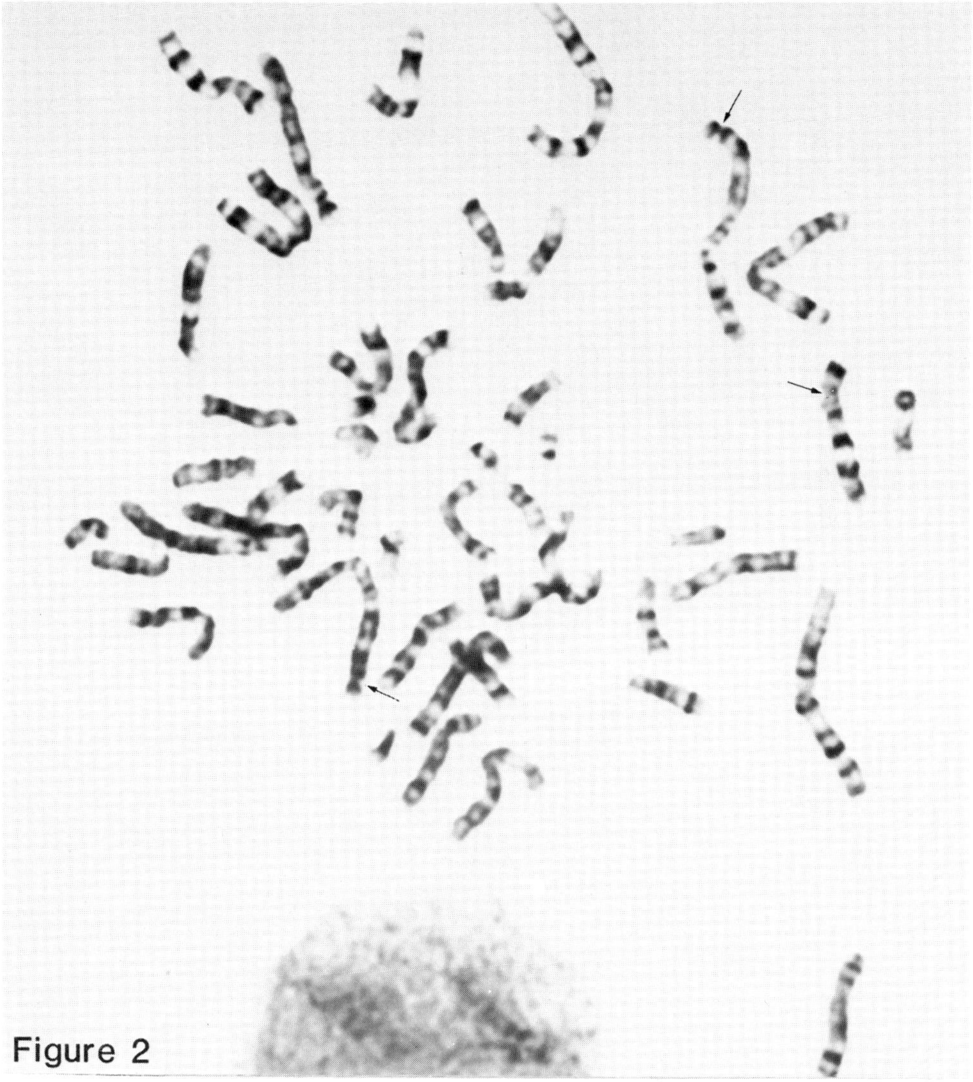

Figure 2

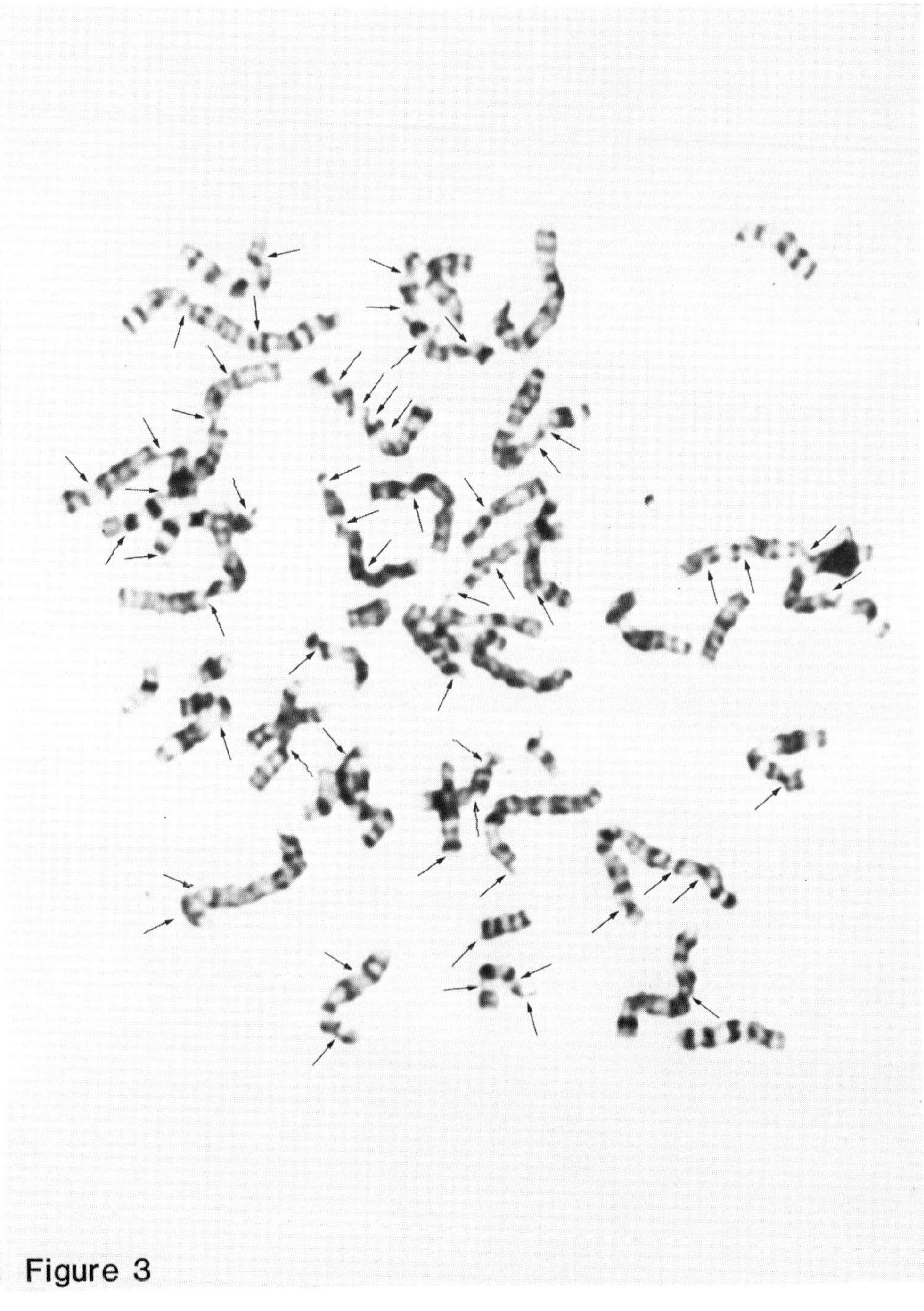

Figure 3

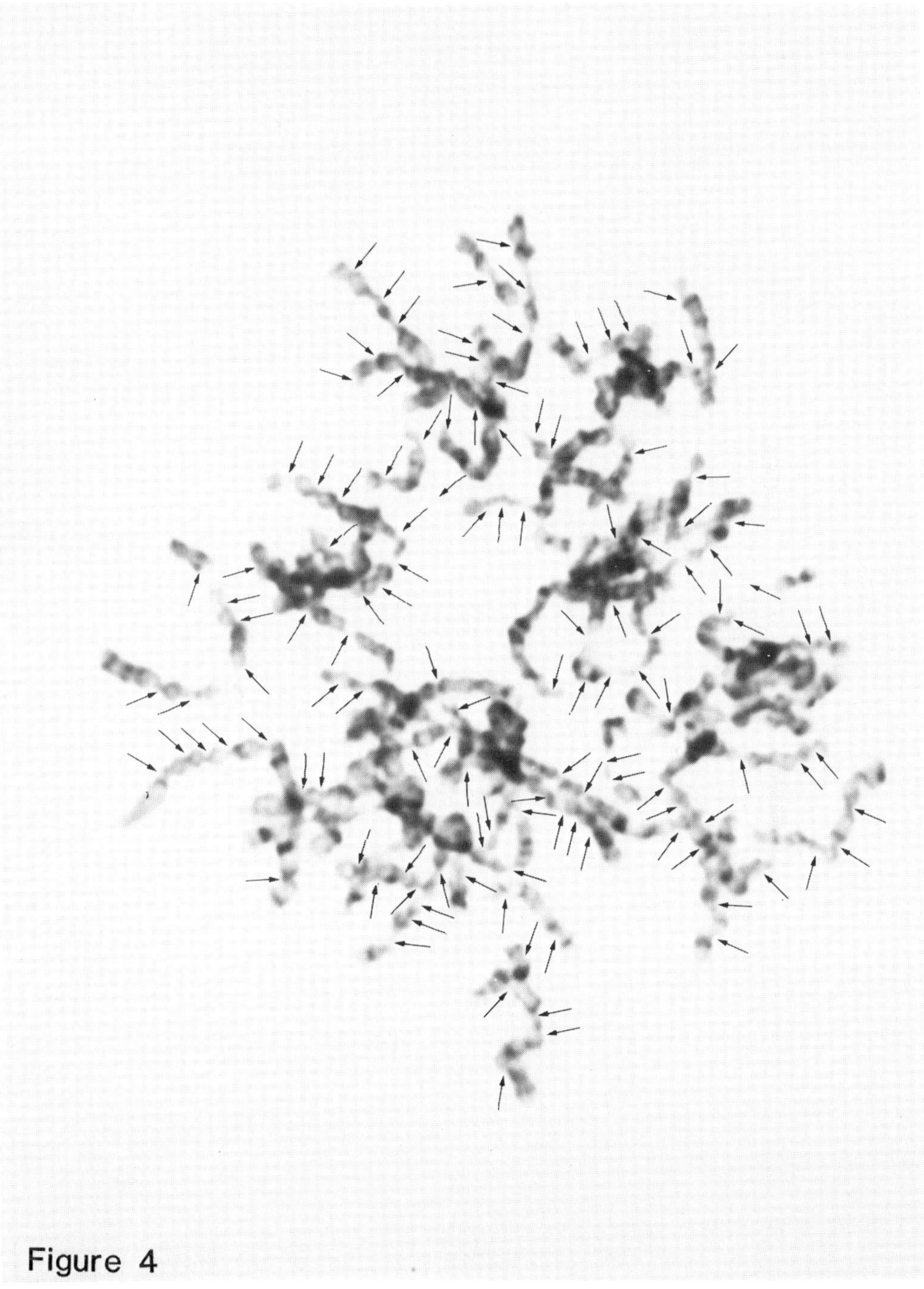

Figure 4

The percentage of breakpoints located at specific sites [oncogene sites, cancer sites, constitutional fragile sites (c-fra), heritable fragile sites (h-fra), and on the X chromosome) are shown in Table III.

Table III

Percentage of Breakpoints at Various Chromosomal Sites

Age Years	Oncogenes	Fragile site breakpoints: Median (range) Cancer Sites	c-fra Sites	h-fra Sites	Sex (X)
21-25	14.3 (8.8-21.9)	39.9 (26.1-47.1)	58.8 (33.2-64.5)	12.3 (7.7-16.67)	3.3 (0.72)
26-50	11.4 (8.6-17.4)	36.9 (38-49)	59.4 (47.83-73)	11.4 (8.36-12.5)	1.33 (0-9.46)
51-62	11.3 (8.0-14.9)	35.65 (22.4-43)	59.2 (49.1-72.4)	10.7 (8.9-14.4)	1.5 (0.3-5)

There are no significant differences between the various age groups. Although no individual had breakpoints on the Y chromosome, every other chromosome demonstrated fragile sites. The 10 most frequent breakpoints for each age group, listed in order of decreasing frequency are shown in Table IV.

Table IV

Ten Most Frequent Breakpoints in Each Age Group
(In Order of Decreasing Frequency)

Age Years	Chromosomal Breakpoint (# Individuals)				
21-25	3p14.2(14)	1q21.3(11)	7q32.3(10)	11q13.3(10)	7q21.2(8)
(15)	1p21.2(8)	3q21.3(7)	11q23.3(7)	1q25.1(6)	7p14.2(5)
26-50	3p14.2(15)	7q32.3(13)	11q13.3(12)	7q21.2(10)	1p21.2(9)
(15)	7q31.2(8)	1q21.3(8)	14q24.11(6)	16q23.2(6)	1q25.1(6)
51-62	3p14.2(14)	1q21.3(11)	7q32.3(10)	1q25.1(8)	14q24.1(7)
(14)	2p13(6)	2q32.12(6)	11q13.3(5)	3q21.3(5)	14q13(5)

Four of the fragile sites are common to all of the agc groups: 3p14.2, 1q21.2, 7q32.3 and 11q13.3. Breakpoints at 7q21.2 and 1q21.2 are common to the young and middle age groups, while the breakpoint at 14q24.1 is common to only the middle and older age groups.

DISCUSSION

While studying families with X-linked mental retardation (fragile X syndrome), Sutherland (6) showed that expression of chromosomal fragile sites in lymphocyte cultures was dependent upon the composition of the tissue culture medium. Requirements for expression of these sites include a medium deficient in folic acid and thymidine. Use of such culture conditions has made it possible to demonstrate the expression of 13 of the 16 known heritable fragile sites (h-fra). Under normal culture conditions, these sites are expressed with a frequency of less than 0.2 percent in the general population and tend to be expressed heterozygously. These fragile sites may represent a mutation of a constitutional fragile site (c-fra) (7).

Yunis and Soreng (1) have reported that expression of the constitutional fragile sites (c-fra) is enhanced when cell cultures are exposed to caffeine during the final 6 hours of incubation. They noted that there were 4 c-fra sites located at the same chromosomal subband as 4 h-fra sites: 9q32, 10q25.2, 11q23.3, and 16q22.1. Twenty of the 51 c-fra sites and 6 of the 16 h-fra sites are located at or near the chromosomal subband of 26 of the 31 cancer sites, while 7 of the 11 oncogenes are located at or near c-fra sites.

Most human cancer appears to be the result of a complex interaction between multiple environmental agents and endogeneous host factors. The development of neoplasia proceeds through multiple steps, i.e. initiation, promotion, and progression. The mechanisms by which translocation or deletion of chromosomes containing the cellular oncogenes might result in malignant transformation of cells is unknown. A spontaneous mutation rate of about one per million per year in mammalian cells renders somatic selection effects and predisposes cells to neoplastic change. In addition, there are a number of endogeneous or genetic factors which place an individual at increased risk for developing leukemia or tumors such as abnormalities found in patients with inherited chromosome instability syndromes (8). One possible explanation for malignant transformation may be found in the discovery by Doolittle et al. (9) of the similarity between the transforming protein derived from a simian sarcoma virus (v-sis) and the human platelet-derived growth factor (PDGF) cellular gene that produces a potent growth factor for human fibroblasts, smooth muscle cells, and glial cells. The abnormal product of the oncogene may lead to uncontrolled growth.

In Burkitt's lymphoma, the three chromosomal rearrangements reported t(8;14), t(2;8), and t(8;22) can be correlated with the type of immunoglobulin produced since the genes for the various types are located at the points of translocation. The gene for production of the kappa light chain is located at 2p13, for the lambda light chain at 22q11-12, and for the heavy chain constant region $\mu(C\mu)$ at 14q34. The kappa light chain immunoglobulin is produced in those tumors with t(2;8), the lambda light chain is produced in tumors with t(8;22), and either of the light chains may be produced in tumors with t(8;14) (10). These immunoglobulin genes are located near known oncogenes, e.g., c-myc is located at

8q24 and c-sis is located at 22q13. Examination of tissue culture cells from an NIH patient (PA682) has shown that the c-sis oncogene was translocated while the c-myc oncogene was not. Although there was no rearrangement of this oncogene, the apparent activation of c-sis is linked circumstantially to its translocation. The correlation of the the type with immunoglobulin secreted in Burkitt's lymphoma or in B-cell acute lymphoblastic leukemia with variant translocations demonstrates that gene expression changes with the chromosome involved and the oncogene translocation site.

Some chromosomal regions have clusters of various types of sites. For example 11q23.3 is a c-fra site, an h-fra site, the site of the oncogene c-ets, and a cancer site [Ewing's sarcoma and peripheral neuroepithelioma: t(11;22)(q24;q12)]. Thus the relationship between the location of chromosome breakpoints, cellular oncogenes, and cancer sites is clearly demonstrated.

In the present study, examination of a large group of normal individuals to determine whether or not there were changes in the frequency of chromosomal aberrations as a person ages showed that there is a significant increase in the degree of hypodiploidy and in the number of minor and major chromosomal aberrations with increasing age for both sexes; this is particularly true if the females were considered separately. There is no significant increase in the amount of hyperdiploidy with in creasing age.

The study of 44 normal volunteers for frequency and location of fragile sites showed that there is an increase in the number of fragile sites per metaphase with increasing age (Table II). However, examination of more individuals, as well as individuals younger than 25 years of age and older than 61 years of age are needed before a statistically significant difference may be demonstrated. The percentage of breakpoints at various locations (oncogenes, cancer sites, c-fra sites, h-fra sites and on the X chromosome) shows no differences among the three age groups. It is of interest that no fragile sites are found on the Y chromosome.

The most frequent breakpoint for each age group is shown in Table IV. Forty-three of the 44 individuals studied have a high frequency of breaks at 3p14.2; three other fragile sites are found in all three age groups: 1q21.3, 7q32.3 and 11q13.3. All of these four locations are also cancer sites. Translocations involving 1q21.2 [t(1;19)(q21-23;p13)] are found in pre-B-ALL. Translocations involving 3p14.2 are found in small cell lung cancer. Numerical and structural abnormalities of this region of chromosome 3 have also been reported in patients with Ph^1 positive CML, AML, ALL, B-prolymphocytic leukemia, ovarian cancer, and other carcinomas and sarcomas (1,11,12). Deletions involving 7q32.3 [del(7)(q31.2q36.3)] are found in acute nonlymphocytic leukemia. Involvement of this region has also been described in cases of CML, AML, polycythemia vera, myeloproliferative disease, malignant lymphoma, ALL, monoclonal gammopathies, carcinomas, and malignant melanoma. Breakpoint 11q13.3 coincides with the location of bcl-1, and with the translocation [t(11;14)(q13.3;q32.3)] found in non-Hodgkin's lymphoma. Two breakpoints are common to both of the younger age groups, 7q21.2 and 1p21.2. 7q21.2 is located at a cancer site and 1p21.2 is near the oncogene N-ras. Only one site is common to the two older age groups, 14q24.1 which is also a cancer site found in ovarian cancer, t(6;14)(q21;q24).

Our findings in the present study show that the natural aging process together with environmental factors such as chemical mutagens, ionizing radiation, or infections with viral or biological organisms may result in an increased frequency of spontaneous chromosomal aberrations and in the expression of fragile sites elicited using fluorodeoxyuridine and caffeine. This may lead to formation of translocations, deletions, and or numerical changes in the chromosomes. The consequences of chromosome rearrangement are unknown but could cause changes in normal constitutional genes, result in transfer of cellular oncogenes, cause alter alterations in gene dosage, or cause activation of a normally quiescent oncogene, and may therefore lead to malignant transformation of normal cells.

REFERENCES

1. Yunis, J.J., and Soreng, A.L. (1984) Science 226, pp. 1199-1205.
2. Whang-Peng, J. Unpublished data
3. Yunis, J.J.(1984) Cancer Genet and Cytogenet 11, pp 125-137.
4. Whang-Peng, J., and Sieber, S.M. (1984) "Risk Factors and Multiple Cancers" (B.A. Stoll, ed.) John Wiley and Sons Ltd. pp. 45-82.
5. Burch, P.R.J. (1964) Annals of the New York Acad of Sci 114, pp. 213-222.
6. Sutherland, G.R. (1977) Science 197, pp. 265-266.
7. Sutherland, G.R., Jacky P.B., Baker, E., and Manuel A. (1983) Am J Hum Genet, p. 432.
8. Hecht, F., and McCaw B.K. (1977) "Genetics of Human Cancer" (J.J. Mulvihill, R.W. Miller, and J.F. Fraumeni, eds), Raven Press, NY, pp. 105-123.
9. Doolittle, R.F., Hunkapiller, M.W., Hoods, L.E., Devare, S.G., Robbins, K.C., Aaronson, S.A., and Antoniades, H.N. (1983) Science 221, pp.275-279.
10. Abe, R., Tebbi, C.K., Yasuda, H., and Sandberg, A.A. (1982) Cancer Genet and Cytogenet 7, pp. 185-195.
11. Mitelman, F. and Levan G. (1981) Hereditas 95, pp. 79-139.
12. Juliusson, G., Robert, K., Ost, A., Frigerg, K., Biberfeld, P., Zech, L, and Gahrton, G. Cancer Genet and Cytogenet (in Press)

Comparison of Human and Rodent Cell Transformation by Known Chemical Carcinogens

J. A. DiPaolo, J. N. Doniger, and N. C. Popescu
National Institutes of Health
National Cancer Institute
Bethesda, Maryland 20205, U.S.A.

ABSTRACT

A few *in vitro* mammalian cell models are available for investigating cellular and molecular mechanisms of chemical carcinogenesis and for determining the potential carcinogenicity of environmental chemicals. Although the results obtained with these models correlate well with results from lifetime studies with experimental animals and human epidemiological data, similar end points have not been obtained with human cells from normal sources or from individuals with inborn errors of metabolism associated with higher risks of malignancy. The results with human cells suggest that a difference in control mechanisms at the target cell level is responsible for "competence" that makes the cell susceptibile to transformation by a carcinogen. Contrary to rodent cells which are transformed by a variety of carcinogens, human cells subjected to the same carcinogens exhibit neither indefinite proliferation nor loss of growth control. Furthermore, the chromosomal defects found in human cells after carcinogen treatment are either not as extensive or lack the specific defect which is associated with indefinite proliferation or loss of growth control usually associated with mammalian cell transformation.

INTRODUCTION

Epidemiological evidence provides the ultimate proof that chemical carcinogens can be carcinogenic agents for humans. Animal studies are confirmatory; when epidemiological evidence is either very weak or lacking, they have often provided the basis for suspecting the carcinogenicity of various species of chemicals. Most carcinogens, particularly chemicals (1,2), can cause transformation that results in malignancy in a few cell culture models. The diverse results from carcinogen treatment of mouse, hamster fetal, guinea pig fetal, and human cells preclude comprehension of the mechanism of competence responsible

B. Pullman et al. (eds.), Interrelationship Among Aging, Cancer and Differentiation, 245–254.

for the susceptibility of target cells to transformation. Although dose-dependent transformation with a variety of chemical carcinogens belonging to different classes can be readily demonstrated in rodent cells, non-quantitative transformation occurs after treatment of guinea pig cells; and with human cells non-progressive growth of the tumor ordinarily develops and/or only a limited extension of cell proliferation occurs. At present, transformation of human cells must be considered empirical, because it occurs only irregularly with physical or chemical carcinogens that are effective with a few non-human cell models or are associated with human cancers (3). Therefore, unknown factors beyond dose-response relationships are responsible for target-cell interactions that result in malignancy of human cells.

Cancer is associated with most organs of the human body. However, the incidence of cancer of different sites varies from one part of the world to another. Therefore, in addition to etiological agents (exogenous factors) some cellular control mechanisms (endogenous factors) must also be responsible for complex interactions that allow the carcinogenic process to proceed through discrete steps. The majority of cancers require the greater proportion of the human life span to develop.

The different cancer sites may have tumors originating from either epithelial or fibroblast cell types. Embryonic and mesenchymal tumors with specific types of leukemias and forms of sarcomas constitute the largest percentage of cancers during the first two decades of life. Subsequently, the frequency of carcinomas increases greatly, and respiratory and digestive system carcinomas increase significantly (4). Probably the simplest explanation for this biphasic response is the alteration in cell cycle which occurs with aging. Although fibroblast cells rarely cycle in adults, continuous but regulated stem cell cycling of epithelial type cells occurs at a rate allowing replacement of epithelial surfaces.

For the last two decades, our laboratory's major emphasis has been the identification of relevant alterations in target cells resulting in malignancy. Several different mammalian cell models have been used, most of which originated in the laboratory. Because of our ability to obtain neoplastic conversion of euploid cells with our transformation models, we are directing our efforts toward understanding the molecular events occurring during transition to malignancy. An *in vitro* human cell model for carcinogenesis does not exist, so animal cell transformation models are being used to elucidate the carcinogenesis process. These cell systems appear to be valid models of carcinogenesis, because they correlate with *in vivo* results concerning the induction of parenchymal liver tumors in the mouse and of tumors of the liver or other organs in the rat and hamster (5). Both established cells lines or early passaged diploid cells that ordinarily become senescent when insulted with a carcinogen become transformed. The life spans of both normal rodent and human cells as reflected in a number of cell doublings are equal or near equal; however, only animal derived cells are capable of undergoing reproducible transformation and progressive changes that lead to malignancy. This paper will review the modulation of the transformation process in two animal diploid cell models and describe how information obtained with them is being used to develop a human model for carcinogenesis studies.

ATTRIBUTES OF CHEMICAL AND PHYSICAL CARCINOGEN TRANSFORMATION OF HAMSTER CELLS IN VITRO

Syrian hamster fetal cells respond to a number of organic and inorganic chemical carcinogens (6-8). A positive correlation of over 90 percent exists between in vivo carcinogenic activity of these compounds and their ability to produce transformation in vitro. Furthermore, the transformed colonies can be isolated and shown to have attributes of neoplastic cells including the production of tumors. The in vitro results correlate with in vivo activity of the carcinogens: potent animal carcinogens produce a high frequency of transformation, weak in vivo carcinogens produce a low rate, and non-carcinogens fail to induce transformation. Thus, the mode of action of the carcinogens can be studied using cells in vitro in place of animals.

However, a fundamental question is whether a carcinogenic chemical transforms or selects pre-existing transformed cells, possibly by the elimination of some normal cells by the toxic effect of the carcinogenic insult. With the Syrian hamster fetal cell (HFC) system, a dose-response relationship and a zero-threshold level were observed. Because each experimental unit can be assumed to have an independent and equal probability of being positive, data obtained with benzo[a]pyrene (BP) alone or with X-irradiation followed by BP were analyzed. Under the assumption of binomial variation, the one-hit curves fitted several sets of experimental data (9). In addition, the total number of cells at risk was estimated with statistical methods developed for use with the number of dishes in an experiment. This statistical analysis provided conclusive proof that transformation is an inductive phenomenon.

Carcinogens which require enzymatic activation are effective only when the proper enzymes are present. For example, BP is ineffective after approximately 30 population doublings of HFC. When HFC are X-irradiated before passage 3, a low transformation frequency is observed, but irradiation of cells seeded for colony formation after the third passage usually does not induce transformation (10 and DiPaolo, unpublished). This phenomenon, which is independent of chromosome aberrations, raises the question of whether there is a transient population that is selected and sensitive to transformation by irradiation or whether cells mature to a more differentiated stage that is resistant to the transforming phenomenon. Although most carcinogens are excellent inducers of sister chromatid changes (SCE), X-irradiation is a weak SCE inducer in both human and third passage HFC (11,12). The very low rate of transformation of primary hamster cells after X-irradiation can be augmented by the addition of a promoting agent, post-irradiation (13). These results suggest that two types of initiation exist: a promoter-dependent and a promoter-independent.

In contrast to X-irradiation, UV treatment results in transformation of late passage cells HFC. The increase is approximately proportional to increased doses in the range of 0.7-6 J/M^2 (14). Analysis of these data indicates that transformation is inductive, because no transformed colony is found without the UV treatment, and the number of transformations increases from zero to 4.5 J/M^2 while cloning efficiency drops to 55 percent relative to controls. This is equivalent to results with a potent chemical carcinogen such as BP.

HFC cells which are intermediate in their excision repair capacity relative to normal (15) and xeroderma pigmentosum (XP) human cell strains are easily transformable; the hamster cell transformation phenomenon, as indicated by altered morphology, is observable after five days of incubation and easily scored after seven days of total incubation. XP cell strains, having no excision repair, cannot be readily transformed. Therefore, excision repair capacity appears to be one factor responsible for the difference between transformability of human and rodent cells.

Although most carcinogens cause DNA damage, the increased transformation frequency found as a result of cocarcinogen treatment is not associated with increased DNA damage or lethality. Furthermore, HFC have the ability to survive and multiply with damage in their parental DNA (15). Therefore, we decided to study how specific genes are involved in carcinogenesis. With oncogenes it should be possible to determine both the changes responsible for their activation and when they occur or are expressed. Because most of the activated oncogenes that have been detected in human tumors and carcinogen-induced mouse and rat tumor cells are members of the ras family, the relationship of the activated HFC transforming gene to ras was determined. After restriction with Hind III, Bam HI, or Eco R1 and agarose electrophoresis, DNA of five lines was independently probed with bas, Ha-ras, and Ki-ras under relaxed hybridization conditions. Under these conditions, proto-oncogenes homologs can be detected in genomic HFC DNA. No hamster ras restriction fragments were detected in the DNA of the NIH/3T3 transformed lines. Furthermore, by similar analysis the HFC activated oncogene(s) is also not related to myc, mos, abl, or src. DNA from transformed HFC formed morphologically transformed foci of NIH 3T3 cells. DNA from normal HFC was ineffective. Isolation of DNA from the foci resulted in second round transformation. This evidence suggests that proto-oncogene(s) were activated during the progression of carcinogen treated cells to neoplasia. Unfortunately, Syrian hamster repeat sequences have a weak homology to 3T3 DNA and middle repeat sequences of the type that exist in human DNA appear rare. Nevertheless, because HFC have between 800 to 950 interstitial "A" particles (retrovirus-like entities) per haploid genome, "A" particle sequences could be detected in some of the transformed 3T3 foci which demonstrated the presence of hamster DNA (16).

ONCOGENES AND TUMORGENICITY OF GUINEA PIG CELLS

The guinea pig transformation model differs from the HFC model because it demonstrates distinct, extended preneoplastic stages similar to <u>in vivo</u> carcinogenesis (17). The guinea pig target cells, similar to hamster and human cells, are diploid and have a finite proliferation capacity. However, transformed cell lines have indefinite proliferation and form progessively growing tumors when injected into syngeneic guinea pigs or nu/nu mice (18). The guinea pig model is being used to study the activation and change in expression of proto-oncogenes as the cells progress from the normal through discrete preneoplastic stages to the neoplastic state.

Transformation of guinea pig cells was initiated with a single carcinogen dose given either <u>in vitro</u> or <u>in utero</u> by transplacental injection (17). Within one month, colonies "morphologically altered" appear. However, the altered morphology is not that of classical transformation with random growth and piling up of cells in a criss-crossed manner. Subsequent to the appearance of morphologic alteration, the cells were again examined for further morphologic changes. Some cultures senesced. In many cultures, however, the cells continued to grow; at different times, depending upon the culture, they progressed to "morphologic transformation". In most cases, morphologic transformation was not associated with anchorage independent growth because no colonies were observed in agar, even after 10^7 cells were seeded; nor were the cells tumorigenic. The cells were subcultured at a ratio of 1 to 10 as before. Some cultures progressed and colonies in soft agar observed and isolated. Progression to anchorage independent growth occurred 6 to 18 months after carcinogen treatment. When anchorage independent cells appeared, the mass cultures were also capable of producing tumors in syngeneic animals. Thus, neoplastic transformation had been achieved. Purified cloned cell lines were established from agar colonies which were also tumorgenic.

The presence of activated oncogenes in seven guinea pig tumorgenic cell lines was assessed by transfecting their DNAs into NIH/3T3 cells to determine whether these DNAs can transform the recipient cells (19,20). Of the seven lines tested, five were positive in the NIH/3T3 transfection assay; DNA from non-carcinogenic treated controls was negative. The transformed foci frequency was in the range of 0.1 focus/microgram of donor DNA. Two other lines were tested once and found to be negative; these may be true or false negatives. Four other agar cloned guinea pig cell lines were also tested in the assay. These were not tumorgenic, although they had progressed through the stages of morphologic alteration, morphologic transformation, and anchorage independent growth. All these anchorage independent non-tumorgenic cell lines were negative in the NIH/3T3 assay, indicating that transformation to anchorage independence does not necessarily involve activated oncogenes detectable in NIH/3T3 cells. Thus, a close association exists between the presence of activated oncogenes and tumorgenicity of transformed guinea pig cells (21). The five tumorgenic lines with activated oncogenes were initiated by four diverse carcinogens: N-methyl-N'-nitro-N-nitrosoguanidine (MNNG) in vitro and in utero, diethylnitrosamine (DEN) in vitro and in utero, 3-methylcholanthrene (3-MCA), and BP.

The nature of the activated oncogenes was further investigated by restriction enzyme analysis (20,21). DNA from cells derived by transfection of NIH/3T3 cells with DNA from each of the five tumorgenic guinea pig cell lines was restricted with Eco R1 and subjected to agarose gel electrophoresis. After Southern blotting, the DNA was probed with P^{32} labeled guinea pig genomic DNA. In all five cases, guinea pig specific DNA sequences were detected in the DNA of these first round NIH/3T3 transformants. NIH/3T3 cells were transfected with this DNA for a second time; second round transformants were observed and isolated. No guinea pig sequences were found in the DNA from these second round transformants, indicating that repeated guinea pig sequences are not closely associated with the activated oncogenes of the tumorgenic guinea pig cells. These DNA were probed with Hras, Kras, myc, amv, rsv, abl, fes, and sis. The only guinea pig sequences detected in the second round transformants were those related to Hras. In guinea pig DNA an 11 KB fragment hybridized under relaxed conditions (20% formamide, 42 C) to v-bas, the mouse Hras, homolog. NIH/3T3 DNA did not have this 11 KB fragment. Thus, the same proto-oncogene, related to Hras, became activated during progression from the normal to the neoplastic state in five independent guinea pig cell lines induced by four diverse chemical carcinogens administered either in vitro or in utero.

Only DNA from neoplastic cells, either the cloned cell lines or the neoplastic mass cultures, induced transformation. However, DNA from the preneoplastic morphologically altered or morphologically transformed cells never gave rise to transformed foci in NIH/3T3 cells. This indicates that activation of the Hras proto-oncogene is closely associated with the acquisition of tumorgenicity (5).

REFRACTIVENESS OF HUMANS CELLS TO NEOPLASTIC TRANSFORMATION

The information gained from the study of rodent cells in vitro is of limited value in application to attempts to transform human cells. Cell survival and metabolic studies indicate that carcinogens are metabolized by human cells in vitro. Chemical carcinogen concentrations effective in inducing transformation in animal cells also increase the frequency of SCE and chromosome aberrations in human cells. Yet, only rarely are normal human cells converted to the malignant state after carcinogen exposure. These results are similar to those from experiments that utilized cells of other mammalian species such as dog, opossum, or monkey. Human cells possess control mechanisms that are responsible for a stable phenotype that cannot be altered readily by one or a series of a few carcinogenic insults. For example, some data indicate that human cells have a capacity to efficiently carry out unscheduled DNA synthesis after exposure to ultraviolet light; other types of cells such as hamster can survive with unrepaired damage. Furthermore, the chromosomal defects found in human cells after carcinogen treatment are either less extensive or lack the specific defects associated either with continued cell proliferation or loss of growth control commonly associated with malignant cells (22-24).

In an attempt to incorporate DNA damage resulting from carcinogen treatment, human foreskin fibroblast or MRC5 cells were blocked in G,S with an amino acid deficient medium, released, and exposed to carcinogen (25). Because carcinogen treatment of human cells did not induce morphologic transformation, the insulted cells were subpassaged and placed in agar. Anchorage independent growth occurred with carcinogens such as 4-nitroquinoline-N-oxide, β-propriolactone, propane sultone, and N-methyl-N'-nitro-N -nitrosoguanidine. The colony size after carcinogen treatment of MRC-5 cells ranged from 50-500 cells and the highest frequency of colony growth in agar was 10^{-3} as a result of ultraviolet irradiation. The addition of 4-nitroquinoline-N-oxide, β-propriolactone, or N-methyl-N'-nitro-N-nitrosoguanidine produced frequencies of colony formation in soft agar of 10^{-5} to 10^{-4}, respectively.

When the colonies were isolated, disaggregated, cultured, and returned to agar colonies were not formed. These results differ from those obtained with animal models in which growth in agar occurs later and once the ability to grow in suspension has been acquired the cells continue to manifest this property. Cytogenetic analysis of carcinogen treated cells indicated relatively few chromosome alterations that consisted primarily of structural changes (26). No specific alteration was observed, because carcinogen treated cells had an extended life span but had not attained indefinite cell proliferation. It is quite possible that the chromosomal changes observed caused the altered growth span. For example cells treated with aflatoxin-B_1 grew more than 60 doublings and produced subcutaneous nodules when injected into nude mice. These cells are aneuploid with alterations of chromosome 1 and 11. Chromosome 1 alterations involved the site of src and fgr (27,28) (feline Gardener Rasheed sarcoma virus) proto-oncogenes. On the short arm of chromosome 11, a change occurs at the site of the Hras proto-oncogene (29). Similarly, a UV initiated line with an extended cell proliferation has an interesting duplication of the segment carrying the sis proto-oncogene. A homology exist between the sis sequence and the gene for the platelett derived growth factor (30,31).

Subcutaneous injection of carcinogen treated cells formed nodules in nude mice which neither regressed nor grew progressively. The tumors were not retransplantable nor did they have an extended proliferation in culture greater than the sister cells which had never been injected. On the other hand, neoplastic mouse, hamster, guinea pig, or human sarcoma cells formed progressive growths that necessitated the killing of the animals. Thus, it is concluded that the human carcinogen treated cells have a high degree of contact inhibition or growth control, a property which is obviously can be responsible for inhibiting invasion and growth in vivo.

In addition to a loss of growth control, another important characteristic of malignancy is the property of indefinite proliferation. This is observed by both spontaneous and induced transformation in animal models and is a characteristic of most cell lines established from spontaneous human tumors, regardless of the organ of origin. It is a very general characteristic and a necessary attribute of neoplasia. Yet this characteristic usually does not occur in cases in which normal human cells were not exposed to chemical carcinogens, and bioassays did not result in progressive tumor growth. The obvious conclusion is that either the characteristics reported after carcinogen insult are not relevant to malignancy or that represent reversible steps. The latter would suggest that the steps leading to transformation of human cells differs from that of rodent models in which aneuploidy and growth in agar correlates with tumorgenicity.

Specific chromosome changes are often found in human cancers. Furthermore, some of the defects are not unique to any one type of neoplasia. The current data would suggest that the experimentally treated human cells do not possess those changes required for the molecular alterations that establish malignancy. Most likely they are responsible for the extension but limited cell proliferation that may be 2-3 times greater than that of nontreated cells. Although the same chemical concentration and radiation energy may cause similar changes in DNA, SCE, cell lethality, etc. in human and rodent cells in vitro, the control mechanisms in human cells obviously differ because only specific rodent cells can be readily transformed. Only the rodent cells acquire loss of growth control and continues proliferation. The former results in invasiveness and the latter perpetuates growth. Both of these properties are necessary to obtain neoplasia.

With the exception of acute childhood leukemia, cancer deaths parallel the overall death rate (32). Dr. Carbone pointed out at this meeting that 50% of cancer occur in people over 65 years of age. This suggests that hereditary factors after the age of reproduction may have a negative effect. Furthermore, the added burden of deleterious life style factors may contribute to cancer induction in that molecular and structural DNA rearrangements occur so that the ultimate result is malignancy.

REFERENCES

1. DiPaolo, J.A., Casto, B.C. (1978): In: "Third Decennial Review Conference: Cell Tissue and Organ Culture Gene Expression and Regulation in Cultured Cells", edited by K. Sanford, Natl. Cancer Inst. Monogr., 48:245-247. United States Printing Office, Washington, D.C.

2. Heidelberger, C. (1975): Ann. Rev. Biochem., 44:79-121.

3. DiPaolo, J.A. (1983): J. Natl. Cancer Inst., 70:3-8.

4. Cutler, S.J. and Young, J.L. (1975): Third National Caner Survey: Incidence Data. Natl. Cancer Inst. Monogr., 41. United States Printing Office, Washington, D.C.

5. Tomatis, J., Partensky, C., and Montesano, R. (1973): Int. J. Cancer, 12:1-20.

6. Berwarld, Y., Sachs, L. (1965): J. Natl. Cancer Inst., 35:641-661.

7. DiPaolo, J.A. (1979): In: "Environmental Carcinogenesis: Occurrence, Risk Evaluation and Mechanisms", edited by P. Emmelot and E. Kriek, pp. 365-380. Amsterdam, Elsevier/North Holland

8. Barrett, J.C., Wong, A., McLachlan, A. (1981): Science, 212:1402-1404.

9. Gart, J.J., DiPaolo, J.A., and Donovan, P.J. (1979): Cancer Res., 39:5069-5075.

10. Borek, C., Sachs, L. (1967): PNAS, 57:1522-1527.

11. Popescu, N.C., Amsbaugh, S.C., and DiPaolo, J.A. (1981): Int. J. Cancer, 28:71-77.

12. Popescu, N.C., DiPaolo, J.A. (1982): In: "Progress and Topics in Cytogenetics", edited by A.A. Sandberg, pp. 425-460. Alan R. Liss, Inc., New York.

13. DiPaolo, J.A., Evans, C.H., DeMarinis, A.J., and Doniger, J. (1984): Cancer Res., 44:1965-1971.

14. DiPaolo, J.A., Donovan, J.P. (1976): Int. J. Radiat. Biol., 30:45-54.

15. Doniger, J., DiPaolo, J.A. (1980): Cancer Res., 40:582-587.

16. Burkhardt, A.L., DiPaolo, J.A., and Doniger, J. (1985): Proc. Amer. Assoc. Cancer Res., 26:69.

17. Evans, C.H., DiPaolo, J.A. (1975): Cancer Res., 35:1035-1044.

18. Evans, C.H., DiPaolo, J.A. (1982): J. Natl. Cancer Inst., 69:1175-1182.

19. Shih, C., Shilo, B.Z., Goldfarb, M., Dannenberg, A., and Weinberg, R.A. (1979): Proc. Natl. Acad. Sci., USA, 76:5714-5718.

20. Sukumar, S., Pulciani, S., Doniger, J., DiPaolo, J.A., Evans, C.H., Zbar, B., and Barbacid, M. (1984): Science, 223:1197-1199.

21. Doniger, J. (1985): In: "Carcinogenesis: Cell Transformation Assays: Application to Studies of Mechanisms of Carcinogenesis and to Carcinogen Testing, edited by C. Barrett, Raven Press, New York. (in press)

22. Sandberg, A.A. (1980): The Chromosomes in Human Cancer and Leukemia, Elsevier Press, Amsterdam.

23. Rowley, J.D. (1984): Cancer Res., 44:3159-3169.

24. Yunis, J.J. (1983): Science 221:227-236.

25. Greiner, J.W., Evans, C.H., and DiPaolo, J.A. (1981): Carcinogenesis, 2:359-362.

26. Popescu, N.C., Amsbaugh, S.C., Milo, G., and DiPaolo, J.A (1985): Proc. Amer. Assoc. Cancer Res., 26:30.

27. Tronick, S.R., Popescu, N.C., Cheah, M.S., Swan, D.C., Amsbaugh, S.A., Lengel, C.R., DiPaolo, J.A., and Robbins, K.S. (1985): Proc. Natl. Acad. Sci., USA. (in press)

28. LeBeau, M.M., Westbrook, C.A., Diaz, M.O., and Rowley, J.D. (1984), Nature, 312:70-71.

29. de Martinville, B., Giacalone, J., Shih, C., Weinberg, R.A., and Francke, U. (1983): Science, 219:498-501

30. Doolittle, R.I., Hunkapiller, M.W., Hood, L.E., Devare, S.G., Robbins, K.C., Aaronson, S.T. and Antoniades, H.A. (1983): Science, 221:275-277.

31. Josephs, S.F., Ratner, L., Clarke, M.F., Westin, E.H., Reitz, M.S. and Wong-Staal, F. (1984): Science, 225:636-639.

32. Knudson, A.G. Jr. (1965): Genetics and Disease. McGraw-Hill, New York.

CRITICAL MOLECULAR EVENTS AND GENE REGULATION IN CARCINOGENESIS, DIFFERENTIATION AND AGING.

Carmia Borek
Radiological Research Laboratory
Departments of Radiology and Pathology
College of Physicians & Surgeons of Columbia University
New York, N.Y.

ABSTRACT

The development of cell culture systems and transformation *in vitro* of rodent and human cells has made it possible to probe into cellular and molecular events and regulatory mechanisms which underlie carcinogenesis differentiation and aging.

Our studies using hamster embryo and C3H 10T1/2 mouse cells show that specific oncogenes are activated in radiogenic transformation and that DNA is the carrier of the radiation induced transforming trait. We find that in hamster and human cells that aging and/or state of differentiation determine cell susceptibility to transformation by radiation. Our studies also show that factors which control cellular metabolic events and regulate gene expression such as thyroid hormones, protective antioxidants and poly(ADP)ribosylation of proteins, play a critical role in radiation and chemically induced neoplastic transformation.

GENETIC AND METABOLIC DETERMINANTS IN GENE EXPRESSION AND CARCINOGENESIS

One of the basic conundrums in biological research evolves from our inability at the present time to distinguish primary events associated with the induction of malignant transformation from those which function as secondary events and to assess their link to differentiation and aging.

Carcinogenesis is a multistep process which has been well studied *in vivo* (3) and in cell cultures under defined conditions free from homeostatic mechanisms (for review see 1,2).

B. Pullman et al. (eds.), Interrelationship Among Aging, Cancer and Differentiation, 255–265.

The initiation of the neoplastic process by physical or chemical agents takes place via irreversible DNA alterations. These may include mutations, methylation gene amplification, oncogene activation or other genetic rearrangements which under permissive physiological conditions may lead to abnormal expression of cellular genes and transformation (for review see 2).

The kinetics of events initiated by various carcinogens differ. Radiation imparts its oncogenic potential within a fraction of a second (1), while the effects of chemicals depend on their presistance in cells and tissues which in turn depends on the type of chemical, the target tissue, dose, period of exposure and the pharmacho-kinetics associated with its metabolism and detoxification. When given to the right target cell and administered in sufficiently high doses, radiation and a variety of chemicals can act alone as complete carcinogens. At low doses a synergistic interaction among various agents may be required for the production of tumors in vivo or transformed colonies in vitro (1,2).

For example, the carcinogenic action of low dose radiation, an agent which damages a variety of macromolecules including DNA is enhanced by interaction with other DNA damaging agents such as chemical carcinogens or with tumor promotors which in themselves are devoid of oncogenic potential (for review see 1,4).

Thus, while at low doses and low toxicity radiation may be deemed a weak carcinogen compared to chemicals, its effectiveness in initiating oncogenic events which may be further amplified by other agents, cannot be ignored(1,4)

Genetic predisposition to cancer as well as age and/or state of differentiation play a significant role in determining cellular susceptibility to transformation by x irradiation (2,4,5) and ultraviolet light (UV) (6-8). Skin fibroblasts from xeroderma pigmentosum patients, a genetic disease with defective DNA repair, and prepondency to cancer (9) are more sensitive to transformation by UVB (5) and UV(A) as compared to normal cells. Sensitivity may vary with propagation in culture. Hamster embryo cell strains become progesssively refractory to transformation by x irradiation as they age and/or differentiate with passage in culture (4). A similar progressive resistance to transformation with passage was found in human embryo transformed by x rays (Borek unpublished) or by ultraviolet light (5).

Once cells mature into adult phase their sensitivity to transformation is altered. Human embryo dermal cells have a higher susceptibility to neoplasic converstion by x rays as compared to adult cells of the same passage (2). Transformation frequency was 3×10^{-6} in the embryonic cells exposed to 200 rad as compared to 8×10^{-7} in the adult cells (2).

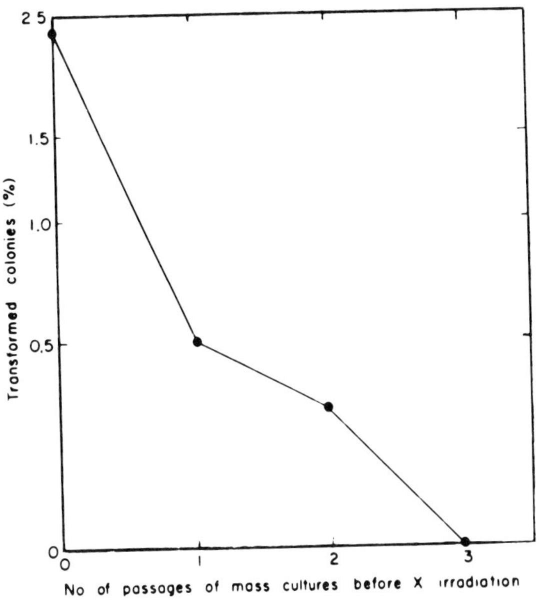

Fig. 1. Decrease in cellular susceptibility of hamster embryo cells to transformation with passage in culture (4).

The exact cellular mechanism underlying these differences in sensitivity to transformation are unclear. DNA damage induced by radiation may be repaired in different ways and although exision repair decreases in diploid cells with passage (11) other types of repair may be retained (10). Cellular metabolism which alters with age/and or differentiation (12) may be playing a crucial role in the differential susceptibility to transformation, between embryonic and adult cells.

Oncogenes in Radiation Induced Transformation

DNA has long been implicated as the target in carcinogenesis, induced by various physical and chemical agents. The more recent developments of cell mediated gene transfer techniques have enabled the identification of cellular transforming genes (c-oncogenes) which are activated following transforming *in vitro* by chemicals (13) and radiation (2,14,15) and confer the neoplastic phenotype on normal cells (Fig. 2).

DNA transfection has made it possible to confirm that DNA is indeed the target in radiation in vitro induced transformation (Table 1) that the transforming activity resides in specific segments of DNA (2,15), and that one type of genetic alteration may be of a mutational nature (16).

Experiments with both hamster embryo cells and C3H 10T1/2 mouse cells indicated that the oncogenes activated following radiogenic transformation in vitro are not of the ras gene family (2,15). These results differ from those in vivo where the Kristen ras gene was shown to be activated in thymic lymphomas induced by radiation (16). Other studies show that specific oncogenes are expressed at a higher level in radiation transformed cells. These include B-lym 1, Abl and Fms (14,15).

While it is apparent that a variety of physical and chemical carcinogens induce the neoplastic process by specific genetic rearrangements leading to abnormal expression of cellular genes certain cellular regulatory processes serve as determinants and appear to act as permissive or protective factors (17). These may vary with age and the differentiated state of a cell and differ within tissues (18).

Hormones

Thyroid hormones which are important in cellular differentiation serve as critical permissive factors in the initiation of cell transformation by a variety of oncogenic agents (14-21) in tumor promotion (22).

Table 1. Transfection of DNA from Radiation In Vitro Transformed Cell C3H 10T½ and Hamster Embryo Cells onto NIH 3T3 and C3H 10T½ Cells.

DNA Source	No foci/μg DNA On NIH/3T3	No foci/μg DNA On C3H 10T½
Normal hamster embryo	0	0
X-ray transformed hamster embryo	0.27	0.20
Normal C3H 10T½	0	0
X-ray transformed C3H 10T½	0.25	0.19
NIH 3T3	0	0
C3H 10T½	0	0

Under hypothyroid conditions cells are resistant to transformation by radiation (10) and chemicals (20) and are in part resistant to the oncogenic effect of the Kirsten murine sarcoma viruse (21). When the hormone triiodothyronine (T_3) is added at physiological doses of 10^{-12} - 10^{-10}M cellular transformation by these various agents is potentiated and becomes T_3 dose dependent. The enhancement of radiogenic transformation by the tumor promotors TPA and teleocidin also requires the presence of thyroid hormones (21).

Some of the data suggest that thyroid hormone is involved in the synthesis of a "cellular protein associated with transformation," (19,20). The effect of T_3 are abolished by treatment with cyclohexamide and we find a similarity between the T_3 dose dependence of transformation and the synthesis of an enzyme Na/K ATPase (19,20) (Fig. 3). However other evidence suggests that the hormone may have a wider scope of action (22). Since thyroid hormones serve as determinants of basal cellular respiration and metabolism (23). they may modulate the oncogenic process by altering the oxidative state of the cell (22).

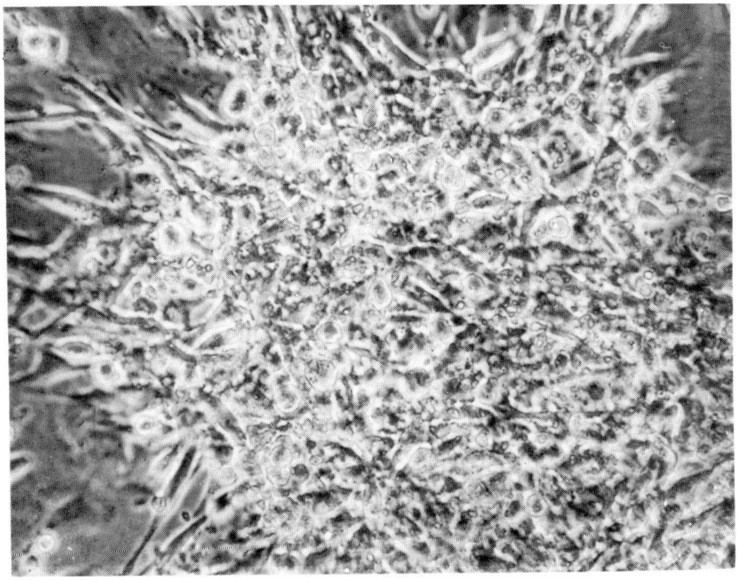

Fig. 2 A focus of transformed NIH/3T3 cells transferred by DNA from hamster embryo cells transformed <u>in vitro</u> by x rays (14,15).

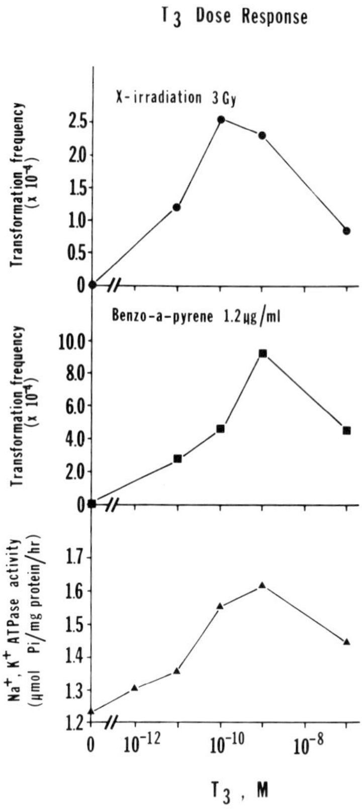

Fig. 3. The synthesis of the enzyme Na/K ATPase and initiation of cell transformation by x rays and benzo(a)pyrene are thyroid hormone dose dependent (19,20).

Free Radicals Mediated Processes and Antioxidants

Our work in vitro has shown that free oxygen radicals play a role in the initiation of cell transformation by radiation and chemicals and in tumor promotion (18). Their oncogenic action is highly dependent on the level of inherent protective scavanging systems that prevail in the cell such as superoxide dismutase (SOD) (1,8) catalase (1,8) and glutathione peroxidase (24,25). These factors serve as antioxidants and may differ among cells and species (18).

Free radicals are ubiquitous in living systems and are inevitable by products of aerobic processes. These reactive oxygen derivatives such as the superoxide anion radical O^-_2 are enhanced in living cells by the action of radiation and a variety of chemicals (including tumor promotors. Superoxides go on to form the reactive hydrogen peroxide (H_2O_2) which is toxic in itself but can also interact with another superoxides to form the reactive hydroxyl radical (OH•). One consequence of free radicals is their oxidative power and the production of lipid peroxides. These can decompose to yield a variety of products such as aldehydes which cross link cellular macromolecules including DNA (26). The results are damage to the structural and functional integrity of the cell and a plausible role in the carcinogenic process (27) as well in aging where free radical processes have been associated with the accumulation of aging pigments (27,28).

During cellular exposure to radiation or some chemicals the inherent cellular protective systems are insufficient to cope with oxidant stress imparted by these agents. These protective systems whose role in aging must still be defined can be enhanced by external means. These factors are nutritional factors, such as retinoids (29,30), selenium (24,25,27) ascorbic acid and β carotene (for review see 27), or enzyme such as SOD and catalase (18).

SOD (18) and a variety of retinoids which show a dramatic ability to inhibit cell transformation by radiation and chemicals and suppress the effect of tumor promotors (18,29). Selenium, a micronutrient in our diet exerts an inhibitory action on transformation by inducing glutathione peroxidase, a selenium dependent antioxidant, as well as catalase which detoxifies peroxide and glutathione, an important scavanger and a crucial factor for maintaining the cellular redox potential (28). Changes in glutathione level would alter cellular NAD^+-NADH pools which serve as cofactors for maintaining its reduced state (30) and plays a role in gene regulation via its role in poly(ADP)ribosylation (31-34).

Table 2. Transformation, Glutathione Peroxidase (GSH), Catalase and Nonprotein Thiols (NPSH) in Selenium Pre-treated and untreated C3H 10T½ cells

	Untreated	Selenium Treated (0.1 M Na_2SeO_3)
Transformation 400 rad X-ray	1.2×10^{-3}	6.1×10^{-4}
Transformation by B(a)P 1.2μg/ml	1.1×10^{-3}	2.2×10^{-4}
GSH px*	5.2	10.0
Catalase*	4.3	6.0
NPSH+	1.0	2.1

*N moles H_2O_2 reduced/min/mg protein.

+N moles/mg protein.

Poly(ADP)Ribose Modifyer of Gene Expression.

Covalent modification of proteins is important in regulating cellular processes. One type of modification which has been associated with DNA damage, differentiation and carcinogenesis (31-34) is the adenosine diphosphate (ADP) ribosylation of specific nuclear proteins. The process is catalysed by poly(ADP)ribose synthetase which is located in the cell nucleus and links the (ADP) ribosyl moiety of NAD to produce chairs of variable lengths which covalently bind to Histone and none Histone proteins (31). Poly(ADP)ribose synthesis is triggered by DNA damaging agents including radiation and alkylating chemicals, which produce oxidant stress. (31,33,34). The stimulation of poly(ADP)ribosylation represents a cellular response to DNA strand breaks by which a large array of proteins becomes covalently modified while in transient fashion with a resulting decrease in NAD pools (31-34).

Studies on the role of poly(ADP)ribose in biological systems have taken advantage of Benzamides, a family of compounds which at low concentration specifically inhibits poly(ADP)riboxe synthetase (31-34).

Using these compounds we have found that inhibition of poly(ADP)ribose results in a marked suppression of neoplastic transformation induced by x rays UV light and methylating chemicals (34). The enhancement of transformation by the tumor promotor TPA was also inhibited by 3AB (22). No effect on UV or x ray induced DNA breaks was observed (34).

Table 3. Inhibition of initiation and promotion by 3AB+ in C3H 10T½ cells (22,34).

Damaging Agent	Drug	Transformation rate x 10^{-4}
X rays (400 rad)	none	0.3 ± 2.1
X rays (")	3AB	4.9 ± 0.6
UV (8.6J/m^2)	none	0.7 ± 2.2
UV (")	3AB	2.4 ± 1.0
MNNG (3.4μM)	none	10.9 ± 2.0
MNNG (")	3AB	2.8 ± 1.0
X ray + TPA*		19.9 ± 1.1
X ray + TPA + 3AB		5.1 ± 0.9

+3AB added at the time of cell plating
*TPA added after x ray exposure

The results indicate an important role for poly(ADP)ribose in carcinogenesis one that is mediated via poly(ADP)ribosylation of specific proteins and altered gene expression rather than via changes in DNA repair (34).

While the exact role of poly(ADP)ribose in cellular regulation of transformation is unknown one could speculate the following as one possibility. Depletion of NAD pools by poly(ADP)ribosylation would remove an important cellular protective system closely associated with glutathione (Fig 4) which in turn would alter the cellular redox potential favoring condition for initiation and promotion (22) of transformation.

The role of poly(ADP)ribose in cellular aging has yet to be defined however, any events which contribute to changes in gene regulation and altered gene expression would clearly affect the metabolic state of the cell and its path towards senescence.

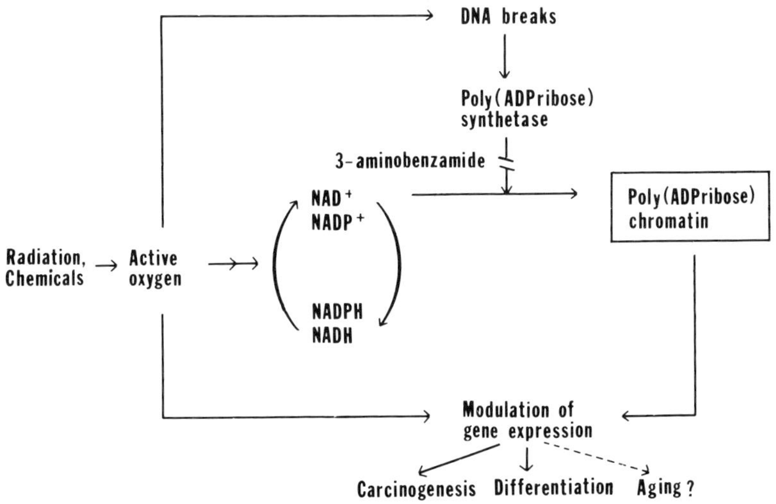

Fig. 4 One possible mechanism for the role of poly(ADP)ribosylation in cellular gene expression as related to oncogeneis, differentiation and aging.

This work was supported by a contract from the National Foundation for Cancer Research and grant no. CA 12536 from the National Cancer Institute.

REFERENCES

1. Borek, C.: 1982, Adv. Cancer Res. 37, pp. 159-232.
2. Borek, C. J.: 1985, J. Pharmachol. Ther. (in press)
3. Berenblum, I.: 1982, in Cancer, A Comprehensive Treatise. F. F. Becker (ed), pp.451-484.
4. Borek, C. and Sachs, L.: 1982. Proc. Nat. Acad. Sci. pp. 1522-1527.
5. Borek, C.: 1980, Nature 283, pp. 776-778.
6. Borek, C. and Andrews, A.: 1983, In: Human Carcinogenesis, Harris, C. C. and Atrup, H. (eds.) pp.519-541.
7. Sutherland, B. et al.: 1981, Cancer Res. 41, pp. 2211-2214.
8. Maher, V. M. et al.: 1982, Proc. Nat. Acad. Sci. (USA) 79, 2613-2617.
9. Cleaver, J. E. and Bootsma, D.: 1975, A. Rev. Genct. 9, pp. 10-38.
10. Cleaver, J. E.: 1978, Bioch. Biophys. Acta. 516, pp. 489-516.
11. Hart, R. W.: 1976, Aging Carcinogenesis and Radiation Biology. Smith K. C. (ed), Plenum Press, NY.
12. Aging, Cancer and Cell Membrane.: 1980, Adv. in Pathobiol. 7, Borek, C. et al (eds) Thieme Stratton, Inc. NY.
13. Shilo, B. Z. and Weinberg, R. A.: 1981, Nature 289, pp. 607-609.
14. Borek. C. et al: 1984, Proc. Am. Assoc. Cancer Res. 25, pp. 100.
15. Borek, C. and Ong, A. Submitted.
16. Guerrero, I. et al.: 1984, Proc. Nat. Acad. Sci.(USA)
17. Borek, C.: 1984, in: The Biochemical Basis of Chemical Carcinogenesis Greim, H. et al. (eds), Raven Press, NY pp. 175-188
18. Borek, C. Troll, W.: 1983, Proc.Nat. Acad. Sci. USA 80, pp. 5749-5752.
19. Guernsey, D. L., Borek, C. and Edelman, I. S.: 1981, Proc. Nat. Acad. Sci. USA 78, pp. 5708-5711.
20. Borek, C., Guernsey, D. L., Ong, A. and Edelman, I. S.: 1983, 80, pp. 5749-5752.
21. Borek, C., Ong, A. and Rhim, J. S.: 1985, Cancer Res. pp. 1702-1706.
22. Borek, C., Cleaver, J. E. and Fujiki, H.: 1984, in: Cellular Interactions by Environm,ental Tumor Promotors. Fujiki et al (eds) Japan Sci. Soc. Press, Tokyo, Japan pp. 195-206.
23. Edelman, I. S.: 1984, N. Engl. J. Med. 290, pp. 1305.
24. Borek et al.: 1984, Proc. Am. Assoc. Cancer Res. 25, pp. 125.
25. Borek et al. Proc. Nat. Sci. (in press).
26. Lesko, S. A., Lorentzen, R. J. and Ts'o P. O. P.: 1980, Biochemistry 19, pp. 3023-3028.
27. Ames, B. N.:1983 Sciencc 22, pp. 1256-1263.

28. Leibovitz, B. E. and Siegal, B.V.: 1980, J. Geront. 35, pp. 45-56.
29. Borek, C.: 1982, In: Molecular Interrelation of Nutrition and Cancer M. S. Arnot et al (eds), Raven Press, NY pp. 337-350.
30. Oxygen Radicals in Biological Systems.: 1984, L. Packer (ed). Methods in enzymology 105, Acad. Press. NY.
31. Sugimura, T. et al.: 1980, Adv. Enzyme Kegal 18, pp. 195-220.
32. Kimura N. et al.: 1983, DNA 2, pp. 195-203.
33. Shall, S.: 1984, Adv. Rad. Biol. II, pp. 1-69.
34. Borek, C., Morgan, W. F., Ong, A. and Cleaver, J. E.: 1984, Proc. Nat. Acad. Sci. (USA) 81, pp. 243-247.

MONOCLONAL ANTIBODIES DIRECTED AGAINST ALKYL-DEOXYNUCLEOSIDES AND CELL TYPE- AND DIFFERENTIATION STAGE-SPECIFIC CELL SURFACE DETERMINANTS IN THE STUDY OF ETHYLNITROSOUREA-INDUCED CARCINOGENESIS IN THE DEVELOPING RAT BRAIN

M. F. Rajewsky, J. Adamkiewicz, N. Huh,
A. Kindler-Röhrborn, U. Langenberg,
R. Minwegen, and P. Nehls
Institut für Zellbiologie (Tumorforschung),
Universität Essen (GH), Hufelandstrasse 55,
D-4300 Essen 1, Federal Republic of Germany

ABSTRACT. In this report we describe some properties and applications of two different groups of monoclonal antibodies (Mab) produced in our laboratory for the analysis of molecular and cellular aspects of the process of carcinogenesis induced by N-ethyl-N-nitrosourea (EtNU) in the developing rat brain. Mab of the first group are directed against defined alkylation products in the DNA of cells exposed to alkylating N-nitroso compounds. In conjunction with sensitive immunoanalytical methods, these Mab are used for quantitation of alkyl-deoxynucleosides in the DNA of tissues and individual cells, for studies of DNA repair, and for analyses of the frequency and distribution of specific alkyldeoxynucleosides in the DNA of different chromatin fractions and in individual DNA molecules. Mab of the second group recognize and bind to cell type- and developmental stage-specific cell surface determinants (CSD) of rat brain cells (BC). These Mab are presently applied for the separation by fluorescence-activated cell sorting (and subsequent in vitro culture) of distinct BC subpopulations previously exposed to EtNU in vivo, as well as for comparative CSD analyses of defined subpopulations of BC and their malignant counterparts.

1. INTRODUCTION

Mutagens and most chemical carcinogens cause structural alterations in cellular DNA (1, 2). Such DNA lesions are likely to be of critical importance in carcinogenesis; their accumulation in somatic cells may also contribute to the ageing process (3, 4). Among the possible genetic consequences of structural changes produced in DNA by exogenous agents, mutations due to misreplication (or misrepair) of chemically modified nucleotides in DNA may be of particular relevance with respect to the modification or inactivation of genes involved in the multi-step process of malignant transformation (1, 2, 5). Various carcinogen-DNA adducts have already been characterized in terms of their molecular structure, notably those caused by the interaction of DNA with alkylating N-nitroso compounds (1, 2, 5, 6, 9, 10). When

B. Pullman et al. (eds.), Interrelationship Among Aging, Cancer and Differentiation, 267–275.

applied *in vivo*, the latter class of carcinogens are converted into highly reactive, electrophilic derivatives either enzymatically or via rapid heterolytic decomposition. The resulting covalent binding of alkyl residues to about a dozen nucleophilic N and O atoms in cellular DNA occurs with very fast reaction times. The mutagenic and carcinogenic potency of different N-nitroso compounds correlates positively with their respective O/N alkylation ratios in DNA (1, 9, 11). A case in point is the highly carcinogenic nitrosamide N-ethyl-N-nitrosourea (EtNU) which, under *in vivo* conditions, undergoes rapid non-enzymatic decomposition (half-life < 8 min; 12) to an ethyldiazonium ion and exhibits one of the highest known O/N alkylation ratios ($\sim$0.7; 9, 10, 11, 13). Among the ethylated purines and pyrimidines produced in DNA exposed to EtNU, the miscoding DNA lesions O^6-ethyl-2'-deoxyguanosine (O^6-EtdGuo) and O^4-ethyl-2'-deoxythymidine (O^4-EtdThd) are considered to be principally responsible for mutagenesis and carcinogenesis by ethylating N-nitroso compounds (1, 2, 5, 6, 7, 9, 11, 14, 15). The absolute and relative frequencies of the ethylation products formed in the DNA of EtNU-exposed target cells (i.e., the accessibility of electron-rich atoms in DNA to the electrophilic ethyldiazonium ion generated from EtNU) are dependent on chromatin structure and ionic strength and on nucleotide sequence (16, 17, 18, 19).

Upon systemic administration, EtNU ethylates cellular DNA (and other cellular macromolecules) in all tissues to approximately the same extent (5, 11, 12). However, when applied to rats during the perinatal age, a single dose of EtNU results in the formation of almost exclusively neuroectodermal tumors after a dose-dependent latency period; i.e., at the time of death of the animals with tumors of the brain or peripheral nervous system, tumors in non-neural tissues are detected only to a negligible extent (5, 20). An important cellular determinant for this neural tissue tropism of the carcinogenic effect may be the very low capacity (in comparison to other rat tissues) of pre- and postnatal rat brain for enzymatic removal of the miscoding lesion O^6-EtdGuo from their DNA (5, 6, 11, 21). Furthermore, as reflected by the proportion of experimental animals developing neuroectodermal tumors, the number of tumors per tumor-bearing animal, and the median time until death with tumors (latency period) there is a pronounced dependency of the carcinogenic effect on the stage of brain development (differentiation) at the time of the EtNU pulse (5, 6, 20, 22). The maximum neuro-oncogenic effect is observed after exposure to EtNU during the late prenatal or very early postnatal age. The carcinogenic effect decreases strongly not only with increasing postnatal age of the animals at the time of the EtNU pulse, but also when the carcinogen is applied at developmental stages prior to day 15 of gestation. Rats exposed to EtNU transplacentally on prenatal day 11 (number of cells/brain $\sim 10^5$; 21) did not develop any neuroectodermal tumors (23 animals; 20), in spite of the fact that the degree of brain DNA ethylation by EtNU on prenatal day 11 is the same as after exposure to the carcinogen at later stages of development (21). The neuro-oncogenic effect of EtNU is thus positively correlated with the number of (proliferative) neural precursor cells, and subsides with the progression of development (differentiation) of the neural cell system. Lastly, neuro-oncogenesis by EtNU can also be studied under in vitro conditions (23, 24). After an extended "latency period" (reminiscent of the process of carcinogenesis in vivo) clones of tumorigenic neuroectodermal cells develop in populations of prenatal rat brain cells (BC) transferred to cell culture after transplacental exposure to EtNU in vivo. The appearance of malignant cell clones (with invasive properties; 25) is preceded by a characteristic sequence of phenotypic changes in the cultured BC populations, including an increased rate of cell proliferation, changes in cell morphology, and a decrease in substrate adhesiveness.

2. ANTI-(ALKYL-DEOXYNUCLEOSIDE) MONOCLONAL ANTIBODIES

The detection and quantitation of specifically altered structural components of genomic DNA, especially at the low levels of DNA modification usually resulting from carcinogen-DNA interactions, and in small numbers of target cells, require particularly sensitive analytical methodology. Conventional radiochromatographic procedures have a number of shortcomings, the most important of which is that they require application of radioactively labeled carcinogens. The sensitivity of radiochromatography is, therefore, limited by the specific ^{3}H or ^{14}C radioactivity of the respective compounds, with the further complication that these must be synthesized in the laboratory. Moreover, radiochromatographic methods preclude analyses at the level of individual cells and DNA molecules, and tissues and cells exposed to environmental, non-radioactive agents cannot be studied (except for analyses by ^{32}P "post-labeling" techniques; 26). These difficulties can in many cases be overcome by recently developed immunoanalytical procedures, especially those involving the use of high-affinity monoclonal antibodies (Mab) specific for defined carcinogen/mutagen-DNA adducts.

While other laboratories have focused on the production of antisera, and in some cases of Mab, directed against DNA components structurally altered by reaction with aflatoxin B_1, N-2-acetyl-aminofluorene, benzo(a)pyrene, cis-diamminedichloroplatinum (II), or UV and ionizing radiation (see Refs. 27, 28 and 29 for review), our group has concentrated on the development of Mab specific for alkyl-deoxynucleosides produced in cellular DNA by alkylating N-nitroso compounds (30, 31, 32, 33). Our approach has exploited the exceptional capability of antibodies to recognize subtle alterations of molecular structure, in order to distinguish deoxynucleosides modified by the covalent attachment of single, small alkyl residues (methyl, ethyl, butyl, isopropyl) from their unaltered normal counterparts. A collection of anti-(alkyl-deoxynucleoside) Mab was established (and is being further expanded) by immunization with synthetic alkyl-ribonucleosides as the haptens coupled to a carrier protein (e.g., keyhole limpet hemocyanin, KLH). A number of Mab from this collection are characterized by high affinity (antibody affinity constants of up to 3 x 10^{10} liter/mol) and an extraordinary degree of specificity for the respective alkyl-deoxynucleosides. Application of these anti-(alkyl-deoxynucleoside) Mab in conjunction with sensitive immunoanalytical methods now permits us to analyse much smaller quantities of DNA (e.g., DNA isolated from cell cultures or small tissue samples, or DNA contained in different fractions of chromatin), as well as single cells and DNA molecules. At present our preferred immunoanalytical techniques are (i) the competitive radioimmunoassay (RIA) performed on specific alkyl-deoxynucleoside franctions pre-separated from DNA hydrolysates by high pressure liquid chromatography (HPLC; 31, 32); (ii) "immuno-slot-blot" analysis of DNA denatured after immobilization on nitrocellulose filters (34); (iii) direct immunofluorescence (IF) aided by computer-based image analysis of electronically intensified fluorescence signals (32, 33); and (iv) immuno-electron microscopy (IEM; 17, 32, 33). With the aid of RIA and "immuno-slot-blot" analyses, femtomol to subfemtomol amounts of specific alkyl-deoxynucleosides (e.g., O^6-EtdGuo, O^4-EtdThd) can be detected in microgram-samples of DNA, with alkyl-deoxynucleoside/deoxynucleoside molar ratios in DNA of the order of 10^{-7} and below. When direct IF with specially selected Mab is applied for the visualization of O^6-EtdGuo in the nuclear DNA of individual cells, the detection limit is presently of the order of 10^2 - 10^3 O^6-EtdGuo molecules per diploid genome. Individual cells can thus be monitored for the presence of specific DNA adducts, and with regard to their capacity for enzymatic removal of such adducts from DNA. In conjunction with transmission electron microscopy, Mab permit the direct visualization (via Mab binding sites) of specific carcinogen-DNA adducts in individual DNA molecules. DNA strands of defined nucleotide sequence can be analyzed with this method for the distribution of specific adducts, e.g., for the presence of "hot spots" of specific structural alterations caused by defined carcinogens/mutagens.

Anti-(alkyl-deoxynucleoside) Mab from our collection have permitted us to investigate (i) the frequency and distribution of specific ethylation products (including O^6-EtdGuo) in the DNA of different chromatin fractions and in a particular fraction of chromosomal DNA from fetal rat brain, as well as the ionic strength dependency of the ethylation of O^6-dGuo and other nucleophilic atoms in DNA in vitro (17, 18, 19); (ii) the accessibility of the DNA of embryonic and fetal rat tissues to EtNU applied transplacentally (21); (iii) the capacity of embryonic and fetal rat tissues for enzymatic removal of O^6-EtdGuo from DNA (21); (iv) the capacity of postnatal rat tissues for enzymatic removal of O^4-EtdThd from DNA (15, 35); and (v) the capacity of a panel of malignant neuroectodermal rat cell lines, and of other malignant rodent and human cell lines, for enzymatic removal from DNA of O^6-EtdGuo and O^4-EtdThd (36; N. Huh and M.F. Rajewsky, submitted for publication). The following results from these studies are relevant to the analysis of the process of carcinogenesis by EtNU in the developing rat brain: (i) The DNA of prenatal rat brain and other tissues (as analyzed from prenatal day 11 onwards) are ethylated by EtNU to the same extent as at later stages of pre- and postnatal development. (ii) As during later stages of development, prenatal rat brain ($\geq$ prenatal day 11) exhibits extremely slow, if any, enzymatic removal of O^6-EtdGuo from DNA. (iii) Cloned malignant neuroectodermal rat cell lines (all induced by transplacental exposure of 18th day prenatal rats to EtNU) are capable of enzymatically removing O^6-EtdGuo (but not O^4-EtdThd) from DNA, in a number of cases very rapidly; and the "O^6-EtdGuo repair phenotype" of these tumorigenic cell lines appears to be rather unstable and is possibly modifiable by microenvironmental conditions. (iv) All normal postnatal rat tissues thus far analyzed (i.e., brain, liver, kidney, and lung) lack the capacity for enzymatic removal of O^4-EtdThd from DNA. (v) As analyzed by IEM, a small fraction of the DNA of fetal BC exposed to EtNU *in vivo* (~0.02%) contains DNA molecules with pronounced clustering of O^6-EtdGuo, indicating a highly non-random distribution of O^6-EtdGuo in target cell DNA.

3. MONOCLONAL ANTIBODIES DIRECTED AGAINST CELL SURFACE DETERMINANTS OF PRENATAL RAT BRAIN CELLS

Along the differentiation pathways of mammalian cell lineages specific stages may exist where exposure to a particular carcinogen results in malignant conversion with increased probability. The differential neuro-oncogenic effect of EtNU as a function of rat brain development may thus reflect a cell type- and differentiation stage--dependence of transformation risk (5, 6). In order to compare their relative frequencies of malignant transformation following exposure to EtNU *in vivo*, phenotypically distinct subpopulations of BC must be reproducibly identified and separated for cell culture in a viable state. Using for immunization of female BALB/c mice suspensions of intact rat BC isolated at different stages of prenatal development, we have, therefore, established a collection of Mab directed against cell surface determinants (CSD) expressed on rat BC (A.Kindler-Röhrborn, O. Ahrens, U.Langenberg, and M.F. Rajewsky, submitted for publication). The developmental stage-dependent expression profiles of neural CSD recognized by these Mab were analyzed both on plasma membranes isolated from rat brains at different stages of pre- and postnatal development (with the use of an indirect ^{125}I solid-phase RIA), and on intact prenatal BC using a fluorescence-activated cell sorter (FACS). Several types of expression profiles were obtained; CSD whose expression, increases, decreases, or remains unchanged during rat brain development (see Figure 1), and in some cases CSD whose expression apparently changes transienctly as a function of developmental stage. Moreover, many of the anti-CSD Mab bind to BC subpopulations of different size, and the relative size of BC subpopulations defined by specific Mab varies as a function of brain development.

Anti-CSD Mab will be applied for the characterization, reproducible identification, and isolation of neural cell subpopulations from prenatal rat brain, in order to investigate the risk of malignant conversion by EtNU as a function of cell type and stage development (differentiation). Mab-defined neural CSD are also expressed on premalignant and malignant neuroectodermal rat cell lines (established in culture after *in vivo* exposure to EtNU on

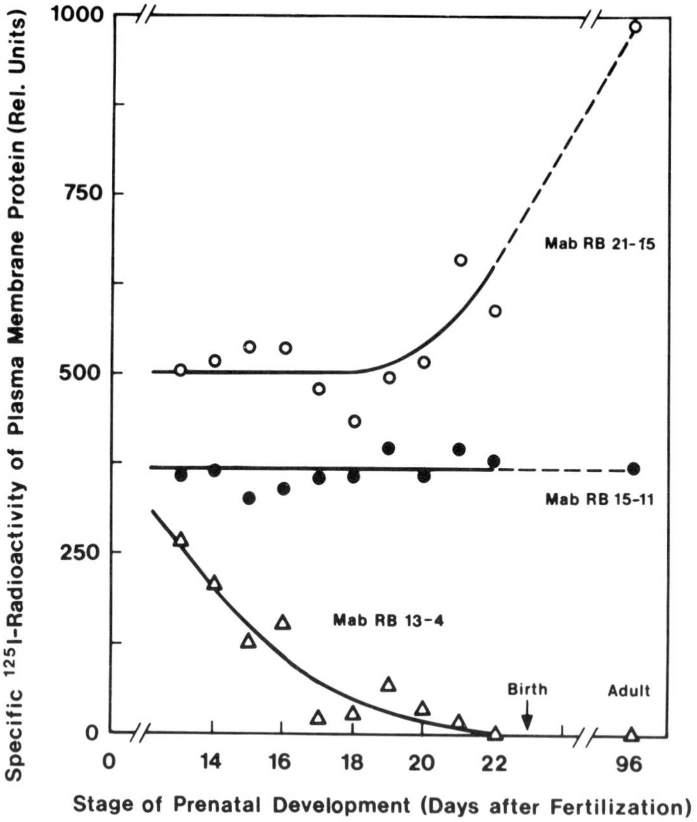

Figure 1

prenatal day 18; see Refs. 23 and 24). Attempts are being made to compare the patterns of Mab-defined CSD expressed on malignant neuroectodermal cell lines with those expressed on the respective normal target cell lineages as a function of prenatal development and at the time of the EtNU pulse.

4. ACKNOWLEDGMENT

The support of this work by the Deutsche Forschungsgemeinschaft (SFB 102/A1 and A9), by the Commission of the European Communities (ENV-544-D B), and by the National Foundation for Cancer Research (Bethesda, Maryland, USA), is gratefully acknowledged.

5. REFERENCES

1. Grover, P.L. (ed.): 1979, Chemical Carcinogens and DNA, Boca Raton, Florida: CRC Press.

2. Singer, B. and Grunberger, D.: 1983, Molecular Biology of Mutagens and Carcinogens. New York, London: Plenum Press.

3. Strehler, B.L.: 1977, Time, Cells, and Aging. New York, San Francisco, London: Academic Press.

4. Ames, B.N.: 1983, Science 221, 1256-1264

5. Rajewsky, M.F., Augenlicht, L.H., Biessmann, H., Goth, R., Hülser, D.F., Laerum, O.D., and Lomakina, L.Ya.: 1977, in Origins of Human Cancer (H.H. Hiatt, J.D. Watson, and J.A. Winsten, eds.), Cold Spring Harbor Conf. on Cell Prolif., 4, pp. 709-726. Cold Spring Harbor, New York: Cold Spring Harbor Laboratory.

6. Rajewsky, M.F.: 1985, in Theories and Models in Cellular Transformation (L. Zardi and L. Santi, eds.), pp. 155-171. London, New York: Academic Press.

7. Loechler, E.L., Green, C.L., and Essigmann, J.M.: 1984, Proc. Natl. Acad. Sci. USA 81, 6271-6275.

8. Zarbl, H., Sukumar, S., Arthur, A.V., Martin-Zanca, D., and Barbacid, M.: 1985, Nature (Lond.) 315, 382-385

9. Rajewsky, M.F.: 1980, in Molecular and Cellular Aspects of Carcinogen Screening Tests (R. Montesano, H. Bartsch, and L. Tomatis, eds.), IARC Scientific Publ. No. 27, pp. 41-54. Lyon: International Agency for Research on Cancer.

10. Singer, B., Bodell, W.J., Cleaver, J., Thomas, G.H., Rajewsky,M.F., and Thon, W.: 1978, Nature (Lond.) 276, 85-88.

11. Goth, R. and Rajewsky, M.F.: 1974, Z. Krebsforsch. 82, 37-64.

12. Goth, R. and Rajewsky, M.F.: 1972, Cancer Res. 32, 1501-1505.

13. Beranek, D.T., Weis, C.C., and Swenson, D.H.: 1980, Carcinogenesis 1, 595-606.

14. Loveless, A.: 1979, Nature (Lond.) 223, 206-207

15. Swenberg, J.A., Dyroff, M.C., Bedell, M.A., Popp, J.A., Huh, N., Kirstein, U., and Rajewsky, M.F.: 1984, Proc. Natl. Acad. Sci. USA 81, 1692-1695.

16. Briscoe, W.T. and Cotter, L.E.: 1984, Chem. Biol. Interact. 52, 103-110.

17. Nehls, P., Rajewsky, M.F., Spiess, E., and Werner, D.: 1984, EMBO J. 3, 327-332.

18. Nehls, P. and Rajewsky, M.F.: 1985, Cancer Res. 45, 1378-1383.

19. Nehls, P. and Rajewsky, M.F.: 1985, Mutation Res. (in press).

20. Ivankovic, S. and Druckrey, H.: 1968, Z. Krebsforsch. 71, 320-360.

21. Müller, R. and Rajewsky, M.F.: 1983, Cancer Res. 43, 2897-2904.

22. Druckrey, H., Schagen, B., and Ivankovic S.: 1970, Z. Krebsforsch. 74, 141-161.

23. Laerum, O.D. and Rajewsky, M.F.: 1975, J. Natl. Cancer Inst. 55, 1177-1187.

24. Laerum, O.D., Haugen, A., and Rajewsky, M.F.: 1979, in Neoplastic Transformation in Differentiated Epithelial Cell Systems (L.M. Franks and C.B. Wigley, eds.), pp. 190-201. London, New York: Academic Press.

25. Laerum, O.D., Rajewsky, M.F., and de Ridder, L.: 1982, Ann. N.Y. Acad. Sci. 381, 264-273.

26. Reddy, M.V., Gupta, R.C., Randerath, E., and Randerath, K.: 1984, Carcinogenesis 5, 231-243.

27. Poirier, M.C.: 1981, J. Natl. Cancer Inst. 67, 515-519.

28. Müller, R., Adamkiewicz, J., and Rajewsky, M.F.: 1982, in Host Factors in Human Carcinogenesis (H. Bartsch and B. Armstrong, eds.), IARC Scientific Publ. No. 39, pp. 463-479. Lyon: International Agency for Research on Cancer.

29. Strickland, P. and Boyle, J.M.: 1984, Progr. Nucleic Acid Res. Mol. Biol. 31, 1-58.

30. Rajewsky, M.F., Müller R., Adamkiewicz, J., and Drosdziok, W.: 1980, in Carcinogenesis: Fundamental Mechanisms and Environmental Effects (B. Pullman, P.O.P. Ts'o, and H. Gelboin, eds.), pp. 207-218. Dordrecht, Boston: D. Reidel Publ. Co.

31. Adamkiewicz, J., Drosdziok, W., Eberhardt, W., Langenberg, U., and Rajewsky, M.F.: 1982, in Indicators of Genotoxic Exposure (B.A. Bridges, B.E. Butterworth, and I.B. Weistein, eds.), Banbury Report 13, pp. 265-276. Cold Spring Harbor, New York:

Cold Spring Harbor Laboratory.

32. Adamkiewicz, J., Ahrens, O., Huh, N., Nehls, P., Spiess, E., and Rajewsky, M.F.: 1984, in N-Nitroso Compounds: Occurence, Biological Effects and Relevance to Human Cancer (I.K. O'Neill, R.C. von Borstel, C.T. Miller, J. Long, and H. Bartsch, eds.), IARC Scientific Publ. No. 57, pp. 581-587. Lyon: International Agency for Research on Cancer.

33. Adamkiewicz, J., Eberle, G., Huh, N., Nehls, P., and Rajewsky, M.F.: 1985, J. Env. Health Perspect. 62 (in press).

34. Nehls, P., Adamkiewicz, J., and Rajewsky, M.F.: 1984, J. Cancer Res. Clin. Oncol. 108, 23-29.

35. Müller, R. and Rajewsky, M.F.: 1983, Z. Naturforsch. 38c, 1023-1029

36. Rajewsky, M.F., and Huh, N.: 1984, Recent Results in Cancer Res. 96, 18-29.

STRUCTURE AND BIOLOGY OF SCRAPIE PRIONS

Stanley B. Prusiner
Departments of Neurology and of Biochemistry and Biophysics,
University of California, San Francisco, California 94143 USA

Scrapie and Creutzfeldt-Jakob disease (CJD) are caused by prions which appear to be different from both viruses and viroids. Prions contain protein which is required for infectivity, but no nucleic acid has been found within them. Prion proteins are encoded by a cellular gene and not by a nucleic acid within the infectious prion particle. A cellular homologue of the prion protein has been identified. The role of this homologue in metabolism is unknown. Prion proteins, but not the cellular homologue, aggregate into rod-shaped particles that are histochemically and ultrastructurally identical to amyloid. Extracellular collections of prion proteins form amyloid plaques in scrapie- and CJD-infected rodent brains. Within the plaques, prion proteins assemble to form amyloid filaments. Elucidating the molecular differences between the prion protein and its cellular homologue should lead to an understanding of the chemical structure of prions.

INTRODUCTION

In recent years there has been recognition amongst physicians and medical scientists that aging of the nervous system is not necessarily accompanied by dysfunction. For many decades, older patients were thought to show an inevitable decline of intellectual function, not as a consequence of disease, but merely due to their old age.

That the majority of older individuals do not lose their intellect has forced physicians to reconsider the causes of senility. The term 'senility' has been used as a wastebasket to describe older individuals with diminished intellectual functions. Epidemiologic studies show that after the age of 60, the number of people who become demented rises with increasing age. By the age of 85, some studies suggest that as many as one in four individuals have some form of senile dementia. Other studies place this number at one in six. In any case, it is a very large number; indeed, senile dementia does not seem to be a consequence of the aging process. On the other hand, aging does seem to predispose or facilitate the development of diseases that lead to loss of intellect, diminished memory and poor judgment.

B. Pullman et al. (eds.), Interrelationship Among Aging, Cancer and Differentiation, 277–287.

There are two major causes of senile dementia: Alzheimer's disease and cerebral vascular disease. Alzheimer's disease is a degenerative disease; its cause is unknown, but many hypotheses have been offered. There are multiple causes of cerebral vascular disease, but the mechanisms responsible for cerebral vascular disease in individual patients are generally poorly defined.

In attempting to develop models for studying degenerative neurologic diseases, we have focused our attention on the study of scrapie, a degenerative neurologic disorder of sheep and goats. Scrapie occurs after a prolonged incubation period and is readily transmissible to laboratory rodents. The disease is characterized by a progressive neurologic disorder due to degenerative changes of the CNS. These changes include proliferation of glial cells, vacuolation of neurons and deposition of extracellular amyloid in the form of plaques.

Three degenerative human diseases similar to scrapie have been described: kuru, Creutzfeldt-Jakob disease (CJD) and Gerstmann-Sträussler syndrome (GSS). All three of these human degenerative disorders are transmissible to experimental animals. Kuru presents as a cerebellar ataxia (1). Cases with incubation periods as long as 30 years have been described. Kuru appears to have been transmitted during ritualistic cannibalism amongst New Guinea natives. Creutzfeldt-Jakob disease occurs at a rate of one per million population throughout the world. It is a dementing disorder much like Alzheimer's disease, but in addition myoclonus is frequently seen. Gerstmann-Sträussler syndrome presents very much like kuru as a cerebellar ataxia initially (2). Later, dementia becomes a prominent symptom of GSS prior to death. Gerstmann-Sträussler syndrome is very rare and generally is confined to families; it is apparently inherited as an autosomal dominant trait. In all three of these human diseases, as well as scrapie, the host remains afebrile, and there is no inflammatory response despite overwhelming and fatal CNS 'slow infection'.

PURIFICATION OF SCRAPIE PRIONS

Progress in purification of the infectious particles causing scrapie is beginning to lead to an understanding of their chemical structure. Numerous attempts have been made to purify the scrapie agent over the past three decades (3-9). Few advances in this area of investigation were made until a relatively rapid and economical bioassay was developed (10, 11). Over a period spanning nearly a decade, our investigations of the molecular properties of the scrapie agent have been oriented toward developing effective procedures for purification. We began our studies by determining the sedimentation properties of the scrapie agent in fixed angle rotors and sucrose gradients (12-14). Subsequent work extended those findings and demonstrated the efficacy of nuclease and protease digestions as well as sodium dodecyl sarcosinate gel electrophoresis in the development of purification protocols (15, 16). Once a 100-fold purification was achieved, convincing evidence demonstrating that a protein is required for infectivity was obtained (17, 18).

Even before the scrapie protein was identified, we began an intensive search for the putative nucleic acid genome of the scrapie agent. To date, we have failed to find this elusive nucleic acid (19-22); indeed, our results are consistent with those reported by Alper and her colleagues nearly two decades earlier (23-25). The requirement of a protein for infectivity and the extraordinary resistance of the scrapie agent to inactivation by procedures that modify or hydrolyze nucleic acids led to the introduction of the term 'prion' to denote these infectious particles (19).

SCRAPIE PRIONS CONTAIN A SIALOGLYCOPROTEIN

In our search for a scrapie-specific protein, it became necessary to substitute discontinuous sucrose gradients in vertical rotors for gel electrophoresis (26). The resulting purification scheme led to the first identification of a macromolecule within the scrapie prion (27-31). This molecule is a sialoglycoprotein designated PrP $27\text{-}30^{Sc}$ (32). Hydrolysis or selective chemical modification of PrP $27\text{-}30^{Sc}$ resulted in a loss of scrapie infectivity. The development of a large-scale purification protocol has allowed us to determine the N-terminal sequence of PrP $27\text{-}30^{Sc}$ and to raise antibodies against the protein (30, 33, 34). Other investigators using purification steps similar to those developed by us seem to have demonstrated the presence of this protein in their preparations (35, 36).

SEARCH FOR A PRION GENOME

The size of the smallest infectious unit remains controversial, largely because of the extreme heterogeneity and apparent hydrophobicity of the scrapie prion (19, 20, 37, 38). Early studies suggested a molecular weight of 60,000 to 150,000 (23). While an alternate interpretation of that data has been proposed (38), there is no firm evidence to suggest that these molecular weight calculations are incorrect. In fact, sucrose gradient sedimentation, molecular sieve chromatography and membrane filtration studies all suggest that a significant portion of the infectious particles may be considerably smaller than the smallest known viruses (19, 38). However, the propensity of the scrapie agent to aggregate makes molecular weight determinations by each of these methods subject to artefact.

To date, no experimental data has been accumulated which indicates that scrapie infectivity depends upon a nucleic acid within the particle. Attempts to inactivate scrapie prions with nucleases, ultraviolet irradiation at 254 nm, Zn^{++} catalyzed hydrolysis, psoralen photoinactivation and chemical modification by hydroxylamine have all been negative (19, 21, 22) even using preparations which contain one major protein as determined by amino acid sequencing (Bellinger, C. G., Cleaver, J. E., Diener, T. O., and Prusiner, S. B., in preparation). While these negative results do not establish the absence of a nucleic acid genome within the prion, they make this a possibility worthy of consideration. Attempts to identify a nucleic acid in purified prion preparations by silver staining and [^{32}P]-end-labeling have been unsuccessful to date (Bellinger, C. G., McKinley, M. P., Meyer, R. K., and Prusiner, S. B., in preparation).

Recently, a cDNA encoding PrP 27-30 has been cloned. The cloned PrP cDNA has been used to search for a complementary nucleic acid within purified preparations of prions. To date, we have failed to identify either a PrP-related DNA or RNA molecule in these preparations. Assuming PrP is a component of the infectious scrapie particle, then our observations demonstrate that scrapie prions are not typical but elusive viruses. As noted, we still cannot eliminate the possibility of a small, nongenomic nucleic acid which lies highly protected within the prion core.

CELLULAR GENE ENCODES PRION PROTEINS

A cDNA encoding PrP 27-30^{Sc} has been used as a probe to show that the gene for PrP is found in healthy uninfected hamsters (39). In contrast to most viruses where their major proteins are encoded within the viral genome, prions contain no nucleic acid which encodes PrP 27-30, as described above. These studies have led to the discovery of a cellular homologue of the prion protein designated PrP 33-35^{C}. This protein can be distinguished from PrP 33-35^{Sc} by its sensitivity to proteases, its inability to polymerize into amyloid filaments, and its subcellular localization in the smooth endoplasmic reticulum (Meyer, R. K., McKinley, M. P., Barry, R. A., and Prusiner, S. B., in preparation). The role of PrP 33-35^{C} in cellular metabolism is unknown. In contrast, the scrapie prion protein, PrP 33-35^{Sc} accumulates to high levels in infected brains, forms amyloid rods and filaments, and is found largely in the rough endoplasmic reticulum. Proteolytic digestion converts PrP 33-35^{Sc} to PrP 27-30^{Sc} which is resistant to further degradation (39). Both PrP 33-35^{C} and PrP 33-35^{Sc} appear to be integral membrane proteins (Meyer, R. K., McKinley, M. P., Barry, R. A., and Prusiner, S. B., in preparation). How the accumulation of PrP 33-35^{Sc} disrupts cellular metabolism and nervous system function is unknown. It is of interest that PrP mRNA within brain is largely confined to neurons (Kretzschmar, H., DeArmond, S. J., Stowring, L. E., and Prusiner, S. B., in preparation). The same mRNA is found at lower levels in organs outside the CNS (39).

Whether or not prions contain macromolecules other than PrP 33-35^{Sc} remains to be established. The biologic properties of prions argue in favor of a small nucleic acid, but there is no physical or biochemical evidence for such a molecule (19, 20). How prions replicate is unknown. Understanding the chemical differences between PrP 33-35^{C} and PrP 33-35^{Sc} will be important to unraveling the mechanism of prion biosynthesis.

The discovery of PrP 33-35^{C} explains one of the most interesting yet perplexing features of scrapie. This slow infection progresses in the absence of any detectable immune response (40, 41). Since PrP 33-35^{C} and PrP 33-35^{Sc} share epitopes, PrP 33-35^{C} probably renders the host tolerant to PrP 33-35^{Sc} (39).

ULTRASTRUCTURAL IDENTIFICATION OF PRION AGGREGATES

Many investigators have used the electron microscope to search for a scrapie-specific particle. Spheres, rods, fibrils and tubules have been described in scrapie, kuru and CJD-infected brain tissue (42-49). Notable amongst the early studies are reports of filamentous, virus-like particles in human CJD brain measuring 15 nm in diameter (47), and rod-shaped particles in sheep, rat and mouse scrapie brain measuring 15-26 nm in diameter and 60-75 nm in length (48, 49). Studies with ruthenium red and lanthanum nitrate suggested that the rod-shaped particles possessed polysaccharides on their surface; these findings are of special interest since PrP 27-30 has been shown to be a sialoglycoprotein (32).

In purified fractions prepared from scrapie-infected brains, rod-shaped particles were found measuring 10-20 nm in diameter and 100-200 nm in length (26, 29). Although no unit morphologic structure could be identified, most of the rods exhibited a relatively uniform diameter and appeared as flattened cylinders. Some of the rods had a twisted structure suggesting that they might be composed of protofilaments. In the fractions containing rods, one major protein (PrP 27-30^{Sc}) and ~$10^{9.5}$ ID_{50} units of prions per ml were also found. The high degree of purity of our preparations demonstrated by radiolabeling and sodium dodecyl sulphate polyacrylamide gel electrophoresis allowed us to establish that the rods are composed of PrP 27-30^{Sc} molecules. Since PrP 27-30^{Sc} had already been shown to be required for, and inseparable from, infectivity (28), we concluded that the rods must be a form of the prion (29). In earlier studies with less purified fractions, we could not determine whether the rods were a pathologic product of infection or an aggregate of prions (26). Subsequently, others faced the same dilemma because their preparations lacked sufficient purity due to protein contaminants (35). Recent immunoelectron microscopic studies using antibodies raised against PrP 27-30^{Sc} have confirmed that the rods are composed of PrP 27-30^{Sc} molecules (50). Sonication of the prion rods reduced their mean length to 60 nm and generated many spherical particles without altering infectivity titers (McKinley, M. P., Braunfeld, M. B., and Prusiner, S. B., unpublished results). In contrast, fragmentation of M-13 filamentous bacteriophage by brief sonication reduced infectivity significantly (Bellinger, C. G., McKinley, M. P., and Prusiner, S. B., in preparation).

PRION MORPHOLOGY

It seems doubtful whether electron microscopic studies to date have been able to demonstrate the smallest infectious unit or fundamental particle of the scrapie prion: certainly, the morphology of the unit structure has not been defined. The extreme morphologic heterogeneity of the rods is inconsistent with the recently advanced hypothesis that prions are filamentous viruses (51).

Spherical particles have been found within postsynaptic evaginations of the brains of scrapie-infected sheep and mice as well as CJD-infected humans and chimpanzees (42-45); these particles measured 23-35 nm in diameter. Since sonication fragmented prion rods and generated spheres measuring 10-30 nm in diameter, the question arises of whether or not the spherical particles in brain tissue are related to the sonicated spheres.

PRION RODS AND FILAMENTS ARE AMYLOID

The ultrastructure of the prion rods is indistinguishable from many purified amyloids (29). Histochemical studies with Congo red dye have extended this analogy in purified preparations of prions (29) as well as in scrapie-infected brain where amyloid plaques have been shown to stain with antibodies to PrP $27\text{-}30^{Sc}$ (33). In addition, PrP $27\text{-}30^{Sc}$ has been found to stain with periodic acid Schiff reagent (32); amyloid plaques in tissue sections readily bind this reagent. Amyloid plaques have also been found in three transmissible disorders similar to scrapie and CJD: kuru and GSS of humans as well as a chronic wasting disease of mule, deer and elk (52-54). These findings raise the possibility that prion-like molecules might play a causative role in the pathogenesis of nontransmissible disorders such as Alzheimer's disease (55). Amyloid proteins are prevalent in Alzheimer's disease, but for many decades these proteins have been considered a consequence rather than a possible cause of the disease.

Immunocytochemical studies with antibodies to PrP $27\text{-}30^{Sc}$ have shown that filaments measuring approximately 16 nm in diameter and up to 1,500 nm in length within amyloid plaques of scrapie-infected hamster brain are composed of prion proteins (56). The antibodies to PrP $27\text{-}30^{Sc}$ did not react with neurofilaments, glial filaments, microtubules and microfilaments in brain tissue. The prion filaments have a relatively uniform diameter, rarely show narrowings and possess all the morphologic features of amyloid. Except for their length, the prion filaments appear to be identical ultrastructurally to the rods which are found in purified fractions of prions.

In extracts of scrapie-infected rodent brains, abnormal structures were found by electron microscopy and labeled scrapie-associated fibrils (57). These abnormal fibrils were distinguished from other filamentous structures by their characteristic and well-defined morphology. Published electron micrographs of the scrapie-associated fibrils consistently show helically wound structures measuring 300-800 nm in length. Based on their ultrastructural characteristics, the fibrils have been reported repeatedly to be different from amyloid (57, 58). Attempts to stain scrapie-associated fibrils with Congo red dye have yielded negative results; however, even a positive result would have been impossible to interpret because of impurities in the extracts.

No structures with the ultrastructural morphology of scrapie-associated fibrils have been found in thin sections of scrapie-infected brain specimens. If scrapie-associated fibrils in brain extracts are eventually found to be composed of PrP $27\text{-}30^{Sc}$ molecules, then the possibility that these fibrils are an artefact of the preparative extraction procedure must be entertained. Some investigators have attempted to adapt the term 'scrapie-associated fibrils' (35, 59) to describe the rod-shaped particles found in purified preparations of prions (26, 29). This revision of the terminology seems neither appropriate nor useful in view of the following observations: 1) filaments within scrapie-infected brain are composed of PrP $27\text{-}30^{Sc}$ molecules, 2) these filaments have a uniform diameter and rarely twist, 3) they are morphologically and histochemically identical to amyloid, and 4) they possess the same ultrastructural and antigenic characteristics as the rods found in purified fractions of prions except for length. Clearly, both the prion filaments and rods are indistinguishable from amyloids, but can be readily differentiated morphologically from scrapie-associated fibrils.

CREUTZFELDT-JAKOB DISEASE PRIONS

Investigations of scrapie prions have recently been extended to studies on CJD. The CJD agent has been partially purified using procedures developed for scrapie prions (34, 60). The CJD agents from humans, mice and guinea pigs contain protease-resistant proteins that exhibit cross-immunoreactivity with PrP 27-30 antisera. Electron microscopy reveals that the CJD preparations contain rod-shaped particles of similar dimensions to those found in scrapie prion preparations. Furthermore, the CJD prion rods stain with Congo red dye and exhibit green-gold birefringence. It is noteworthy that long, helically twisted fibrils have been reported in extracts from human, mouse and guinea pig CJD brains and called scrapie-associated fibrils (61); however, our results with purified preparations of CJD prions show that structures with the morphology of these fibrils are not required for infectivity.

Recent studies have shown that PrP antisera stain amyloid plaques in rodent brains from animals with experimental CJD (Kitamoto, T., Tateishi, J., Tashima, T., Takeshita, I., Barry, R. A., and Prusiner, S. B., in preparation).

THE PRION HYPOTHESIS

New knowledge about the molecular structure of scrapie prions is beginning to accumulate rapidly. If prions are viruses, they they should contain a genomic nucleic acid which encodes PrP 27-30^{Sc}. They do not! Prions may contain a small, nongenomic nucleic acid which does not encode PrP 27-30^{Sc}. There is no chemical or physical evidence to indicate the presence of such a nucleic acid, but the apparent biological diversity of prions could readily be explained by such a model. Alternatively, prions may be devoid of nucleic acid. Information for the synthesis of new PrP 27-30^{Sc} molecules is encoded within the host genome. A cloned PrP cDNA as well as antibodies to the protein provide new tools with which to extend our investigation of the chemical structure of prions.

Once it is determined whether or not prions contain other macromolecules besides glycoproteins, chemical studies to determine the molecular mechanisms by which prions reproduce and cause disease should become possible. Indeed, efforts to purify and characterize the infectious particles causing scrapie and CJD have yielded important new information about the structure and biology of prions.

ACKNOWLEDGMENTS

Portions of this article were adapted from a recent review in Microbiological Sciences 2, pp. 33-39, 1985. Important contributions from Drs. M. McKinley, R. Barry, C. Bellinger, R. Meyer, S. DeArmond, D. Kingsbury, D. Stites and G. Lewis are gratefully acknowledged. Collaborative studies with Drs. L. Hood, C. Weissmann, S. Kent, T. Diener, J. Cleaver and W. Hadlow have been important to the progress of these studies. The author thanks D. Groth, K. Bowman, M. Braunfeld, P. Cochran and L. Pierce for technical assistance as well as L. Gallagher, J. Sleath and F. Elvin for editorial and administrative assistance. This work was supported by research grants from the National Institutes of Health (AG02132 and NS14069) as well as by gifts from R. J. Reynolds Industries, Inc., Sherman Fairchild Foundation and Koret Foundation.

REFERENCES

1. Gajdusek, D.C.: 1977, Science 197, pp. 943-960.
2. Masters, C.L., Gajdusek, D.C., Gibbs, C.J., Jr.: 1981, Brain 104, pp. 559-588.
3. Hunter, G.D.: 1972, J. Infect. Dis. 125, pp. 427-440.
4. Millson, G.C., Hunter, G.D., and Kimberlin, R.H.: 1976, in Slow Virus Diseases of Animals and Man, Kimberlin, R.H. (ed.). American Elsevier Publishing, New York, pp. 243-266.
5. Siakotos, A.N., Gajdusek, D.C., Gibbs, C.J., Jr., Traub, R.D., and Bucana, C.: 1976, Virology 70, pp. 230-237.
6. Mould, D.L., Smith, W., and Dawson, A.M.: 1965, J. Gen. Microbiol. 40, pp. 71-79.

7. Diringer, H., Hilmert, H., Simon, D., Werner, E., and Ehlers, B.: 1983, Eur. J. Biochem. 134, pp. 555-560.
8. Marsh, R.F., Dees, C., Castle, B.E., Wade, W.F., and German, T.L.: 1984, J. Gen. Virol. 65, pp. 415-421.
9. Brown, P., Green, E.M., and Gajdusek, D.C.: 1978, Proc. Soc. Exp. Biol. Med. 158, pp. 513-516.
10. Prusiner, S.B., Groth, D.F., Cochran, S.P., Masiarz, F.R., McKinley, M.P., and Martinez, H.M.: 1980, Biochemistry 19, pp. 4883-4891.
11. Prusiner, S.B., Cochran, S.P., Groth, D.F., Downey, D.E., Bowman, K.A., and Martinez, H.M.: 1982, Ann. Neurol. 11, pp. 353-358.
12. Prusiner, S.B., Hadlow, W.J., Eklund, C.M., and Race, R.E.: 1977, Proc. Natl. Acad. Sci. USA 74, pp. 4656-4660.
13. Prusiner, S.B., Hadlow, W.J., Eklund, C.M., Race, R.E., and Cochran, S.P.: 1978, Biochemistry 17, pp. 4987-4992.
14. Prusiner, S.B., Hadlow, W.J., Garfin, D.E., Cochran, S.P., Baringer, J.R., Race, R.E., and Eklund, C.M.: 1978, Biochemistry 17, pp. 4993-4999.
15. Prusiner, S.B., Groth, D.F., Bildstein, C., Masiarz, F.R., McKinley, M.P., and Cochran, S.P.: 1980, Proc. Natl. Acad. Sci. USA 77, pp. 2984-2988.
16. Prusiner, S.B., Groth, D.F., Cochran, S.P., McKinley, M.P., and Masiarz, F.R.: 1980, Biochemistry 19, pp. 4892-4898.
17. Prusiner, S.B., McKinley, M.P., Groth, D.F., Bowman, K.A., Mock, N.I., Cochran, S.P., and Masiarz, F.R.: 1981, Proc. Natl. Acad. Sci. USA 78, pp. 6675-6679.
18. McKinley, M.P., Masiarz, F.R., and Prusiner, S.B.: 1981, Science 214, pp. 1259-1261.
19. Prusiner, S.B.: 1982, Science 216, pp. 136-144.
20. Prusiner, S.B.: 1984, in Advances in Virus Research, Vol. 29, Lauffer, M.A., and Maramorosch, K. (eds.). Academic Press, New York, pp. 1-56.
21. Diener, T.O., McKinley, M.P., and Prusiner, S.B.: 1982, Proc. Natl. Acad. Sci. USA 79, pp. 5220-5224.
22. McKinley, M.P., Masiarz, F.R., Isaacs, S.T., Hearst, J.E., and Prusiner, S.B.: 1983, Photochem. Photobiol. 37, pp. 539-545.
23. Alper, T., Haig, D.A., and Clarke, M.C.: 1966, Biochem. Biophys. Res. Commun. 22, pp. 278-284.
24. Alper, T., Cramp, W.A., Haig, D.A., and Clarke, M.C.: 1967, Nature 214, pp. 764-766.
25. Alper, T., Haig, D.A., and Clarke, M.C.: 1978, J. Gen. Virol. 41, pp. 503-516.
26. Prusiner, S.B., Bolton, D.C., Groth, D.F., Bowman, K.A., Cochran, S.P., and McKinley, M.P.: 1982, Biochemistry 21, pp. 6942-6950.
27. Bolton, D.C., McKinley, M.P., and Prusiner, S.B.: 1982, Science 218, pp. 1309-1311.
28. McKinley, M.P., Bolton, D.C., and Prusiner, S.B.: 1983, Cell 35, pp. 57-62.
29. Prusiner, S.B., McKinley, M.P., Bowman, K.A., Bolton, D.C., Bendheim, P.E., Groth, D.C., and Glenner, G.G.: 1983, Cell 35, pp. 349-358.

30. Prusiner, S.B., Groth, D.F., Bolton, D.C., Kent, S.B., and Hood, L.E.: 1984, Cell 38, pp. 127-134.
31. Bolton, D.C., McKinley, M.P., and Prusiner, S.B.: 1984, Biochemistry 23, pp. 5898-5905.
32. Bolton, D.C., Meyer, R.K., and Prusiner, S.B.: 1985, J. Virol. 53, pp. 596-606.
33. Bendheim, P.E., Barry, R.A., DeArmond, S.J., Stites, D.P., and Prusiner, S.B.: 1984, Nature 310, pp. 418-421.
34. Bendheim, P.E., Bockman, J.M., McKinley, M.P., Kingsbury, D.T., and Prusiner, S.B.: 1985, Proc. Natl. Acad. Sci. USA 82, pp. 997-1001.
35. Diringer, H., Gelderblom, H., Hilmert, H., Özel, M., Edelbluth, C., and Kimberlin, R.H.: 1983, Nature 306, pp. 476-478.
36. Hilmert, H., and Diringer, H.: 1984, Biosci. Rep. 4, pp. 165-170.
37. Diringer, H., and Kimberlin, R.H.: 1983, Biosci. Rep. 3, pp. 563-568.
38. Rohwer, R.G.: 1984, Nature 308, pp. 658-662.
39. Oesch, B., Westaway, D., Wälchli, M., McKinley, M.P., Kent, S.B.H., Aebersold, R., Barry, R.A., Tempst, P., Teplow, D.B., Hood, L.E., Prusiner, S.B., and Weissmann, C.: 1985, Cell 40, pp. 735-746.
40. McFarlin, D.E., Raff, M.C., Simpson, E., and Nehlsen, S.: 1971, Nature 233, p. 336.
41. Kasper, K.C., Bowman, K., Stites, D.P., and Prusiner, S.B.: 1981, in Hamster Immune Responses in Infectious and Oncologic Diseases, Streilein, J.W., Hart, D.A., Stein-Streilein, J., Duncan, W.R., and Billingham, R.E. (eds.). Plenum Press, New York, pp. 401-413.
42. David-Ferreira, J.F., David-Ferreira, K.L., Gibbs, C.J., Jr., and Morris, J.A.: 1968, Proc. Soc. Exp. Biol. Med. 127, pp. 313-320.
43. Bignami, A., and Parry, H.B.: 1971, Science 171, pp. 389-399.
44. Lampert, P.W., Gadjusek, D.C., and Gibbs, C.J., Jr.: 1971, J. Neuropathol. Exp. Neurol. 30, pp. 20-32.
45. Baringer, J.R., and Prusiner, S.B.: 1978, Ann. Neurol. 4, pp. 205-211.
46. Field, E.J., Mathews, J.D., and Raine, C.S.: 1969, J. Neurol. Sci. 8, pp. 209-224.
47. Vernon, M.L., Horta-Barbosa, L., Fuccillo, D.A., Sever, J.L., Baringer, J.R., and Birnbaum, G.: 1970, Lancet 1, pp. 964-966.
48. Field, E.J., and Narang, H.K.: 1972, J. Neurol. Sci. 17, pp. 347-364.
49. Narang, H.K.: 1974, Acta Neuropathol. (Berl.) 28, pp. 317-329.
50. Barry, R.A., McKinley, M.P., Bendheim, P. E., Lewis, G.K., DeArmond, S.J., and Prusiner, S.B.: 1985, J. Immunol., in press.
51. Merz, P.A., Rohwer, R.G., Kascsak, R., Wisniewski, H.M., Somerville, R.A., Gibbs, C.J., Jr., and Gajdusek, D.C.: 1984, Science 225, pp. 437-440.
52. Klatzo, I., Gajdusek, D.C., Zigas, V.: 1959, Lab. Invest. 8, pp. 799-847.

53. Masters, C.L., Gajdusek, D.C., and Gibbs, C.J., Jr.: 1981, Brain 104, pp. 559-588.
54. Bahmanyar, S., Williams, E.S., Johnson, F.B., Young, S., and Gajdusek, D.C.: 1985, J. Comp. Pathol. 95, pp. 1-5.
55. Prusiner, S.B.: 1984, N. Engl. J. Med. 310, pp. 661-663.
56. DeArmond, S.J., McKinley, M.P., Barry, R.A., Braunfeld, M.B., McColloch, J.R., and Prusiner, S.B.: 1985, Cell 41, pp. 221-235.
57. Merz, P.A., Somerville, R.A., Wisniewski, H.M., and Iqbal, K.: 1981, Acta Neuropathol. (Berl.) 54, pp. 63-74.
58. Merz, P.A., Wisniewski, H.M., Somerville, R.A., Bobin, S.A., Masters, C.L., and Iqbal, K.: 1983, Acta Neuropathol. (Berl.) 60, pp. 113-124.
59. Kimberlin, R.H.: 1984, Trends Neurosci. 7, pp. 312-316.
60. Bockman, J.M., Kingsbury, D.T., McKinley, M.P., Bendheim, P.E., and Prusiner, S.B.: 1985, N. Engl. J. Med. 312, pp. 73-78.
61. Merz, P.A., Somerville, R.A., Wisniewski, H.M., Manuelidis, L., and Manuelidis, E.E.: 1983, Nature 306, pp. 474-476.

DISEASES OF AGING: VIRAL GENES AND PERTURBATION OF DIFFERENTIATED FUNCTIONS IN PERSISTENT INFECTION

Michael B. A. Oldstone
Scripps Clinic and Research Foundation, La Jolla, CA, USA

ABSTRACT

Here we present evidence that viruses alter the differentiated (luxury) function of a cell without affecting the cell's vital (housekeeping) activity. The outcome may be disturbed homeostasis and, thereby, disease. These events can occur with neither altered cell morphology nor cell destruction. Hence, viruses can induce disease by "silently" infecting cells in the absence of ordinarily expected footprints of viral infection. These events may foment the fatal physiologic breakdowns characteristic of aging.

I. INTRODUCTION

Of the many diseases to which the organs and systems of an aging population are particularly susceptible, most are of unknown origin and often reflect the faltering of a differentiated function. Examples include sufficient loss of neurotransmittors to disorder cognitive thought, intellectualization, personality or locomotion[1,2]. Within this spectrum are diseases from Alzheimers to Parkinsonism. Other examples abound for dysfunctions of endocrine and immune systems; some of the former are diminished hormone synthesis and/or utilization by hormonal receptors as in type II (adult onset) diabetes, thyroid disease and myasthenia gravis[1,3]. The latter includes imbalances among subsets of immunocompetent cells (lymphocytes, macrophages), leading to unsuccessful or misdirected regulation of immune responses[1,4]. An ineffectual response can lead to immunosuppression with concomitant reactivation of latent microbial agents and/or escape from immunologic surveillance of transformed cells. On the other hand aberrant or heightened responses can lead to autoimmunity.

Are these diseases with definable etiologies and pathogenic mechanisms? Alternatively, are these manifestations of a natural "aging" process, <u>i.e.</u>, the winding down of finite functions[5]? Specific replacement therapy may reasonably restore lost differentiated function

B. Pullman et al. (eds.), Interrelationship Among Aging, Cancer and Differentiation, 289–299.

in both instances; if environmental factors are the cause (viruses, bacteria, etc.), preventive measures might also be utilized.

II. HYPOTHESIS

We have postulated that several diseases notably occurring in the aged population and currently of unknown origin may be caused directly by lingering infectious agents and/or immune responses against such agents[6-9]. These could be introduced either early or late in the host's life, since the disease need not become manifest until accumulated damage from the defect is expressed with advanced age.

III. OBSERVATIONS

Support for the infectious etiology hypothesis comes from three sources--first, multiple observations of differentiated, cultured cells during acute and persistent virus infections; second, many studies of animals during natural infection or of animal models during experimental infections; and third, a modest but definitive number of human diseases. Such human illnesses include virus induced diabetes mellitus[10], virus associated Parkinson's disease[11], virus induced disorders of cognition or intelligence and/or balance and locomotion[12] and virus induced dysfunction of the immune system[9,13,14]. Interestingly, these known virus induced disorders of humans have similar counterparts in animal diseases and models. Thus, it is reasonable that the converse may be so. The abundance of examples showing virus induced chronic and degenerative diseases in animals should mirror disorders with the same or similar causes in man.

A. Cytopathology in the Presence of Cell Death

Viruses cause human diseases by killing cells with critical and specific life functions. For example, Coxsackie B4 virus that destroys beta cells has been isolated from the islets of Langerhans in a few patients with juvenile diabetes. Transfer of the recovered virus to susceptible animals induces destruction of beta cells within the islets[10]. As beta cells die, the synthesis of insulin decreases followed by hypoinsulinemia, hyperglycemia and acute diabetes. Similarly the 1918-1920 influenza pandemic has been associated with destruction of nerve cells in the basal ganglia and a clinical picture of Parkinson's disease[11]. Spongioform encephalopathy viruses cause Jakob-Creutzfeldt disease worldwide or Parkinson's-dementia complex, primarily in Guam[12]. Human T cell leukemia virus III (HTLV-III) is implicated in causing acquired immunodeficiency syndrome. The former two diseases are associated with destruction of central nervous system (CNS) cells and the latter with the OKT4 (helper function primarily) subset of lymphocytes. However, recent observations suggest that some of these infectious agents, especially HTLV-I and -III, instead of killing, can persist in lymphoid and some CNS cells.

Viruses cause cell death in two general ways. The first is by a direct effect on cells. Included are virus alterations of cells' transcriptional or translational machineries or disturbances of cells' lysosomal, nuclear or plasma membranes (reviewed 8, 15). The second is indirect, as infected cells express virus proteins[8,15] on their surfaces, calling forth an immune response that kills the cells. Further, viruses or their proteins bind to antiviral antibodies in the circulation and other body fluids to form immune complexes. Once trapped in capillary beds and arteries, such complexes cause glomerulonephritis, arteritis and choroiditis[16]. In any of these instances, cell destruction and inflammatory infiltration is the final event and frequently occurs at sites of injury. The unique tropisms of the virus for selected cells or cell subsets, and the number of cells killed, reflect the clinical symptoms, signs and severity of related illness. Mechanisms of cell destruction or perversion by viruses are illustrated in Figure 1.

B. Cytopathology in the Absence of Cell Death: in vitro Findings

Current knowledge about viruses and the diseases they cause is conditioned by the expectation of finding concurrent cell death and inflammatory lesions[1-4,8,11,26]. Yet many diseases occurring in the aged do so in the absence of major cell necrosis or inflammation but with significant cell dysfunction. Included are adult onset diabetes, disorders of cognitive function or personality changes and autoimmunity. To account for this circumstance, it was necessary to record events in which noncytolytic viruses could involve differentiated cells and alter their functions without causing their destruction.

The initial evidence that this occurred came from observations with cultured cells[17-22]. These experiments uniformly showed that a variety of viruses could alter differentiation (luxury) functions but not housekeeping (vital) functions. Hence, all such dysfunction occurred in the absence of cell death. Oldstone et al.[17,18] showed that cultured murine neuroblastoma cells infected with lymphocytic choriomeningitis virus (LCMV) become deficient in the production of enzymes needed to make or degrade acetylcholine (Table 1). LCMV is a noncytolytic virus that replicates with ease in the majority of cells it infects in vitro or in vivo and persistent infection usually ensues. During both acute or persistent infection, no anatomic or morphologic alteration of involved cells or tissues occurs and growth rates, cloning efficiencies and functions of vital enzymes are not impaired[17,18]. Interestingly, neuroblastoma cells infected at multiplicities of 0.1 to 3 develop viral antigens within three days, and continue to do so, but no deficiencies in esterase or transferase of acetylcholine are measurable till several days later (Table 1). Similar results occur with some but not all antibody producing murine hybridoma cell lines infected with LCMV. Usually no significant decrease in antibody synthesis is noted until 10 days after the initial infection. The synthesis of antibody may then decrease during the next 150 days after which the defect remains constant. Again, cloning efficiencies and growth

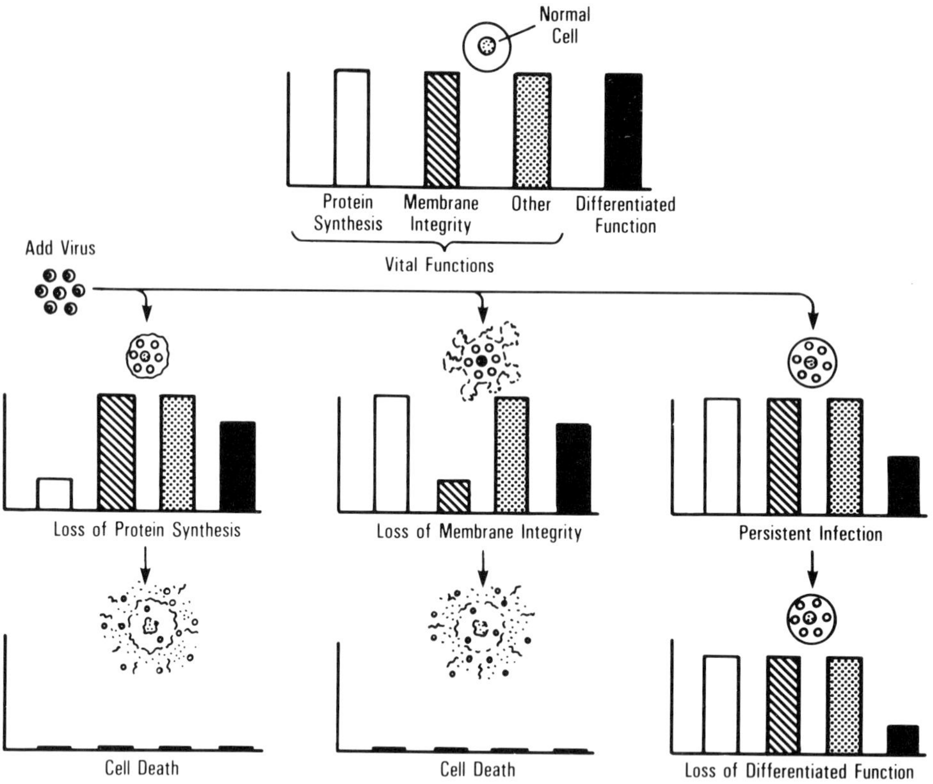

FIGURE 1. Cartoon of mechanisms by which a virus infection alters cells' functions, disturbs homeostasis and leads to disease. The first mechanism depicts the accompanying loss of vital or housekeeping functions, i.e., cell death due to either abrogation of a cell's protein synthesis or alteration of its membrane integrity. The second mechanism depicts a more subtle effect by which the housekeeping or vital functions are not impaired, cell anatomy and morphology remain normal but alterations in differentiated functions occur. See text for biomedical implications.

rates of infected cells are no different from those of matched controls. Holzer and his associates[19,20] found corresponding results with differentiated chick cells infected with Rous sarcoma virus (RSV). RSV-infected chondroblasts failed to manufacture sulfated-proteoglycans, and similarly infected myotubes or melaninoblasts did not make the heavy or light chains of myosin or melanin, respectively. Using a temperature sensitive mutant of RSV, the authors showed that these differentiation products could be cycled by altering the temperature (high temperature: nonpermissive state for virus = normal synthesis of differentiation products; low temperature: permissive state for virus =

TABLE 1: Virus induced alteration of a differentiation product *in vitro*: deficiency of the acetylase and esterase of acetylcholine in cultured murine neuroblastoma cells persistently infected with virus*

Day of LCMV infection	Acetylase**		Esterase**	
	uninf	virus inf	uninf	virus inf
pre-infection	86	--	242,600	--
5	84	81	294,300	263,183
30	89	42	279,870	140,600
120	87	44	239,820	143,620
180	83	48	144,294	98,135
360	81	40	212,491	126,329

* N115 neuroblastoma cells were infected at a multiplicity of 0.1 with LCMV Armstrong strain. Within 3 days, 95% of cells expressed viral nucleoprotein and continued to do so throughout the observation period. The cell viability and cloning efficiencies of infected and uninfected cells were equivalent. See references 17, 18 and 36 for details.
** p moles/mg cell protein/minute at 37°C.

diminished synthesis of differentiated products). Recently researchers in our laboratory[21,22] extended these findings to virus infected human cells. Infection of lymphocytes with RNA (measles, influenza) or DNA (cytomegalovirus) viruses altered several specific immune functions of these differentiated cells (Figure 2). Like murine neuroblastoma and hybridoma cells infected with LCMV and chick cells infected with RSV, these infected human lymphocytes displayed viabilities and growth equivalent to those of uninfected matched control cells despite their pronounced loss of function.

C. Cytopathology in the Absence of Cell Death: *in vivo* Findings

Do similar events occur *in vivo* whereby viruses can alter differentiation functions in the absence of cytomorphic injury? Further, could such alterations lead to a significant enough imbalance in homeostasis to cause clinical disease?

The answer recently came from observations of C3H/St mice infected at birth with the Armstrong strain of LCMV[23,Table 2]. In the pituitary gland, this virus showed tropism only for cells synthesizing growth hormone. The biochemical consequence was diminished synthesis of growth hormone. Production of other hormones like prolactin, ACTH, cortisol and insulin was unimpaired. Anatomically, the pituitary gland was normal and histologically its virus infected cells revealed no morphologic abnormalities by either light or electron microscopy[24]. Clinically, the decreased growth hormone synthesis led to severe growth retardation and hypoglycemia. Reconstitution of persistently infected mice with growth hormone via transplants of hormone producing cells corrected both the growth retardation and the glucose metabolism[25, Table 2].

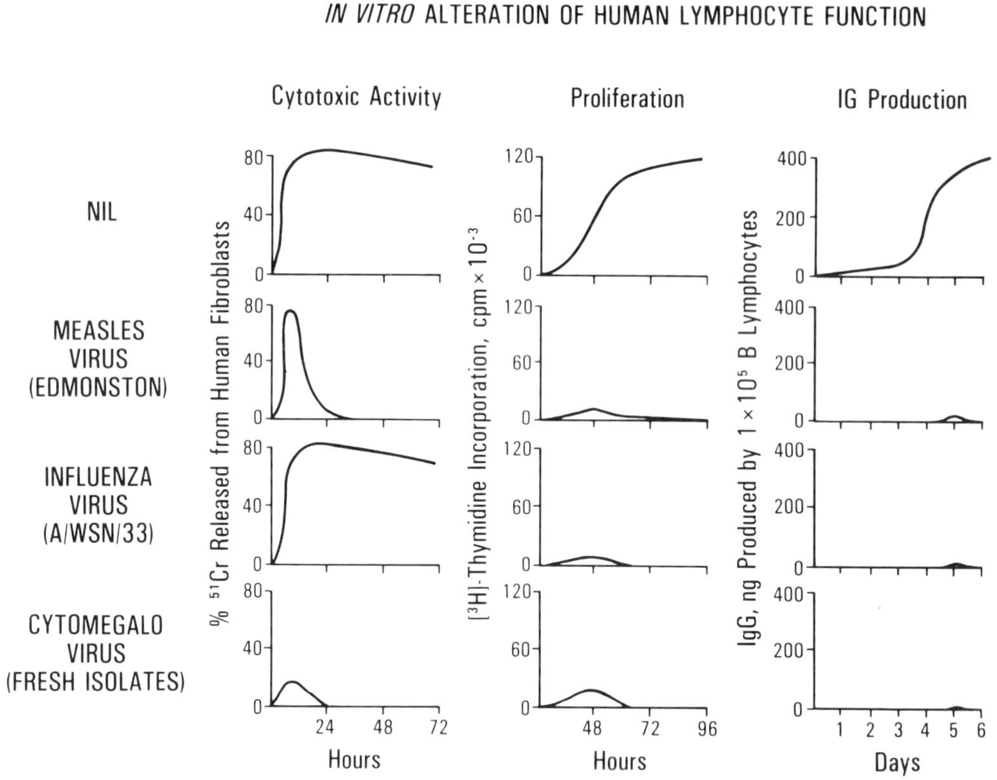

FIGURE 2. Schematic representation from several experiments in which viruses were shown to disrupt lymphocyte functions. Assays were: killing by either natural killer or virus specific-HLA restricted T lymphocytes, proliferation in response to antigen or mitogen, and immunoglobulin (Ig) secretion by B lymphocytes in the presence of T lymphocytes and pokeweed mitogen. See references 21 and 22 for details.

These studies indicate that virus infection of specialized cells in vivo can disorder an important differentiation product, thereby unbalancing homeostasis sufficiently to cause disease. Further, cytopathology occurred in the absence of cytolysis. Thus, routine histopathologic study would yield no clue to the cause of abnormal growth hormone synthesis and, by convention, the absence of cell lysis and inflammatory infiltration rules out participation by a virus[1,11,26]. Recent studies indicate a block in transcription of growth hormone messenger RNA; that is, significantly less message is made by virus infected pituitaries than by uninfected pituitaries (A. Valsamakis and M. B. A. Oldstone, unpublished observations). For the purpose of investigating the host genes and virus genes involved, LCMV-resistant and -susceptible mice were cross-bred. So far, the results involve many host genes and no strong linkage to the major histocompatibility locus. Recently virulent and avirulent strains of LCMV with respect

TABLE 2: Virus induced alteration of a differentiation product *in vivo*: growth hormone deficiency, growth retardation and hypoglycemia associated with persistent virus infection*

Experimental Group	Growth (wgt/gms) 15 days	Growth (wgt/gms) 50 days	Blood Sugar(dl/ml) 15 days	Blood Sugar(dl/ml) 50 days
uninfected control	7.7±.1	19.5±.2	144±20	134±8
virus infected	4.1±.1	<1% ** survive	43 ±5	<1% ** survive
virus infected reconstituted with growth hormone	7.0±.2	19.0±.2	128 ±4	128 ±5

* Mice infected with LCMV Armstrong strain at birth (60 plaque forming units) and maintaining persistent infection thereafter. All contained replicating virus in cells making growth hormone in the anterior lobe of the pituitary and diminished levels of growth hormone in the pituitary gland. Weights (wgt) and blood sugar levels determined 15 and 50 days after initiating infection. See references 23 and 25 for details.
** Data limited owing to death from hypoglycemia.

to capacity for retarding growth of C3H/St mice have been noted[27]. Since the LCMV genome consists of two nucleic acid segments, intertypic reassortants among virulent and avirulent viruses could be generated[28]. After using a number of reassortants, the S RNA (4Kb segment) of LCMV and the genes it encodes have been identified as causing this disease[29].

With other strains of mice and of LCMV, persistent infection of beta cells of the islets of Langerhans has been observed[30]. The result of this infection is hyperglycemia, normal or moderately low levels in blood insulin and abnormal glucose tolerance tests. These findings resemble those of adult onset diabetes. Further, as in adult onset diabetes, minimal morphologic injury to beta cells or islets occurs[3,26,30,31].

Other studies have been directed towards understanding how LCMV or other viruses persist *in vivo*. It was noted that LCMV is tropic for a small proportion of lymphocytes, frequently less than 1% of the total circulating population[32]. As a result, too few virus specific H-2 restricted cytotoxic T lymphocytes (CTL) are generated to eliminate LCMV infected cells. Thus LCMV persists *in vivo* in multiple cell types and tissues, in large part because the generation of virus specific CTL is hindered[33,34]. Reconstitution of such CTL, by using splenic lymphocytes from LCMV immune, syngeneic donors clears the infectious virus, viral nuclei and sequences and proteins from tissues of persistently infected mice[34,Figure 3].

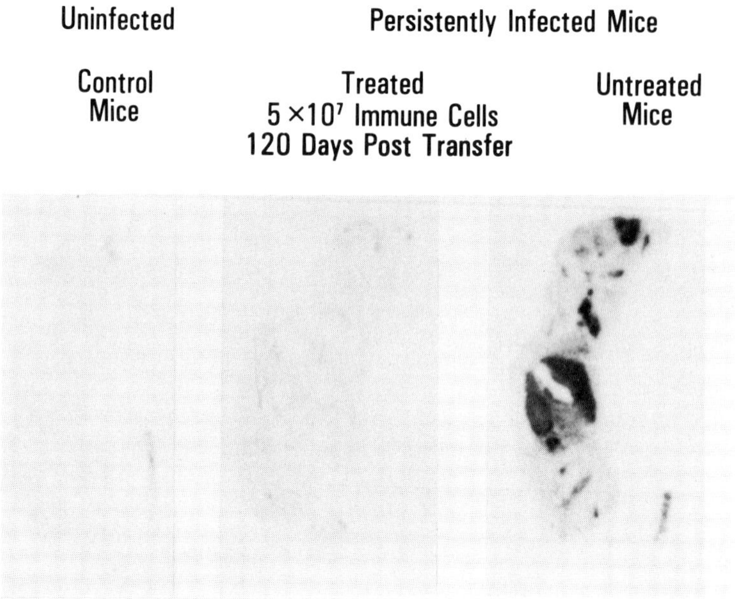

Tissue	LCMV titer (log_{10} PFU/organ or ml)	
Serum	<1.6	4.5
Spleen	<1.6	5.8
Liver	<1.6	5.3
Lung	<1.6	5.3
Brain	1.9	5.2
Kidney	3.9	6.5

FIGURE 3. Mice persistently infected with LCMV were adoptively transferred with splenic lymphocytes from immune, syngeneic donors. The amount of infectious virus was then quantitated by plaque assay (PFU). Data represent mean values from four independent observations. The presence of LCMV nucleic acid sequences was observed *in situ* after hybridization to whole mouse sections (30 micron) and use of a 32p labeled cDNA probe specific for the glycoprotein 2 gene of LCMV Armstrong strain. See reference 35 for details.

IV. CONCLUSION

Increasing amounts of evidence suggest that viruses can cause injury and disease by subtly disordering the function of specialized cells. Depending on the differentiated cell(s) infected and the function it performs or product it makes, an imbalance of homeostasis may result. Often the consequence is a chronic and progressive disease that appears in the later stages of life. Such disease can occur in the absence of cell lysis or inflammatory infiltration. Involvement of nervous, endocrine and immune systems by such mechanisms has been described in several experimental models. Currently, probes are being generated and strategies planned to evaluate human diseases of unknown etiologies for similar abnormalities. Perhaps we will learn how that wonderful one hoss shay of the deacon did break down.

> "Have you heard of the wonderful one hoss shay,
> That was built in such a logical way
> It ran a hundred years to a day,
> And then, of a sudden, it ----- ah, but stay,
> I'll tell you what happened without delay...
>
> "...in building of chaises, I tell you what,
> There is always somewhere a weakest spot...
>
> "...And that's the reason, beyond a doubt,
> That a chaise breaks down, but doesn't wear out."[5]

ACKNOWLEDGEMENTS

This is Publication No. 3961-IMM from the Department of Immunology, Scripps Clinic and Research Foundation, La Jolla, California 92037.

This research was supported by U.S.P.H.S. grants AG-04342, NS-12428 and AI-09484, and Biomedical Research Support Grant RR0-5514.

The author wishes to thank Ruth Danielle and Phyllis Minick for their help in manuscript preparation.

REFERENCES

1. Isselbacher, K.J., et al. (eds.): 1982, HARRISON'S PRINCIPLES OF INTERNAL MEDICINE, McGraw Hill, New York.
2. Bloom, F.: 1981, Sci. Amer. 245, pp. 148-168.
3. DeGroot, L.J., et al. (ed.): 1979, ENDOCRINOLOGY, Grune & Stratton, New York.
4. Paul, W.E. (ed.): 1984, FUNDAMENTAL IMMUNOLOGY, Raven Press, New York.
5. Holmes, O.W.: "The Deacon's Masterpiece: or, The Wonderful 'One-Hoss Shay'" in POEMS FROM THE AUTOCRAT OF THE BREAKFAST TABLE.
6. Oldstone, M.B.A.: 1971, "Autoimmunity and viruses--fact or fiction: Persistent LCM viral infection, anti-LCM viral immune response and tissue injury", Amer. J. Clin. Pathol. 56, pp. 299-302.
7. Oldstone, M.B.A. & Dixon, F.J.: 1974, "Aging and chronic virus infection: Is there a relationship?", Panel of Immunopathology of Aging, FASEB 33, pp. 2057-2059.
8. Oldstone, M.B.A.: 1982, "Immunopathology of persistent viral infections", Hosp. Prac. 17, pp. 61-72.

9. Notkins, A.L. & Oldstone, M.B.A. (eds.): 1984, CONCEPTS IN VIRAL PATHOGENESIS, Springer-Verlag, New York.
10. Yoon, J., et al: 1979, New Eng. J. Med. 300, pp. 1173-1179.
11. Blackwood, W. & Corsellis, J. (eds.): 1976, GREENFIELD'S NEURO-PATHOLOGY, Year Book Publishers, Chicago.
12. Gajdusek, D.C.: 1978, Harvey Lectures 72, p. 283.
13. von Pirquet, C.E.: 1908, "Das Verhalten der kautanen Tuberculin-reaktion wahrend der Masern", Dtsch. Med. Wochenschr. 34, p. 1297.
14. Reviewed Marx, J.: 1983, Science 220, pp. 806-809; Black, P. & Levy, E.: 1983, N. Eng. J. Med. 309, p. 856.
15. Oldstone, M.B.A. & Southern, P.: 1985, "Medical consequences of persistent virus infections", N. Eng. J. Med., submitted.
16. Oldstone, M.B.A.: 1975, "Virus neutralization and virus-induced immune complex disease: Virus-antibody union resulting in immuno-protection or immunologic injury--Two sides of the same coin" in Melnick, J.L. (ed): PROGRESS IN MEDICAL VIROLOGY, Vol. 19, S. Karger, Basel, pp. 84-119.
17. Oldstone, M.B.A., Holmstoen, J. & Welsh, R.M.: 1977, "Alterations of acetylcholine enzymes in neuroblastoma cells persistently in-fected with lymphocytic choriomeningitis virus", J. Cell Physiol. 91, pp. 459-472.
18. Oldstone, M.B.A, Welsh, R.M. & Joseph, B.S.: 1975, "Pathogenic mechanisms of tissue injury in persistent viral infections", Annals N.Y. Acad. Sci. 256, pp. 65-72.
19. Holtzer, H., Pacifici, M., Tapscott, S., Bennett, G., Payette, R. & Dlugosz, A.: 1982, "Lineages in cell differentiation and in cell transformation" in Revoltella, R.F. (ed.): EXPRESSION OF DIFFEREN-TIATED FUNCTIONS IN CANCER CELLS, Raven Press, New York, p. 169.
20. Holtzer, H., Biehl, J., Yeoh, G., Meganathan, R. & Kaji, A.: 1975, "Effect of oncogenic virus on muscle differentiation", Proc. Natl. Acad. Sci. USA 72, pp. 4051-4055.
21. Casali, P., Rice, G.P.A. & Oldstone, M.B.A.: 1984, "Viruses dis-rupt functions of human lymphocytes: Effects of measles virus and influenza virus on lymphocyte-mediated killing and antibody pro-duction", J. Exp. Med. 159, pp. 1322-1337.
22. Rice, G.P.A., Schrier, R.D. & Oldstone, M.B.A.: 1984, "Cytomega-lovirus infects human lymphocytes and monocytes: Virus expression is restricted to immediate-early gene products", Proc. Natl. Acad. Sci. USA 81, pp. 6134-6138.
23. Oldstone, M.B.A., Sinha, Y.N., Blount, P., Tishon, A., Rodriguez, M., von Wedel, R. & Lampert, P.W.: 1982, "Virus-induced alterations in homeostasis: Alterations in differentiated functions of infected cells in vivo", Science 218, pp. 1125-1127.
24. Rodriguez, R., von Wedel, R.J., Garrett, R.S., Lampert, P.W. & Oldstone, M.B.A.: 1983, "Pituitary dwarfism in mice persistently infected with lymphocytic choriomeningitis virus", Lab. Invest. 49, pp. 48-53.
25. Oldstone, M.B.A., Rodriguez, M., Daughaday, W.H. & Lampert, P.W.: 1984, "Viral perturbation of endocrine function: Disorder of cell function leading to disturbed homeostasis and disease", Nature 307, pp. 278-280.

26. Robbins, S., Angell, M. & Kumar, V. (eds.): 1981, BASIC PATHOLOGY, Saunders Co., Philadelphia.
27. Oldstone, M.B.A., Ahmed, R., Buchmeier, M.J., Blount, P. & Tishon, A.: 1985, "Perturbation of differentiated functions during viral infection in vivo. I. Relationship of LCMV and host strains to growth hormone deficiency", Virology 142, pp. 158-174.
28. Riviere, Y., Ahmed, R., Southern, P.J., Buchmeier, M.J., Dutko, F.J. & Oldstone, M.B.A.: 1985, "The S RNA segment of lymphocytic choriomeningitis virus codes for the nucleoprotein and glycoproteins 1 and 2", J. Virol. 53, pp. 966-968.
29. Riviere, Y., Ahmed, R., Southern, P. & Oldstone, M.B.A.: 1985, "Perturbation of differentiated functions during viral infection in vivo. II. Viral reassortants map growth hormone defect to the S RNA of the lymphocytic choriomeningitis virus genome", Virology 142, pp. 175-182.
30. Oldstone, M.B.A., Southern, P., Rodriguez, M. & Lampert, P.: 1984, "Virus persists in beta cells of islets of Langerhans and is associated with chemical manifestations of diabetes", Science 224, pp. 1440-1443.
31. Rodriguez, M., Garrett, R.S., Raitt, M., Lampert, P.W. & Oldstone, M.B.A.: 1985, "Virus persists in the beta cells of islets of Langerhans and infection is associated with chemical manifestations of diabetes. II. Morphologic observations", Amer. J. Path., submitted.
32. Doyle, M.V. & Oldstone, M.B.A.: 1978, "Interactions between viruses and lymphocytes. I. In vivo replication of lymphocytic choriomeningitis virus in mononuclear cells during both chronic and acute viral infections", J. Immunol. 121, pp. 1262-1269.
33. Ahmed, R., Salmi, A., Butler, L.D., Chiller, J.M. & Oldstone, M.B.A.: 1984, "Selection of genetic variants of lymphocytic choriomeningitis virus in spleens of persistently infected mice: Role in in suppression of cytotoxic T lymphocyte response and viral persistence", J. Exp. Med. 60, pp. 521-540.
34. Ahmed, R., Southern, P., Blount, P., Byrne, J. & Oldstone, M.B.A.: 1985, "Viral genes, cytotoxic T lymphocytes, and immunity" in Lerner, R.A., Chanock, R.M. & Brown, F. (eds.): VACCINES 85: MOLECULAR AND CHEMICAL BASIS OF RESISTANCE TO PARASITIC, BACTERIAL, AND VIRAL DISEASES, Cold Spring Harbor Laboratory, New York, pp. 125-132.
35. Southern, P.J., Blount, P. & Oldstone, M.B.A.: 1984, "Analysis of persistent virus infections by in situ hybridization to whole-mouse sections", Nature 312, pp. 555-558.
36. Welsh, R.M. & Oldstone, M.B.A.: 1977, "Inhibition of immunologic injury of cultured cells infected with lymphocytic choriomeningitis virus: Role of defective interfering virus in regulating viral antigenic expression", J. Exp. Med. 145, pp. 1449-1468.

AGEING AND CANCER. A COMMON FREE RADICAL MECHANISM?

Lawrence H. Piette
Department of Chemistry and Biochemistry, Utah State University

Abstract
An attempt is made to present evidence that there is a close relationship at the molecular level between biological events associated with oxidative metabolism that can lead to cell ageing and ultimate tumorigenesis. It is suggested that the common thread between these two gross biological events is the production of and lack of inhibition of labile oxygen radicals $\cdot O_2^-$ and $\cdot OH$. Evidence is presented that OH radicals are responsible for the initiation of lipid peroxidation which is associated with cell ageing and these same radicals may be responsible for carcinogen activation or direct mutagenesis leading ultimately to tumorigenesis.

Cancer epidemiological data has clearly shown the association between cancer mortality and cancer incidence with age. This dependence for certain malignancies may even suggest that cancer is in fact a natural cause of death in man. If one compares mortality rates for all causes of death as a function of age with those associated with cardiovascular disease there is a strong similarity in the two curves suggesting that cardiovascular disease is probably the most important natural cause of death. Denham Harman (6) approximately 30 years ago suggested a commonality in mechanism between the ageing process and cardiovascular disease particularly arteriosclerosis. He suggested the two were due to an accumulation of toxic oxygen radicals. Harman suggested that oxygen radicals such as $\cdot O_2^-$ and $\cdot OH$ which are normal intermediates in oxidative metabolism under certain circumstances could induce lipid peroxidation which in turn is responsible for both plaque formation in cardiovascular disease and cellular ageing in the general ageing process. I should like to suggest this other age dependent disease namely cancer, may lend itself to exactly the same type of mechanism as suggested by Harman for cardiovascular disease and life expectancy.

John Totter (14) has suggested that if one looks at the total spontaneous cancer rates for man throughout the world and assumes a common origin for the cause of these cancers, namely a somatic muta-

B. Pullman et al. (eds.), Interrelationship Among Aging, Cancer and Differentiation, 301–312.

tion of DNA, then one can calculate from what is known from human radiation dose effects and cancer that this number of spontaneous cancers would require about 450 to 2100 rads of total exposure or approximately 0.3 to 1.3 single strand breaks per cell DNA per day. Totter further suggests that such equivalent DNA breaks can easily be obtained from the production of oxygen radicals that occur naturally through the reduction of oxygen during normal oxidative metabolism in the absence of any radical inhibition or defense system.

This explanation as to the cause of spontaneous cancer is clearly an age or dose dependent effect, i.e. the accumulation in this case of endogenous potential carcinogenic radicals, however, this is probably not the only mechanism. Labile radicals can also participate in the general ageing process that can result in biological changes that can promote carcinogenesis and tumor development. Which of these are dependent mechanisms is the correct one remains to be seen.

Animal studies on this subject are inconclusive for many reasons. Variations in effects due to species and strain. No control for rates of metabolism. Insufficient numbers of truly aged animals used in the studies. There is, however, a growing knowledge that suggests oxidative metabolism rates may play a common role in both processes, i.e. ageing and cancer. It has been well established that those animal species with very high oxidative metabolism rates age much more rapidly and have shorter life expectancies than those with slower rates, i.e. man versus the mouse. Similarly when inhibitors of free radical oxidative processes are present ageing is slowed and life expectancy increases. As there is an inverse correlation between life expectancy and oxidative metabolism rates there is a direct correlation between levels of antioxidants such as vitamin E, C, and perhaps A with life expectancy.

It has also been suggested (11) that these same antioxidants are also effective in inhibiting or preventing certain types of neoplasms. The common mechanism that may link the two together, i.e. ageing or life expectancy and cancer is the inhibition of oxidatively generated free radicals. Both $\cdot O_2^-$ and $\cdot OH$ radicals are naturally ocurring by-products of oxidative metabolism, i.e. the cytochrome P450 system. Both radical species are potential gene modifying agents. The OH radical is considered to be the strongest oxidant known. Current mechanisms suggested for the action of the antibiotic type antitumor agents is the generation of labile OH radicals through a Fenton type redox coupling of the drug bound to DNA which results in unrepaired nicking of the genetic material. Since it is also known that most antitumor drugs are also carcinogenic a similar mechanism has been proposed for carcinogenesis.

Under normal circumstances these labile radicals are effectively controlled by the naturally occurring antioxidants vitamin E, A, and C and the oxidative enzymes such as catalase, superoxide dismutase and glutathione peroxidase.

In addition to altering the genetic material and thus inducing the mutation that may be responsible for tumorigenesis these oxidative radicals are also capable of initiating cell membrane lipid peroxidation, a process that is definitely associated with cell degeneration and ageing.

It has been thought for some time that lipid peroxidation is initiated by either labile superoxide $\cdot O_2^-$ radicals or $\cdot OH$ radicals; however, up until now there has been no definitive detection of either of these radicals in the initiation step leading to lipid peroxidation. Electron spin resonance lends itself very nicely to the detection of free radicals; however, very labile free radicals can only be detected by this technique if they are observed in the steady state, i.e. if their production rate is sufficiently high to compensate for their rapid disappearance. In a biological system in which lipid peroxidation is taking place, the steady state levels of the initiating radical levels or those involved in propagation are too small to be detected by most standard ESR methods.

A new method has been developed called spin trapping (5, 8) which has the potential of allowing one to detect labile radicals such as those that may be involved in lipid peroxidation. The method uses a variety of spin traps derived from the reactive alkyl nitrones (7). These nitrones have the advantage that when reacted with a labile radical they form a stable nitroxide that is easily detected by ESR, the general reaction being as follows.

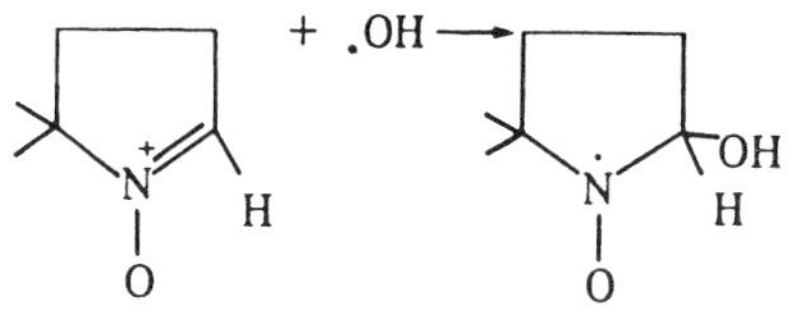

The radical adduct adding to the nitrone can usually be identified by the characteristic splitting observed with the nitroxide produced. Thus, for example, when .OH radicals add to 5,5'-1-Pyrroline-1-Oxide (DMPO) one obtains a characteristic five line spectrum in which the nitrogen and Beta H hyperfine couplings are comparable $a^N\ a^H_\beta = 15.0$ gauss whereas when $.HO_2$ radicals add, one gets twelve lines in which $a^N = 14.3$, $a^H_\beta = 11.7$ and $a^H_\gamma = 1.25$ gauss. Carbon centered radical adducts, on the other hand, give totally different splittings.

The liver microsomal membrane system is an excellent model for studying lipid peroxidation. Hochstein and Ernster (7) have demonstrated an induced NADPH-dependent enzymic lipid peroxidation in liver microsomes. The extent of peroxidation can be altered if the animals from which the microsomes have been isolated are given large doses of antioxidants.

Repeating Hochstein's experiment in Figure 1, an increase is seen in the production of malondialdehyde, the end product of lipid peroxidation in the microsomal system as a function of added NADPH. If we monitor instead of malondialdehyde production the amount of free radicals that can be trapped by DMPO, we get in this same system during induced lipid peroxidation an identical curve as shown in Figure 1. The spectrum shown in Figure 2a for the trapped adduct in this system is characteristic of the OH radical. The splitting and the hyperfine coupling constants are identical with those obtained by independent generation and trapping of the OH radical during U.V. photolysis of hydrogen peroxide or as the product of the Fenton reaction as shown in Figure 3a and b.

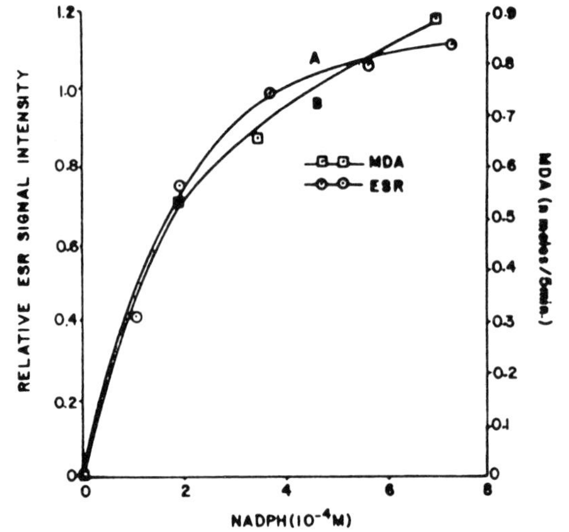

FIGURE 1. *Correlation between DMPO-OH radical adduct and malondialdehyde production in liver microsomes as a function of NADPH. A) —o—o—o— the relative ESR signal intensities of the DMPO-OH adduct obtained when the reaction mixtures containing 2.1 mg/ml microsomes, 2.2 x 10^{-5}M Fe^{2+}, 2.2 x 10^{-5}M EDTA, 5.6mM DMPO and 0.15 M KCl in 1mM phosphate buffer, pH 7.4 were mixed with various amounts of NADPH. B) —□—□—□— the same samples as in A except DMPO is omitted and MDA production was monitored.*

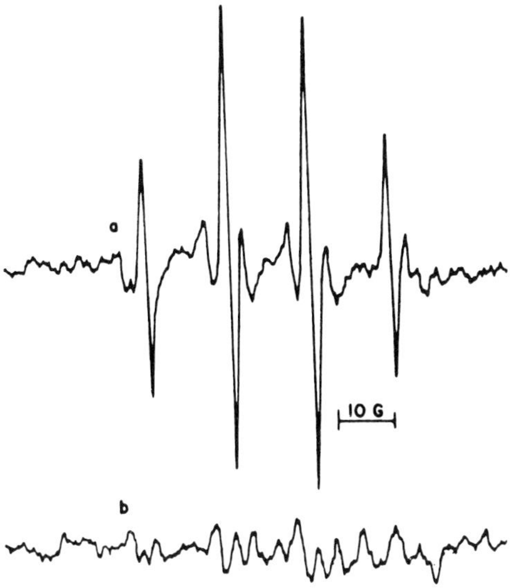

FIGURE 2. *ESR spectrum of the DMPO-OH radical adduct obtained from liver microsomes in the presence of DMPO. The reaction mixture containing 1.8 mg/ml microsomes, 2.2 x 10^{-5}M Fe^{2+}, 4.4 x 10^{-5}M EDTA, 7.0 mM DMPO at pH 7.4 in 0.15 M KCl solution was mixed (a) with 0.74 mM NADPH, (b) without NADPH. $a^N = a^H_\beta$ = 15.0 G and g = 2.0062.*

Further evidence that the trapped radical in this system is due to OH radicals was from inhibition studies. Ethanol is a known inhibitor of OH radicals. If ethanol is added to the microsomal system and lipid peroxidation induced in the presence of DMPO, the spectrum in Figure 4 is obtained. If the same experiment is repeated using a model .OH generating system such as the Fenton reaction instead of the microsomal system, a similar spectrum is obtained. The hyperfine splittings clearly identify the adduct as a carbon centered radical, most likely CH_3-CH-OH.

One can also establish a model lipid peroxidation system using the Fenton reaction and various unsaturated fatty acids. If one adds Fe^{++} to H_2O_2 (Fenton reaction) in the presence of linoleic acid and monitors malondialdehyde formation as a function of added DMPO, the results in Figure 5 are obtained. We see that the addition of DMPO to the system produces an inhibition in malondialdehyde formation and the production of the DMPOH adduct suggesting that like ethanol DMPO selectively traps OH radicals and thus inhibits the initiation of lipid peroxidation. Thus, the evidence is quite convincing that .OH radicals are produced during NADPH induced lipid peroxidation but the question remains as to the source of the .OH radical.

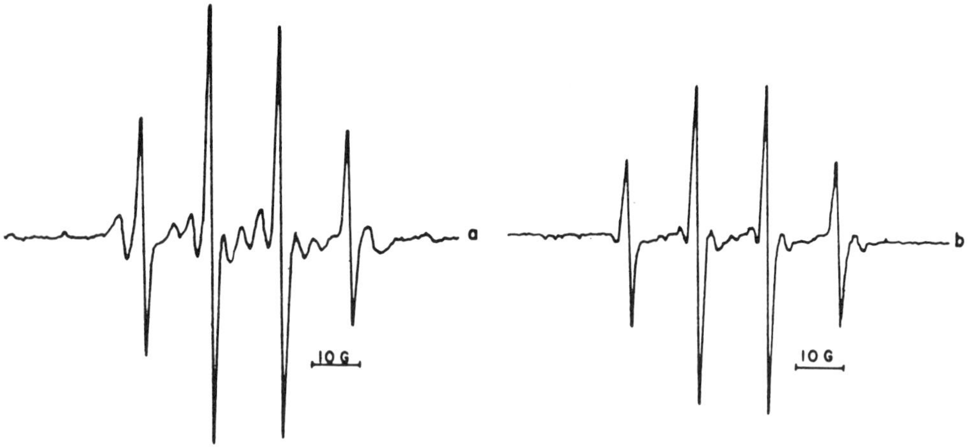

FIGURE 3. *a) ESR spectrum of the DMPO-OH adduct obtained upon U. V. irradiation of H_2O_2 in the presence of DMPO. b) ESR spectrum of the DMPO-OH adduct obtained in the Fenton reaction in the presence of DMPO. A 6.2 mM DMPO solution was added to the reaction mixture containing 2.0 x 10^{-4}M Fe^{2+}, 4.0 x 10^{-4}M EDTA, 0.5% H_2O_2 in in 0.15 M KCl solution. $a^N = a^H_\beta = 15.0$ gauss and g = 2.0062.*

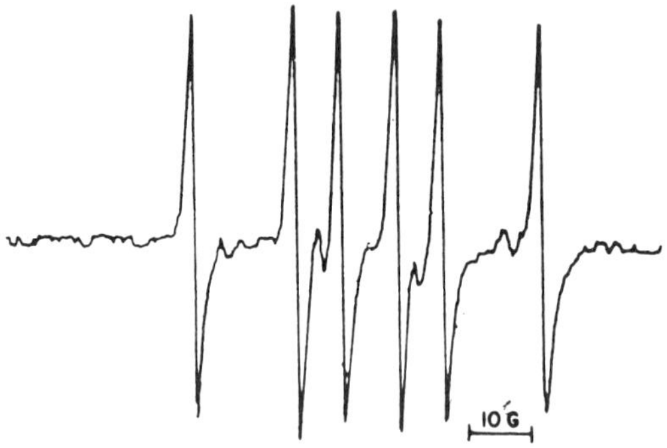

FIGURE 4. *ESR spectrum of the DMPO-ethanol adduct obtained when a 2% ethanol solution is added to a mixture containing 2.2 mg/ml microsomes, 100 mM DMPO, 2 x 10^{-4}M Fe^{2+}, 1 x 10^{-4}M EDTA, 1 mM phosphate, pH 7.2 and 1 mM NADPH.*

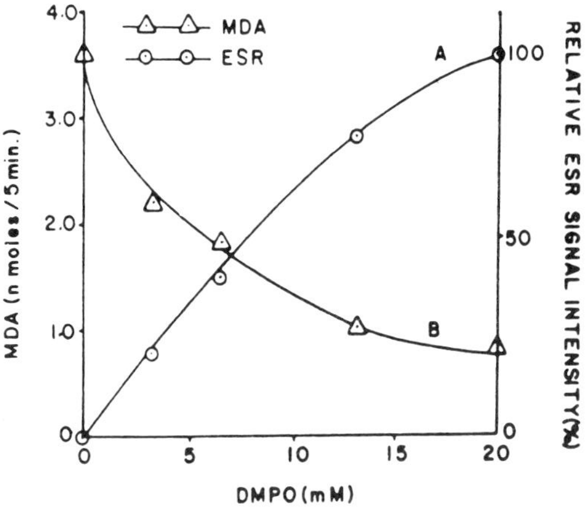

FIGURE 5. *Relationship between the DMPO-OH radical adduct and malondialdehyde production in the presence of linolenic as a function of DMPO. A) —○—○—○— Relative ESR signal intensities of the DMPO-OH radical adduct. Reaction mixtures contained 4.9 x 10^{-5}M Fe^{2+}, 4.9 x 10^{-5}M EDTA, 1.0 mM linolenic acid, 1.6 x 10^{-4}M H_2O_2, and 0.15 M KCl in 1 mM phosphate buffer, pH 7.4 and DMPO (0-20 mM) B) —△—△—△—△— A set of samples identical in composition to those in A were used to measure MDA production.*

It is well known that certain flavin proteins upon reduction are capable of reducing oxygen to $.O_2^-$ or $.HO_2$, the most notable of these proteins being xanthine oxidase (3). In rat liver microsomes, the mixed function oxidase system also has a flavin protein, namely cytochrome P450 reductase. Thus, it would not be unreasonable to assume that this NADPH dependent enzyme upon reduction is also capable of reducing oxygen to $.O_2^-$. In Figure 6a we see the results of the incubation of a purified preparation of cytochrome P450 reductase with DMPO and NADPH. The spectrum is quite different from that of the .OH radical adduct but identical to that obtained from the oxidation of xanthine by xanthine oxidase Figure 6b. This radical is identified as the DMPOOH adduct from the addition of $.O_2^-$ which yields twelve lines with splittings of a^{N2} = 14.3, a^H = 11.7 and a^H = 1.25 gauss.

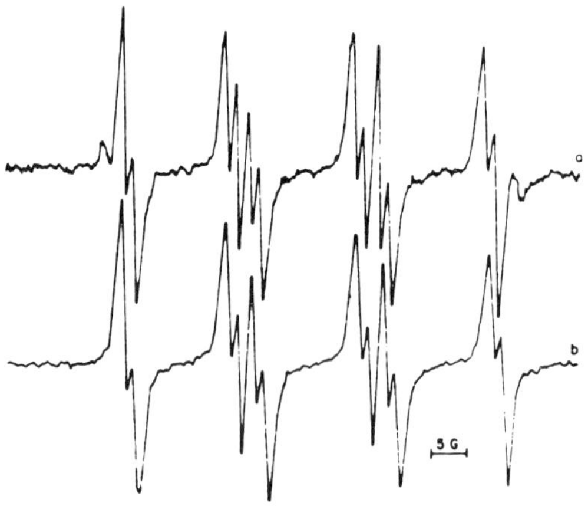

FIGURE 6. *ESR spectra of the DMPO-OOH adduct. a) spectrum obtained from the reaction of a mixture containing 56 mg/ml purified cytochrome P450 reductase with 1 mM NADPH, 1 mM phosphate buffer, 1 mM DETAPAC (diethylenetriamine penta--aceticacid) and 50 mM DMPO. b) reaction of 0.2 mg of xanthine oxidase with 5×10^{-4} M xanthine in 1 mM detapac and 50 mM DMPO, 1 mM phosphate buffer, pH 7.4 Scan rate is one half that of other spectra.*

No adduct formation takes place if superoxide dismutase (SOD) is added to the incubation. If, on the other hand, a small amount of free Fe^{+2} is added to the incubation, both enzyme reactions yield the same spectrum shown in Figure 2a, suggesting that in the presence of endogenous iron $.O_2-$ or its product of dismutation H_2O_2 reacts rapidly in a Fenton or iron dependent Haber Weiss reaction to yield .OH. In the microsomal system the reaction scheme would be as follows:

Oxidized P450 reductase +NADPH $\rightarrow$ Red P450 reductase + NADP+ (1)

Reduced P450 reductase $+O_2$ $\rightarrow$ Ox P450 reductase + $.O_2-$ (2)

$$.O_2^- + H^+ \rightleftarrows HO_2^{\bullet} \quad (3)$$

$$.O_2^- + HO_2^{\bullet} \rightleftarrows O_2 + H_2O_2 + OH^- \quad (4)$$

$$.O_2^- + H_2O_2 \rightleftarrows O_2 + OH^- + .OH \quad (5)$$

$$Fe^{2+} + .O_2^- + 2H^+ \rightleftarrows Fe^{3+} + H_2O_2 \quad (6)$$

$$Fe^{2+} + H_2O_2 \rightleftarrows Fe^{3+} + .OH + OH^- \quad (7)$$

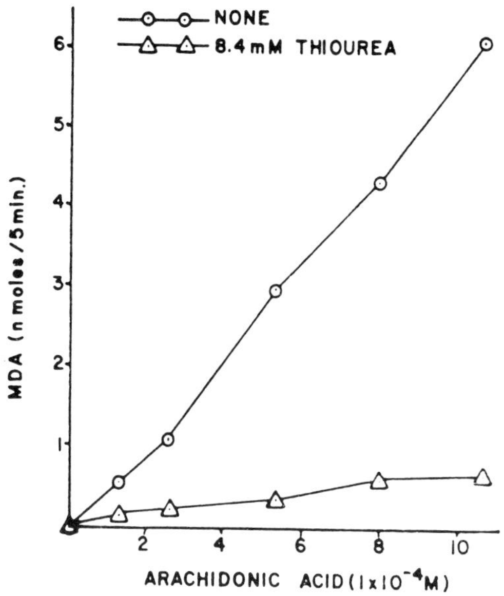

FIGURE 7. *Malondialdehyde production in the Fenton reaction as a function of arachidonic acid. The amounts of malondialdehyde production were measured when reaction mixtures containing 5.3 x 10^{-5}M Fe^{2+}, 5.3 x 10^{-5}M EDTA, 1.8 x 10^{-4}M H_2O_2, 0.15 M KCl in 1 mM-phosphate buffer, pH 7.4 were mixed with various concentrations of arachidonic acid without thiourea, a potent ·OH radical inhibitor, and with 8.4 mM thiourea.*

The .OH radical so produced in this system can then initiate lipid peroxidation as shown in Figure 7 where in the Fenton reaction the .OH induces the formation of malondialdehyde. The .OH radical can also be scavenged by other organic substrates including carcinogens to yield activated carcinogens or modified cells leading eventually to tumorigenesis in a scheme such as that illustrated in Figure 8.

This scheme suggests that the same system responsible for lipid peroxidation, a known cause of ageing can possibly lead to carcinogenesis as well.

Up until now, the accepted mechanism for nitrosoamine activation involved oxidative demethylation by a liver demethylase enzyme resulting in the formation of a reactive carbonium ion that serves as the primary alkylating agent. Spin-trapping studies, however, suggest that an alternate mechanism may be the production of a labile radical that can also act as an alkylating agent.

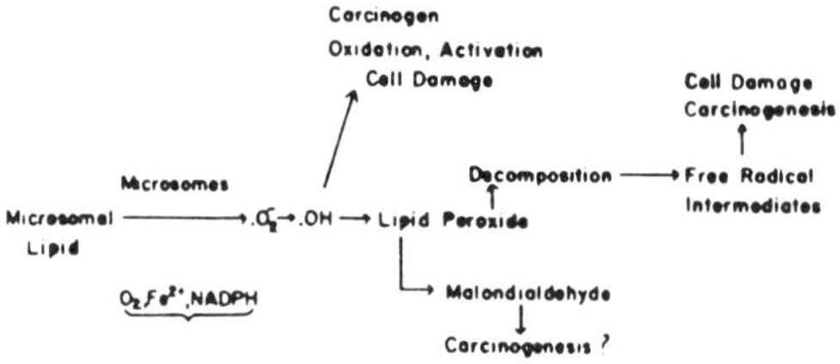

FIGURE 8. *A proposed scheme for carcinogenesis involving free radical activation via the mixed function oxidase system and lipid peroxidation.*

As suggested earlier in Figure 8, any kind of reactive free radical that can damage surrounding proteins or the genetic material of a cell could be involved in tumorigenesis. Experiments conducted in a number of laboratories but most particularly those of N. M. Emanuel and A. N. Saprin of the U.S.S.R. have attempted to correlate the presence of reactive free radicals with tumor development in animals. These investigators have shown that during tumorigenesis or prior to the onset of tumor formation by ESR one can detect paramagnetic species, presumably free radicals, generated in the liver of animals that are either being fed or injected with chemical carcinogens such as dimethylazobenzene (2, 13). They observed a steady increase in the free radical level for approximately four months or until tumor formation could be observed, at which time the free radical level decreases and eventually disappears. This increase in free radical formation is parallel to an increase in chemiluminescence from the same tissue, suggesting peroxide decomposition as both the source of chemiluminescence and free radical formation.

Another interesting and important result that these investigators observed was that as the free radical level in the liver increased, there was a parallel decrease in the level of cytochrome P450 in the tissue as measured by ESR.

In Figure 9, it is seen in liver microsome experiments where lipid peroxidation is induced and followed by malondialdehyde production that there is also a decrease in the cytochrome P450 level measured spectrophotometrically. If lipid peroxidation is inhibited by the addition of a spin trap as PBN, the P450 levels are not affected and no malondialdehyde is formed. Thus, it would seem that the animal experiments of Emanuel and Saprin could possibly be associated with lipid peroxidation induced free radical tumorigenesis.

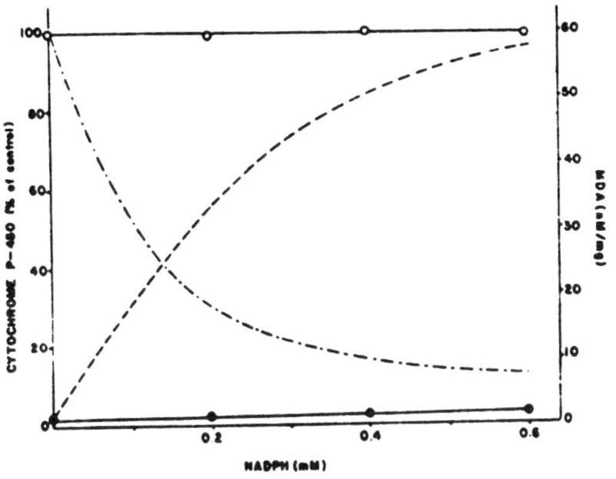

Figure 9. *Comparison of the effect of lipid peroxidation and PBN on cytochrome P450 levels in microsomes. P450 levels were measured on aliquots incubated with varying amounts of NADPH. (—·—·—·—·—) represents the change in P450 content, (-----) the change in MDA, (—o—o—o—o—o—) the change in P450 content when PBN is also present in the systems (—●—●—●) the change in MDA levels in the presence of PNB. PBN concentration is 0.15 M, microsomes 2 mg/ml, 1.2 x 10^{-5}M Fe^{2+}, 2.0 x 10^{-4}M pyrophosphate, 0.05 tris, pH 7.4.*

The mixed function oxidases of which cytochrome P450 is a component are responsible not only for activating certain carcinogens, but also detoxifying other potentially cell damaging xenobiotics. Thus, any perturbation of this system could very easily lead to the loss "optimum potential function" (ageing) of an organ or the induction of uncontrolled growth (cancer).

References

1. ANISIMOV, V. N. and TURUSOV, V. S.: Modifying effect of ageing on chemical carcinogenesis. A Review Mechanism of Ageing and Development 15, 399, 1981.

2. EMANUAL, N. M.: Free radicals on the appearance and growth of tumors. In Free Radicals and Cancer. R. A. Floyd, editor. Marcel Dekker, Inc., New York (in press).

3. FRANKFURT, O. S., LYPCHINU, L. P., BUNTO, T. V. and EMANUAL, N. M.: Inhibition of carcinogenesis with antioxidants BHT. Exp. Biol. Med. 8, 163, 1967.

4. FRIDOVITCH, I.: Free radicals in biology, Vol. I. W. A. Pryor, editor. Academic Press, New York, 1976, pp. 239-277.

5. HAMBOUR, J. R. and BOLTON, J. R.: Superoxide formation in spinach chloroplasts: ESR detection by spin trapping. Biochem. Biophys. Res. Commun. 64, 803-807, 1975.

6. HARMON, D. and PIETTE, L. H.: Free radical theory of ageing: Free radical reactions in serum. J. Gerontol. 21, No. 4, 560, 1966.

7. HOCHSTEIN, P. and ERNSTER, L.: APP-activated lipid peroxidation coupled to the TPNH oxidase system of microsomes. Biochem. Biophys. Res. Commun., 12, 388-394, 1963.

8. JANZEN, E. G. and BLACKBURN, B. J.: Detection and identification of shortlived free radicals by electron spin resonance trapping. Accts. Chem. Res. 4, 31-40, 1981.

9. KOMEDA, K. ONO, T. and IMAI, Y.: Participation of superoxide and OH radicals in NADPH-cytochrome P450 reductase catalyzed peroxidation of methyl linolenate. Biochem. Biophys. Acta. 572, 77-82, 1979.

10. LAI, C. S. and PIETTE, L. H.: Spin-trapping studies of OH radical production involved in lipid peroxidation. Arch. Biochem. Biophys. 190, 27-38, 1978.

11. PETO, R., DOLL, R., BUCKLEY, J. D., SPORN, M. B.: Nature (London) 290, 201, 1980.

12. SHAMBERGER, R. J., TYTKO, S. A. and WILLIS, C. E.: Malondialdehyde is a carcinogen. Fed. Proc. 34, 827, 1975.

13. SHULYAKOVSKAYA, T. A., SAPRIN, A. N. and EMANUAL, N. M.: Effects of chemical compounds with varying toxicity on the content of free radicals and microsomal system of non-specific oxidases in the liver of animals. Dokl. Akad. Nauk., U.S.S.R. 210,221-223, 1973.

14. TOTTER, J. R.: Spontaneous cancer and its possible relationship to oxygen metabolism. Proc. Natl. Acad. Sci. USA, 77, 1763-1767, 1980.

CANCER IN THE ELDERLY: CLINICAL AND BIOLOGIC CONSIDERATIONS

Paul P. Carbone, M.D.
University of Wisconsin Clinical Cancer Center, Madison, WI
Colin Begg, Ph.D.
Harvard Medical School, Boston, MA
Jeanne Moorman, M.S.
University of Wisconsin Clinical Cancer Center, Madison, WI
Supported by grants: CA21076, CA23318, CN75348 and CA36591

Cancer is a disease of the elderly in that most cancers occur in persons over age 65. Cancer increases in incidence with age, however, not all cancers have that characteristic and some cancers decline in incidence in the very aged (over 80). Most physicians who treat patients with cancer tend to offer the elderly patient fewer options for therapy. There is a feeling that the older patient is less able to tolerate the stresses of adjunctive therapy, whereas the treatment in the elderly is as likely to involve surgery. There is no evidence that elderly patients have more aggressive disease that result in shorter survivals. A careful study of the toxicity of chemotherapy drugs in patients over 70 failed to reveal a generalized increase in severe drug side effects. Certain drugs were associated with increased toxicity but these are resultant to the known physiologic changes that occur in the elderly, namely, decreases in renal clearance and renal blood flow. Finally, the theory that cancer in the elderly is associated with loss of immune competence and therefore fosters a more aggressive disease cannot be substantiated. It is postulated that the measures of early detection and prevention are just as applicable to the elderly as they are to the younger patient.

"Old age is the harbor of all ills." Quote from Bion: 280 BC: Diogenes Laertius Book IV.

INCIDENCE AND AGE

Most cancers occur in the people over 65 that make up only 11.5% of the population. There is a progressive increase in cancer incidence with age. However, this relationship is not always a direct one. Some cancers peak in childhood (ALL, Wilms, retinoblastoma, Ewings), while others, like breast cancer, increase in age but there is a change in the slope of the incidence rate at about age 50 (the average time of menopause). In geographic areas where breast cancer incidence rates are lower, the decrease is primarily in the postmenopausal variety. The estrogen receptor content of the tumors increases with age (1).

B. Pullman et al. (eds.), Interrelationship Among Aging, Cancer and Differentiation, 313–323.

These facts demonstrate the complexity of neoplastic diseases as it relates to age and incidence.

Furthermore, data from the Wisconsin statewide incidence reporting system shows that cancer incidence increased with age and doubled every decade except for the following types: corpus uteri, cervix, and lung cancer (Chart 1). In the same study, cancer of the cervix peaked at age 35-39 years. Likewise, in women the rates for corpus uteri and lung cancers peaked at ages 60-64 and ages 65-59 respectively (Chart 2). In men the only site for which the rate declines with age is lung cancer where the maximum occurs at 75 (Chart 3).

Lew (1978) also has shown a similar decline in cancer incidence in the extreme aged "85 and older" (2). He also reported a decrease in cancer death rates for both men and women over the age of 85-90 because of the decline in deaths due to lung, digestive, breast and prostate cancer. Lew reported that in England the decline in mortality rates occurred in people over 85 and in Sweden over 70.

MULTIPLE TUMORS

In addition to the age related incidence, there is a corresponding increase in more than one tumor diagnosis per patient with age. Based on data from the Wisconsin reporting system we found that by age 65, 10% of the patients are reported with multiple tumors (Chart 4). By age 90 the proportion increases to 15%.

TREATMENT AND AGE

From data extracted from over 25,000 records in the tumor reporting system in the State of Wisconsin, we were able to examine the impact of age on the type and frequency of therapy administered. For the years 1978-79, we looked at the frequency and type of treatment given to patients with cancer who either were diagnosed and discharged or who after diagnosis died before discharge (Chart 5). When the patient died in the hospital, there is a clear tendency to have the physician offer more treatment options when the patient is young (43.7% vs. 33.9%). A similar trend is seen for those patients who leave the hospital alive (80.4% vs. 77.1%). A negative association exists between treatments and age for stomach, colon, rectum, pancreas, lung, breast, ovary, and kidney cancers. Prostate cancer is one of the rare tumors where the trend is positive with age.

The tendency to give no treatment seems to be more prevalent in the more advanced stages of the disease rather than in the early stages. Surgery is not likely to be withheld by physicians for age considerations (Chart 6). There seems to be a reluctance to treat the elderly patient with adjunctive treatments. This reluctance to treat as a consequence of age is shown in Chart 7 for breast cancer in

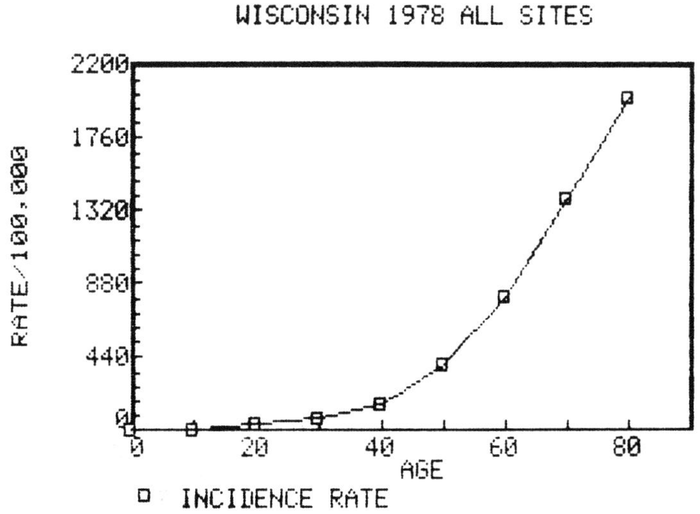

Chart 1.
Incidence of cancer (all sites) by age in Wisconsin. 1978

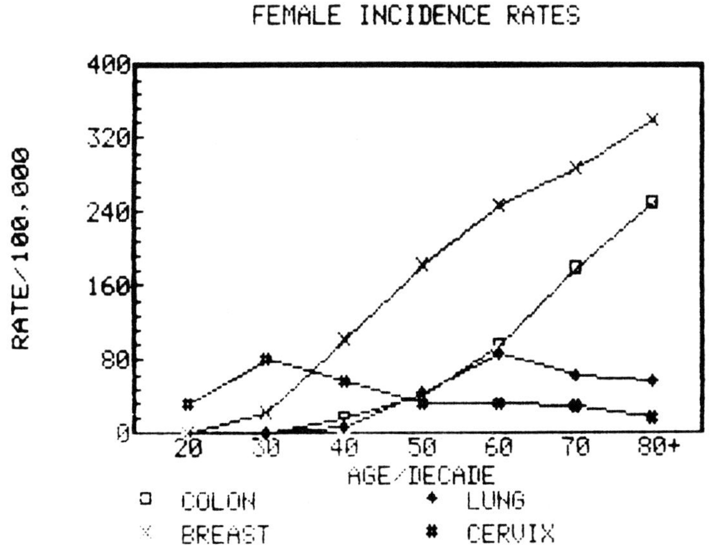

Chart 2.
Incidence of specific cancers in females be decade. 1978

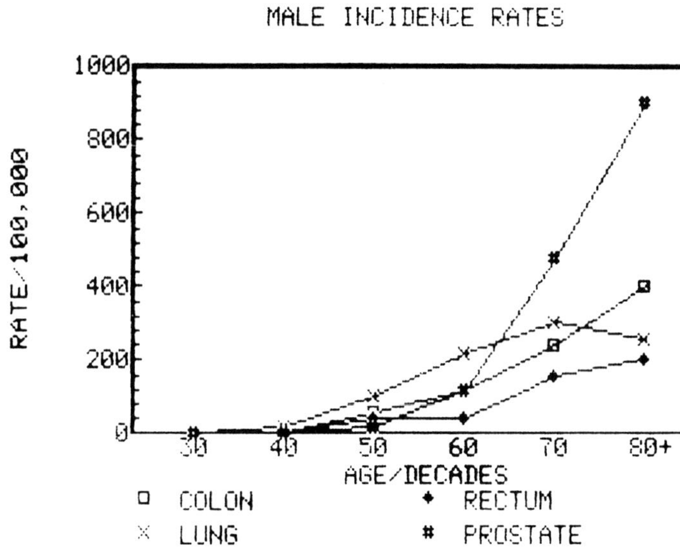

Chart 3. Incidence of specific cancers in males in Wisconsin. 1978

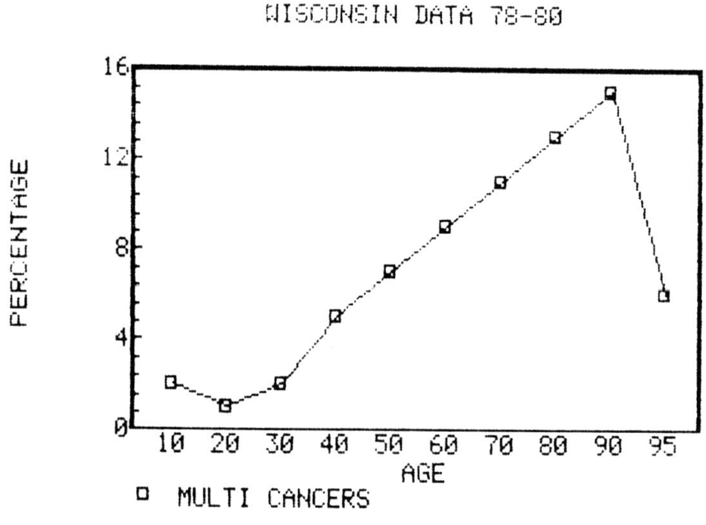

Chart 4. Incidence of multiple tumors in Wisconsin. 1978

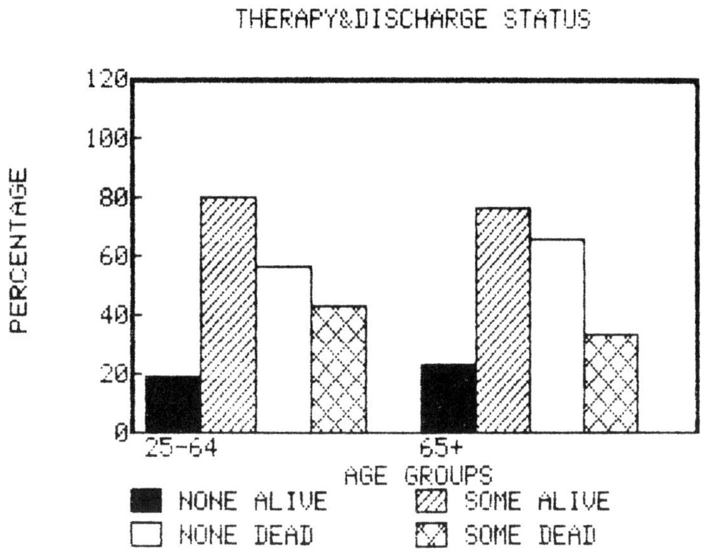

Chart 5. Therapy and discharge status of patients 25-64 and over 65.
Some or None = Some or no treatment
Alive = Left hospital alive
Dead = Died in hospital

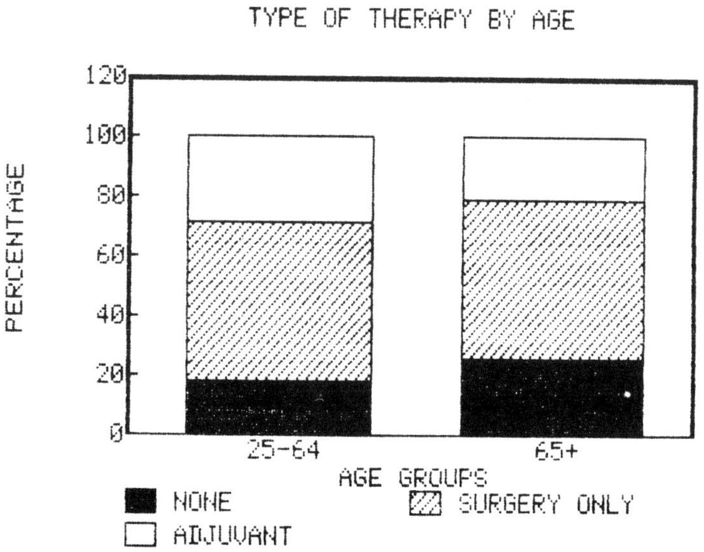

Chart 6. Type of therapy by age.

TABLE 1
BREAST CANCER IN THE ELDERLY*

	AGE GROUP (YEARS)**			
	ALL	<65	>65	>80
TOTAL CASES	874	518	356	81
CLINICAL STAGE				
I-II	717 (81.7)	421 (83.2)	296 (83.1)	66 (81.4)
III	113 (12.9)	74 (14.3)	39 (11.2)	11 (13.6)
IV	38 (4.3)	18 (3.5)	20 (5.3)	4 (4.9)
NODE POSITIVE	324 (43.1)	216 (45.3)	108 (39.4)	13 (35.1)
TREATMENT STAGE I-II				
MASTECTOMY (R,MOD)	625 (92.6)	384 (97.6)	241 (85.5)	35 (53.8)
TUMORECTOMY	13 (1.9)	1 (0.3)	12 (4.2)	11 (16.9)
ADJUNCTIVE THERAPY				
ANY THERAPY	238 (30.0)	185 (38.0)	53 (16.8)	6 (8.5)
RADIOTHERAPY	188 (23.0)	143 (30.0)	45 (14.0)	1 (1.4)
CHEMOTHERAPY	55 (6.8)	53 (10.8)	2 (0.6)	0

* DONNEGAN 1983 (3)
** ( ) are percentages and varying totals are related to differences in number of observations or procedures done.

TABLE 2
BREAST CANCER IN THE ELDERLY FIVE YEAR SURVIVALS*

	AGE GROUPS **		
	<65	>65	>80
ALL CASES	518	356	81
OBSERVED	73%	57%	39%
RELATIVE	75%	67%	61%
MASTECTOMY RADICAL OR MODIFIED WITH NEGATIVE NODES			
TOTAL NUMBER	251	146	17
OBSERVED	79%	80%	86%
RELATIVE	88%	91%	94%
MASTECTOMY RADICAL OR MODIFIED WITH 1-3 POSITIVE NODES			
TOTAL NUMBER	110	47	7
OBSERVED	86%	69%	
RELATIVE	88%	77%	
MASTECTOMY RADICAL OR MODIFIED WITH >3 NODES POSITIVE			
TOTAL NUMBER	92	43	5
OBSERVED	48%	41%	
RELATIVE	48%	47%	

* DONNEGAN 1983 (3)
** ( ) are percentages and varying totals are related to differences in number of observations or procedures done.

Wisconsin.

In another center in Wisconsin Donnegan (3) reported a review of patients treated at a large urban hospital. In this report 874 patients with verified breast cancer treated between years 1967-1979 were reviewed (Table 1). The range of ages of the patients was 22-96 (average age = 60.7 years). The patient groups were analyzed in three groups: <65, <79, >80. Clinical stages (the extent of the disease) were similar among the subsets. Less extensive operations were usually performed in the older women. Segmental mastectomy, a relatively minimal surgical procedure, was done in 1.4% of the patients under 65 and in 13.6% of the patients over 80. Biopsy only was done three times as often in the older group. About 97% of the patients under 65 were treated with a mastectomy (modified or radical) whereas only 40% over 80 were subjected to this more aggressive surgery.

In Donnegan's series overall mortality was higher in the older patients but this mostly was due to an increased frequency of associated morbid conditions (Table 2). A slight difference in the mortality was noted if comparisons for similar pathological stages were made. However, relatively major differences in the kinds of therapy were given to the two populations. Less adjuvant treatment was given to the older patients (<65), 83% vs. 17% (>65). Hormones, a relatively mild therapy, were given to more of the older patients (25% vs. 1%) whereas chemotherapy, a more aggressive treatment, was given more often to the younger patients (23% vs. 0%). Corrected for the same stage and treatment, the survivals were the same.

Donnegan concludes that there was ". . . little support for the assumption that breast cancer is either more malignant or benign when it occurs in a patient at an advanced age."

Patterson, in discussing large bowel cancer, summarizes ". . . no evidence (exists) that colorectal cancer represents a biologically different disease in the elderly. Survivals are the same corrected for non-cancer related mortality." (4)

EASTERN COOPERATIVE ONCOLOGY GROUP (ECOG) STUDIES

Many physicians reduce the doses or fail to offer chemotherapy drugs as part of their treatment plans for older patients. To examine the possible rationale for this decision of withholding treatment with drugs in the elderly, Begg and Carbone analyzed data from 5,459 patients treated in a large American collaborative clinical trial group (5,6). To examine the comparability of the elderly and less elderly patients, we examined several parameters of severity and outcome such as performance status, prior therapy, response, and survival for those patients entered on ECOG trials in the age groups: <70 and >70. There was no evidence that any above the above parameters were different in the two groups, yet correspondingly fewer patients over the age of 70 are put on clinical trials protocols.

Looking at the question of age related toxicity for a variety of drugs revealed that most drugs were not more toxic in the elderly. Two drugs were found to be more toxic for the older patient, namely, methotrexate and a nitrosourea, methyl-ccnu. The main reason for methotrexate's increased hematological toxicity was that this drug is excreted by the kidneys. In the older patient renal blood flow and clearance are diminished thereby increasing the possibility of enhanced drug toxicity. The conclusions from this study were that the older patient should be considered appropriate for drug treatments and that chronologic age is not as important as physiologic age. There is no reason to withhold therapy based on fear of enhanced toxicity if good judgement and understanding about the specific metabolism of the drug are applied.

DISCUSSION

The withholding of treatment of the elderly patient is based on the prejudice of the physician that implies an increased toxicity in the elderly. As mentioned above the older patient receives less aggressive treatment, is less likely to be put on clinical studies (7) and is not treated as frequently with drugs. Some of the problems of increased toxicity in the elderly relate to a lack of appreciation of normal physiological changes in organ function that occur with age. In addition, the older patient is usually being treated for other diseases and is receiving polypharmacy (8,9). Four out of five elderly people have at least one chronic disease. Elderly people are often prescribed between three to ten drugs. Potentially important drug interactions that are clinically important may be possible in half of the elderly hospitalized patients (8). Examples of drugs include ASA, phenobarb, etc. In addition, compliance is a major problem in the elderly. They may not take or they may over medicate themselves.

Drug absorption is probably not different in the older patient as compared to the young. Changes in the blood flow, body composition (more fat), and plasma protein decrease in the elderly patient and may alter drug distribution and metabolism. Kerr and Chabner report that drug metabolism alterations due to advanced age are probably not major factors (9). Well-known is the decrease in creatinine clearance and renal blood flow that occur with aging. Methotrexate and Bleomycin toxicities are increased by renal function impairment. These data are supported by the ECOG studies (5,6).

The fact that the cancers in the elderly are not more aggressive or faster growing suggests that the initial development of malignant cells and/or the progression of the cancer with time is not too different than the same parameters for cancers that occur in other age groups. Likewise, since the cancers do not grow more slowly,

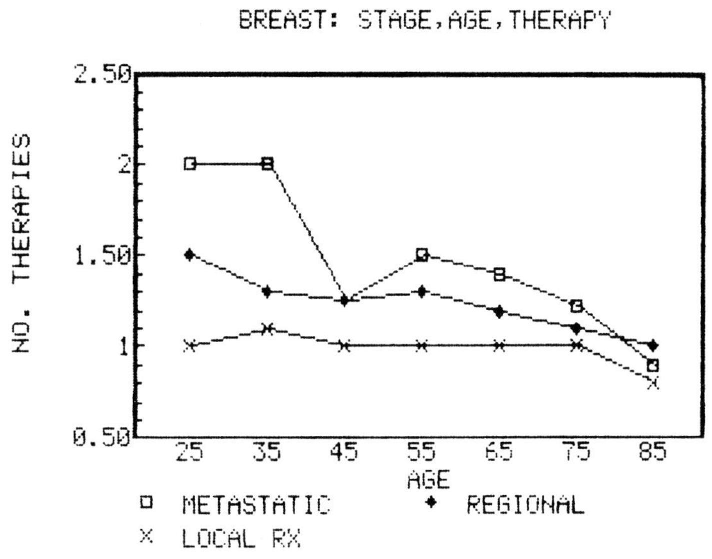

Chart 7. Breast cancer treatment by stage and age.

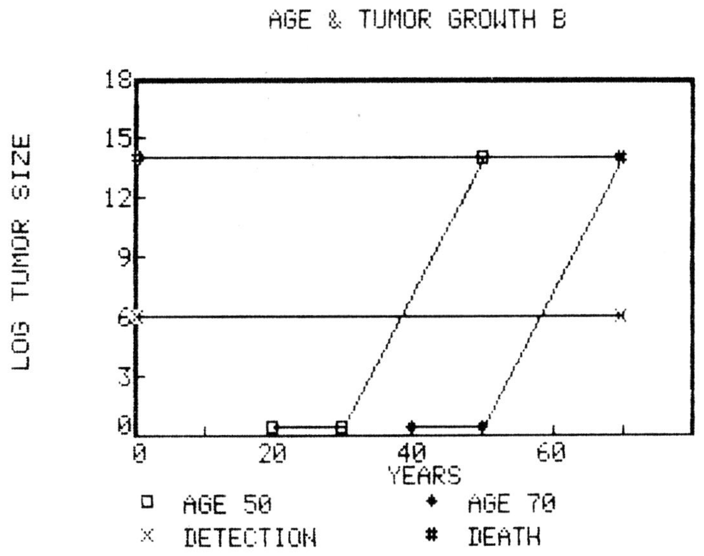

Chart 8. Simulated growth curves for cancer in young and old patients. The growth rates are similar.

suggesting a more benign course, the time from initiation to onset of clinical disease must be similar (Chart 8). I would suggest that the idea of applying either early detection or primary prevention would thus seen as likely in all age groups.

A major theory for cancer development in the elderly involves the concepts embodied in immune surveillance. Immunologic parameters that have been reported to decline with age include thymus size, thymosin, total lymphocytes, natural killer cells, delayed hypersensitivity, etc. (10). Other immune parameters increase with age such as antibody responses and B cell functions. "In experimental animals there is an increase in tumor takes, a reduction in the latent period in tumor cell innoculation, carcinogen administration, virus injection and tumor development...an accelerated tumor growth; and increased resistance to treatment... . These observations are paralleled in humans...patients over 60 usually have a worse prognosis... ." states Hersh. Unfortunately these statements often quoted are not substantiated by facts from other reports or in humans. There is no clear cut relationship between age and disease aggressiveness. Thus, it is difficult to espouse completely the concepts of impaired immune surveillance as the reason for increased tumor development in the elderly.

To conclude: "Senescence and cancer appear at first sight to be related; because cancer is more common in old age, the statement is often made that old people are more susceptible to cancer. Actually, the susceptibility of young and old animals to experimental carcinogens is indistinguishable... . Thus the clustering of cancer in old age is presumably due to a summation of the carcinogenic events that are occurring throughout our lives." writes Cairns (11). This is compatible, according to Cairns, with the multihit hypothesis. However, this too may be an oversimplification and we need to continue to develop and test new hypotheses.

REFERENCES

1. Elwood, K.M. and Godoophin, W.: 1980, Brit. J. Cancer 42, pp. 635-644. (ER and age)

2. Lew, E.A.: 1978, Cancer in old age. Cancer 28, pp. 1-6.

3. Donnegan, W.: Treatment of breast cancer in the elderly. In, Perspectives on Prevention and Treatment of Cancer in the Elderly. Eds.: Yancik, R., Carbone, P.P., Patterson, W.B., Steel, K., and Terry, W.D., Raven Press, New York, pp. 83-96.

4. Patterson, W.B.: 1983, Oncology perspective on colorectal cancer in the geriatric patient. In, Perspectives on Prevention and Treatment of Cancer in the Elderly. Eds.: Yancik, R., Carbone, P.P., Patterson, W.B., Steel, K., Terry, W.D., Raven Press, New York, pp. 105-113.

5. Begg, C. and Carbone, P.P. 1983, Cancer 52, pp. 1986-1992.

6. Carbone, P.P., Begg, C.B., Jensen, L. and Moorman, J.: 1983, In, Perspectives on Prevention and Treatment of Cancer in the Elderly. Eds.: Yancik, R., Carbone, P.P., Patterson, W.B., Steel, K., Terry, W.D., Raven Press, New York, pp. 63-73.

7. Begg, C., Zelen, M., Carbone, P.P., McFadden, E., Brodovsky, H., Engstrom, P., Hatfield, A., Ingles, J., Schwartz, B. and Stohlbach, L.: 1983, Cancer 52, pp. 1760-1767.

8. Ouslander, J.G.: 1981, Ann. Int. Med. 95, pp. 771-722. (Drug therapy in the elderly)

9. Kerr, I.G. and Chabner, B.S.: 1983, The effect of age on the clinical pharmacology of anticancer drugs. In, Perspectives on Prevention and Treatment of Cancer in the Elderly. Eds.: Yancik, R., Carbone, P.P., Patterson, W.B., Steel, K., Terry, W.D., Raven Press, New York, pp. 203-214.

10. Hersh, E.: 1983, Host defense parameters in aging. In, Perspectives on Prevention and Treatment of Cancer in the Elderly. Eds.: Yancik, R., Carbone, P.P., Patterson, W.B., Steel, K., Terry, W.D., Raven Press, New York, pp. 249-259.

11. Cairns, J.: 1982, Aging and cancer as genetic phenomena in research frontiers. In, Aging and Cancer-National Cancer Institute Monograph 60, pp. 237-240.

CELLULAR STUDIES ON THE INTERRELATIONSHIP AMONG CANCER, AGING AND CELLULAR DIFFERENTIATION

Sarah A. Bruce and Paul O. P. Ts'o
Division of Biophysics
School of Hygiene and Public Health
The Johns Hopkins University
Baltimore, Maryland, 21205, USA

ABSTRACT: An experimental animal model and cellular system, based on the Syrian hamster (Mesocricetus auratus), has been developed for the study of the interrelationship among cancer, aging and differentiation. Using this system, we have identified a progenitor-like cell which is present in primary and low passage mesenchymal cell cultures. The loss or absence of these cells is correlated with reduced in vitro proliferative capacity (cellular senescence), reduced response to growth promoting factors, and reduced susceptibility to in vitro neoplastic transformation. Based on these studies we propose a hypothesis on the central role of stem cells and/or progenitor cells in the interrelationship among cancer, aging and cellular differentiation.

INTRODUCTION

This symposium, the third in the Jerusalem Symposium Series devoted to the topic of cancer, has addressed the question of the interrelationship among cancer, aging and cellular differentiation. Each of these three basic biological phenomena is highly complex in itself. Therefore, to understand how they are interrelated becomes a real challenge. However, one key to understanding each phenomenon may be an understanding of how one is modulated by the other two.

A clear example of this modulation is the well established relationship between age and the incidence of carcinomas in humans (1, see also J. Brody, this volume). It is unclear as to whether age is only an extrinsic factor in carcinoma incidence, related to the passage of time and/or the accumulation of exposure to environmental factors (2), or whether there are also intrinsic changes in the target cells leading to altered susceptibility (3). Intrinsic age-related changes in the stem cells, and therefore the renewal capacity, of many organ systems may contribute to the progressive inability to maintain homeostasis associated with aging (4). Further, both a patient's age and the histologically determined degree of differentiation exhibited

B. Pullman et al. (eds.), Interrelationship Among Aging, Cancer and Differentiation, 325–339.

by a tumor influence the prognosis of the cancer.

Another example of this interrelationship is the requirement for a cell to proliferate in order for carcinogen-induced damage to be manifested and for the progeny of the damaged or altered cell to proliferate and form a tumor mass. Carcinogen-induced damage in a post-mitotic (e.g. senescent), terminally differentiated cell is unlikely to result in a tumor. Lastly, carcinogenic transformation of cells leads to their continued proliferation beyond the bounds of normal growth control and precludes, at least in the fully malignant state, their normal differentiation and function within the host organism.

Our laboratory's approach to understanding the interrelationship among cancer, aging and differentiation has been (i) to develop a single experimental system in which all three phenomena can be investigated at multiple levels both in vitro and in vivo and (ii) to focus on less differentiated stem cells or progenitor cells as a point of commonality among these three phenomena.

THE SYRIAN HAMSTER AS AN EXPERIMENTAL ANIMAL MODEL AND CELLULAR SYSTEM FOR STUDIES ON CANCER, AGING AND DIFFERENTIATION

The Syrian hamster (Mesocricetus auratus) is a well established experimental system for both carcinogenesis in vivo (5) and neoplastic transformation in vitro (6). Although less established and less frequently used than mouse and rat systems for studies on aging, there are several age-associated diseases for which hamsters are appropriate animal models (7) and the Syrian hamster has been used for studies of age-related changes in reproductive physiology (8,9) and neurobiology (10-13). In addition, the Syrian hamster has certain advantages for studies of in vitro cellular senescence. These include the following which are described in more detail below: (i) cultures of normal diploid mesenchymal (fibroblastic) cells of Syrian hamster origin exhibit a very low rate of spontaneous establishment into permanent cell lines in contrast to other rodent cell systems (14-17) and similar to human fibroblast cultures which have never been observed to convert spontaneously to permanent cell lines (18,19); (ii) the pattern of senescence of normal Syrian hamster fibroblastic cell cultures is qualitatively similar to that of normal human fibroblastic cell cultures in terms of morphological and proliferative changes; and (iii) both Syrian hamster and human normal cell cultures display an inverse relationship between in vitro proliferative life span and the in vivo age of the donor. Lastly, with regard to cellular differentiation studies, our laboratory has recently shown that several lineages of mesenchymal cell differentiation can be observed and analyzed in Syrian hamster cell cultures. Thus, this one system has the potential for the simultaneous investigation of the three biological phenomena under consideration - cancer, aging and differentiation - both in vivo and in vitro.

If indeed the age and differentiated state of a cell influence the response of the cell to external perturbation, it becomes essential in cell culture studies to define the age and/or developmental stage of the cells being used or at least of the tissue from which the cell culture was derived. Our studies use mesenchymal cell cultures derived from Syrian hamsters at various stages of development and aging, ranging from 9 days gestation to 24 months of age (Table I). Nine day gestation material is used because it is the earliest stage at which the embryo can be clearly distinguished and separated from the maternal tissue. By 9 days gestation in the Syrian hamster, all the major organ systems are developed which suggests that 9-10 days is the end of embryonic development (organogenesis) in this species and that the remainder of the 15.5 days gestation is fetal or maturational development (20,21). In agreement with this conclusion, cell cultures derived from 9 day embryos are very different from those derived from fetal, newborn and adult tissue with regard to the differentiation lineages observed which suggests that there is a significant change in the differentiation potential of the cells *in vivo* between 9 and 13 days gestation. The final time point in our studies is 24 months of age at which time the hamster is considered to be aged. Based on 84 natural deaths in our own Syrian hamster aging colony, the average mean *in vivo* life span in this species (strain LVG) is 18.1 months and the maximum *in vivo* life span is 28.0 months under conventional animal housing conditions (22).

TABLE I. *In Vitro* Cell Systems Analyzed

Cell designation	Source	
	tissue	age
E9	whole embryo	9 days gestation
FC13	fetal carcass	13 days gestation[a,b]
FE13	fetal extremities	13 days gestation[a]
FD13	fetal dermis	13 days gestation[a]
ND3	neonatal dermis	3 days postpartum
AD6	adult dermis	6 months
AD24	adult dermis	24 months[c]

a Gestation in the Syrian hamster is 15.5 days.
b FC13 cells are equivalent to SHE cells used in neoplastic transformation experiments (see ref. 6).
c Average life span and maximum life span of male LVG Syrian hamsters are 18.1 and 28 months respectively.

IN VITRO SENESCENCE OF SYRIAN HAMSTER MESENCHYMAL CELLS

Cell cultures of fetal, neonatal and adult origin

In these studies, we compared the *in vitro* proliferative capacity of normal diploid Syrian hamster (strain LVG) dermal fibroblastic cell cultures derived from 13 day gestation fetal dermis (FD13 cells), 3 day neonatal dermis (ND3 cells), 6 month young adult dermis (AD6 cells) and 24 month aged adult dermis (AD24 cells) (Table I) (22). Freshly isolated cell cultures that had not been stored in liquid nitrogen were used in all experiments. Primary cultures were established by proteolytic digestion of minced tissue with trypsin and, in the case of the adult tissue, collagenase. The resultant single cell suspensions were inoculated into duplicate flasks at $1x10^5$ viable cells/cm^2 in Dulbecco's modified Eagle's medium as further modified by Casto (23) with 20% fetal bovine serum and penicillin and streptomycin. Primary cultures were incubated until confluence. Fetal cells achieved confluence within 1-2 days. All other cell types required longer incubation to achieve confluence (ND3 cells, 3 days; AD6 cells, 5-6 days; and AD24 cells, 7-8 days). Thereafter cells were maintained in medium with 10% serum without antibiotics and were subcultured twice weekly by trypsin dissociation and inoculated into fresh flasks at a constant inoculum throughout their *in vitro* life span. The percentage of inoculated cells that attached to the new culture surface ranged between 76-90% for all cell types throughout their life span. Cultures were judged to be senescent when the harvest density fell below the inoculum for two consecutive subcultures.

Greater than 95% of all replicate Syrian hamster cultures, regardless of the age of the tissue of origin, exhibited cellular senescence characterized by a limited *in vitro* proliferative life span similar to that described by Hayflick and colleagues (18,19) for human diploid fibroblast cell cultures. All Syrian hamster cell cultures exhibited an initial phase of rapid proliferation followed by a progressive reduction in proliferative rate. The overall pattern of senescence of the cells from fetal, neonatal, young adult and aged adult tissue was similar in terms of proliferative changes indicated by a reduction in saturation density, cloning efficiency and ^{3}H-thymidine labeling index and by an increase in population doubling time and cell volume (Table II). However, both the number of days in culture and the maximum cumulative population doubling level (cumPDL) attained at senescence were characteristic for each type of cell culture (Table III). These data show that the *in vitro* proliferative capacity of Syrian hamster fibroblasts is inversely related to the *in vivo* age of the donor tissue in a manner similar to that described for human fibroblasts by Martin, *et al.* (24) and Schneider and Mitsui (25).

TABLE II. Changes in Growth Characteristics of Syrian Hamster Mesenchymal Cell Cultures of Fetal to Aged Adult Origin during *In Vitro* Cellular Senescence

Growth Characteristics	Low PDL[a]	High PDL
Saturation density	$\sim 1\times10^5$ cells/cm^2	$\sim 2\times10^4$ cells/cm^2
Cloning efficiency	3.4%	<0.5%
^3H-Thymidine labeling index	100%	<10%
Population doubling time	~18 hrs	>100 hrs
Cell volume	800-1400μ^3	>6500μ^3

a PDL= population doubling level.

TABLE III. Proliferative Capacity of Syrian Hamster Mesenchymal Cell Cultures of Fetal to Aged Adult Origin

Cell type	Number of days in culture at senescence	Average maximum cumPDL at sensecence
FD13	55	28.6
ND3	42	18.7
AD6	38	13.8
AD24	28	11.1

In addition to the proliferative changes exhibited by normal Syrian hamster fibroblast cell cultures, there are distinct morphological changes associated with senescence of these cells. Primary cultures of fetal dermal cells are morphologically homogeneous and contain small, highly proliferative, fibroblastic cells which do not exhibit strict contact- or density-dependent inhibition of growth (designated as type I cells) (22). In subsequent passages, the morphological homogeneity of the cultures decreases with the emergence of several other cell types (designated type II, III and IV) that exhibit increased size and decreased proliferative capacity (as determined by ^{3}H-thymidine labeling indexes). Type II cells are slightly larger and less refractile than type I cells. They are highly motile and highly proliferative but do exhibit contact-inhibition of cell division. Type III cells are large cells with many stress cables in their cytoplasm. Although some type III cells are binucleate, other type III cells continue to divide. Type IV cells are non-proliferative cells that are grossly enlarged (up to 500-1000 µm in diameter) and frequently multinucleate. The extensive cytoplasm of type IV cells is filled with actin-containing stress cables. Terminally senescent cultures contain only the type III and IV cells. Similar morphologically distinct classes of cells in Syrian hamster fetal lung fibroblast cultures have been described by Raes and Remacle (26) and Raes, et al. (27). Clonal colonies of Syrian hamster cells are generally composed of more than one of the above described cell types indicating that the larger cell types are derived from the type I cells (22). These observations show that there is a clear sequence of morphologically differentiation associated with senescence of Syrian hamster fibroblastic cell cultures.

Cell cultures of embryonic origin

In separate experiments, we have evaluated the proliferative capacity and senescence pattern of 9 day gestation embryonic (E9) cells. Most E9 cell cultures exhibit cellular senescence and the pattern of alterations in morphology and proliferation is similar to fetal and adult cells. However, E9 cells require higher (20%) serum supplementation throughout their in vitro life span, grow more slowly than fetal cells, achieve a lower average maximum cumPDL (15.3) and exhibit a >4-fold higher frequency of spontaneous escape from senescence (20%) relative to fetal and adult cells (<5%).

MESENCHYMAL CELL DIFFERENTIATION IN CULTURED SYRIAN HAMSTER CELLS

Cellular senescence has been proposed to be a form of cellular differentiation (28,29). Thus, if this hypothesis is correct, the phenomenon of senescence, exhibited by Syrian hamster cells of embryonic to aged adult origin may be considered as one example of in vitro cellular differentiation in this system although the corresponding differentiated cell type in vivo has not yet been identified. However, additional examples of differentiation are observed in cultured Syrian hamster cells including adipocyte differentiation and muscle cell differentiation. Notably, these two examples of mesenchymal lineage cell differentiation are only observed in primary and established cell cultures derived from 9 day gestation embryonic tissue (E9 cells) and not in cultures derived from 13 day gestation fetal or older tissue. This observation supports the conclusion that the interface between embryonic development (organogenesis) and fetal or maturational development in the Syrian hamster is at 9-10 days gestation (20,21).

Adipocyte differentiation occurs spontaneously and at high frequency (>50%) in confluent normal diploid E9 cell cultures, which during exponential growth appear fibroblast-like in morphology. The lipid-laden cells which develop have been identified as mature adipocytes by lipid-specific staining with Oil-Red O, electron microscopy and a ≤10-fold acummulation of triglyceride (20,30). A non-proliferative state, resulting from either confluence or senescence, is required for adipocyte differentiation. Clonal analysis of passage 1 E9 cells showed that >50% of the colonies contain adipocytes which indicates that a majority of the cells were preadipocytes or undifferentiated mesenchymal cells that became committed to the adipocyte lineage in culture. Moreover, the presence of multipotent, undifferentiated mesenchymal progenitor cells in primary and low passage E9 cell cultures is indicated by the observation of clonal colonies in these cultures which contain two or more morphologically and/or histochemically (e.g. Oil Red O) distinguished cell types. In addition to adipocyte differentiation, primary and low passage E9 cells also exhibit, at a lower frequency, muscle cell differentiation as indicated by the presence of striated, contractile myotubes and alterations in the creative phosphokinase isoenzyme pattern.

MANIPULATION OF CELLULAR DIFFERENTIATION AND CELLULAR SENESCENCE IN SYRIAN HAMSTER CELL CULTURES

Cellular Differentiation

Both adipocyte and muscle cell differentiation in Syrian hamster primary and established E9 cell cultures are inhibited by treatment with phorbol ester tumor promoters (12-O-tetradecanoyl phorbol 13-acetate (TPA) or phorbol-12,13-didecanoate (PDD)) at 0.1μg/ml in agreement with the numerous reports in the literature on the inhibition by tumor promoters of differentiation of mesenchymal cells including adipocytes (31) and myoblasts (32). Epidermal growth factor (EGF) at 10 ng/ml also inhibits the formation of mature adipocytes and retards the acquisition of contractility by colonies of fused myoblasts. Other treatments such as insulin (0.1 to 10 μg/ml) enhance the differentiation of adipocytes in these cultures.

Cellular Senescence--Effect of Phorbol Ester Tumor Promoters

As part of our study investigating the relationship between senescence and differentiation, we examined the effect of tumor promoters (which are known to inhibit mesenchymal cell differentiation) on the senescence pattern of Syrian hamster cells of fetal to aged adult origin. Exposure to 0.1 μg/ml PDD was initiated a passage 1 and the cells were maintained in the presence of promoter throughout their entire in vitro life span. Such continuous treatment with PDD extended the proliferative life span of fetal and neonatal cells by 60-120% and young adult cells by ~30%, but had minimal effect on aged adult cells (<10%) (33). In addition to increasing the maximum cumPDL of the fetal, neonatal and young adult cells, treatment with PDD retarded both the loss of type I cells and the reduction in proliferative capacity which normally accompany senescence. The continuous presence of PDD was required. If the PDD was removed after the maximum cumPDL of the control had been exceeded, the cells senesced within 1-2 subcultures.

The effects of tumor promoters on E9 cells are distinct from the effects of these compounds on fetal to adult cells. Whereas continuous exposure of fetal to young adult cells to PDD resulted in an extension of in vitro proliferative life span, similar treatment of E9 cells resulted in escape from senescence and conversion to established cell lines which eventually lost their requirement for the presence of tumor promoters for continued proliferation. In contrast to a frequency of 20% escape from senescence in control E9 cultures, 75% of PDD-treated E9 cell cultures escaped senescence. The E9 cell lines derived either spontaneously or by treatment with promoters fall into one of two categories: (i) pre-neoplastic or neoplastic cell lines, or (ii) non-neoplastic, progenitor-like cell lines that have retained the capacity for adipocyte or muscle cell differentiation at frequencies up to 100% in clonal derivatives of these lines (34). All spontaneously established fetal or adult cell lines that have arisen thus far fall only into the first category; the second category appears to be unique to E9 cell lines.

Cellular Senescence --Effect of Fibroblast Growth Factor (FGF) and Epidermal Growth Factor (EGF)

Similar to the effect of tumor promoters on the senescence of E9 cells, continuous treatment from passage 1 of E9 cells with 10 ng/ml EGF or 50 ng/ml FGF increased significantly the frequency of escape from senescence (Table IV). The effect of FGF was slightly greater than that of EGF possibly because of the mesenchymal nature of these cells. FGF-treated E9 cells generally showed an increased growth rate from the beginning of treatment compared to control cells and escaped senescence with a minimal crisis period. Further, those FGF-treated E9 cultures which did senesce exhibited an extension of proliferative capacity indicated by an elevated maximum cumPDL at senescence relative to control cultures. In contrast, EGF neither stimulated the initial growth rate of E9 cells nor extended the in vitro proliferative life span of cultures which did senesce. Further, EGF-treated established cell lines generally exhibited a very long crisis period (30-50 day) characterized by a minimal growth rate. These E9 cell lines derived by treatment with growth factors also fall into one of two catagories (preneoplastic/neoplastic cell lines or non-neoplastic progenitor cell lines) as described above.

TABLE IV. Effect of Growth Factors on the Frequency of Escape from Senescence of Syrian Hamster Embryonic and Fetal Mesenchymal Cell Cultures

Cell type[a]	Percentage of cultures that escaped senescence (number in parenthesis indicates total number of flasks analyzed)		
	Control	EGF (10 ng/ml)	FGF (50 ng/ml)
E9	20% (16)	50% (10)	67% (6)
FC13, FE13	0% (24)[b]	35% (17)	42% (12)

[a] See Table I

[b] No spontaneous escape from senescence was observed among the control flasks from these experiments. We have previously reported a frequency of ~3% among a total of 148 replicate flasks of fetal to aged adult origin (22).

Continuous treatment of fetal cells (FC13 or FE13 cells) with EGF or FGF resulted in a similarly increased frequency of escape from senescence (Table IV). As with E9 cells, FGF is more effective on F13 cells than is EGF in terms of initial growth stimulation and frequency of escape from senescence. All growth factor-treated fetal cell lines appear to be in neoplastic progression (35). They do not exhibit adipocyte or muscle cell differentiation, and with passage in vitro, many have acquired anchorage independent growth and, in some cases, tumorigenicity.

These data show that the in vitro proliferative life span of normal diploid Syrian hamster fibroblast cultures can be extended (either a delay in senescence or escape from senescence) by exogenous factors (e.g. tumor promoters and growth factors) which can inhibit differentiation and/or maintain the cells in a proliferative state. These observations argue against an error accumulation mechanism or a cell division-counting clock mechanism for in vitro cellular senescence, and present evidence to suggest that the loss of proliferative capacity which is characteristic of cellular senescence results from a preprogrammed cellular differentiation.

PROGENITOR CELLS IN SYRIAN HAMSTER CELL CULTURES

As described above, cultures of Syrian hamster E9 cells contain mesenchymal progenitor cells including preadipocytes and myoblasts. However, these progenitor cells are indistinguishable and their presence is only indicated by their differentiation into a recognizable cell type (e.g. mature adipocyte, myotube, etc.). Therefore, a procedure, or phenotypic marker, is required with which the undifferentiated progenitor cells can be identified and isolated. In this regard, we have identified a subpopulation of contact-insensitive (CS^-) cells, which is transiently present in primary and low passage Syrian hamster cell cultures. These CS^- cells, which may be related to the type I cells described above, were originally identified by their lack of post-confluence inhibition of cell division, and can be quantitated by measuring the ability of cells to form colonies on irradiated confluent monolayers of contact-sensitive (CS^+) cells (cell mats) (36). The frequency of CS^- cells in FC13 cell cultures decreases from ~10-20% at passage 1 to less than 0.001% by passage 6 (PDL 16-20). A series of experiments have shown that this loss of CS^- cells cannot be explained by negative selection (36). Further, reanalysis on cell mats of cells from clonal colonies grown on cell mat show that cell mat colonies contain ~ 20% CS^- cells and ~ 80% CS^+ cells (37). We conclude from these and other observations that CS^- cells are lost during serial passage by conversion to CS^+ cells.

The initial frequency of these transient CS^- cells and the rate of their loss appear to be directly related to the proliferative capacity of untreated mass cultures. The cumPDL at which a given primary culture senesces shows a positive correlation to the frequency of CS^- cells present at passage 1 (37). Furthermore, treatments, such as continuous exposure to tumor promoters from passage 1, which extend the _in vitro_ proliferative life span of the mass culture, also retard the rate of loss of the CS^- cells (37). This is true not only for independent preparations of FC13 cell cultures which show some variation in their initial frequency of CS^- cells but also for fetal cell cultures which have a high initial CS^- frequency as compared to adult cell cultures which have a low initial CS^- frequency (33).

The presence of these transient CS^- cells in low passage Syrian hamster cell cultures leads to a consideration of their significance. Based on the above described studies, our working hypothesis is that the CS^- cell represents a less differentiated progenitor-like cell and that the CS^- to CS^+ conversion represents an early stage of differentiation characterized by an altered control of proliferation acquired prior to the appearance of a terminally differentiated phenotype. Current efforts are aimed at the isolation of pure populations of CS^- cells and the precise definition of the role of these cells in cellular senescence and cellular differentiation.

RELATIVE SUSCEPTIBILITY OF THE CS⁻ SUBPOPULATION IN SYRIAN HAMSTER CELL CULTURES TO NEOPLASTIC TRANSFORMATION

To determine whether the above described transient CS^- cells are more sensitive to carcinogenic and/or mutagenic perturbation, the susceptibility to neoplastic transformation and somatic mutation induced by N-methyl-N'-nitro-N-nitrosoguanidine (MNNG) was examined in clonally isolated cell cultures containing various proportions of CS^- cells (4% - 0.02%). The frequencies of MNNG-induced morphological transformation, focus formation and neoplastic transformation were 20-40 fold higher in CS^- enriched cultures (4%) compared to CS^- depleted cultures (0.02%). In contrast, the frequency of MNNG-induced somatic mutation at the Na^+/K^+ ATPase locus was similar among cultures varying in their proportion of CS^- cells. These data show that the transient CS^- cells in primary Syrian hamster cell cultures are more susceptible to neoplastic transformation although equally susceptible to induced point mutation when compared to CS^+ cells (38). By extension of the above stated hypothesis, these observations suggest that less differentiated progenitor-like cells (e.g. CS^- cells) are more susceptible to neoplastic transformation than are their more differentiated progeny (e.g. CS^+ cells).

Preliminary experiments comparing the susceptibility of fetal, young adult and aged adult dermal fibroblast mass cultures to carcinogen-induced neoplastic transformation show an inverse correlation between transformation frequency and donor age which may be related to the lower initial frequency of CS^- cells in adult cell cultures compared to fetal cell cultures. This apparent decrease in susceptibility of mesenchymal cells with increased donor tissue age may be related to the age-related incidence of sarcomas which, in contrast to carcinomas, are most prevalent in childhood (39).

SUMMARY AND CONCLUSION

Our working hypothesis on the interrelationship among cancer, aging and differentiation states that stem cells and progenitor cells play a central role in all three phenomena (Figure 1). The normal pathway for stem cells and progenitor cells to follow is cellular differentiation. Normal differentiation is the progressive development of an undifferentiated slowly proliferating stem cell, into a committed progenitor cell with high proliferative capacity associated with clonal expansion, into a less proliferative differentiating cell, and finally into a fully differentiated cell which performs a necessary function for the host but is generally non-proliferative. In the case of normal development, cell division is coupled to differentiation. In contrast, neoplastic cells appear to have an infinite proliferative capacity and, as a population at least, lack the ability to terminally differentiate although individual tumor cells within the population may cease dividing and

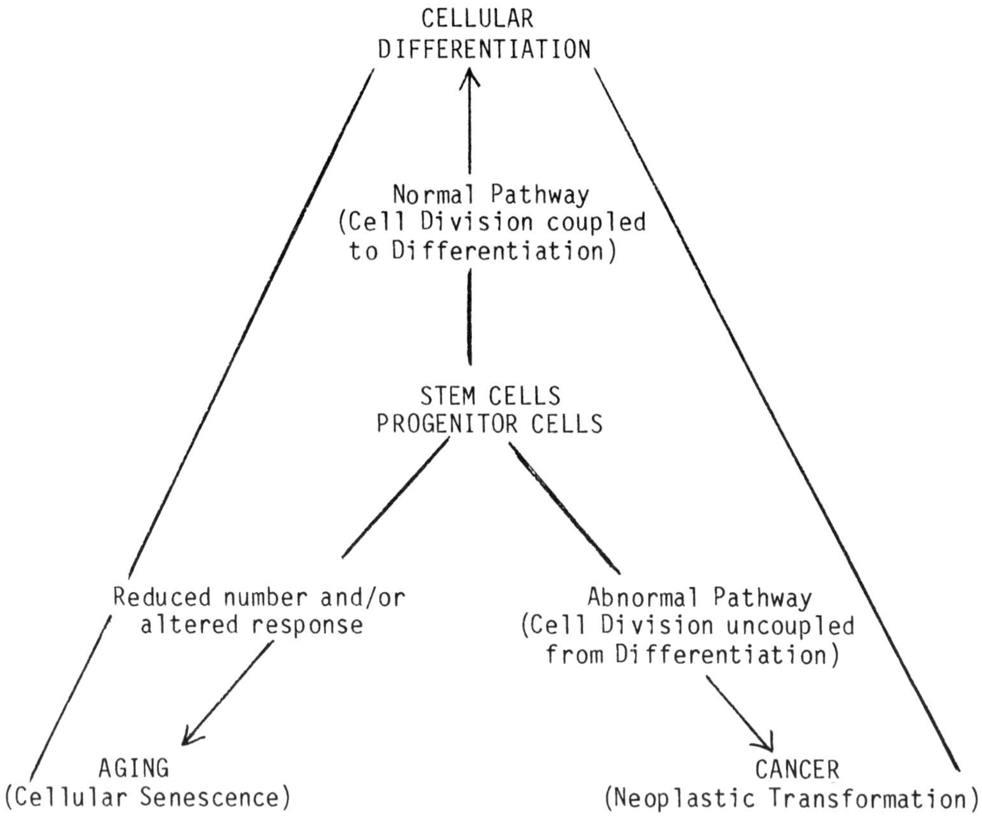

FIGURE 1. Working Hypothesis on the Relationship Between Cancer, Aging and Cellular Differentiation

differentiate or senesce. Neoplastic transformation can therefore, be viewed as the uncoupling of cell division from differentiation (40). Indeed, if we consider cellular senescence as differentiation, escape from senescence, which is a key step in *in vitro* neoplastic transformation, is an example of the uncoupling of cell division and cell differentiation. The involvement of stem cell loss in aging is less well understood. However, in at least some tissues, a reduction in number or response of stem cells leads to a reduced capacity for tissue renewal and homeostasis (4).

Our laboratory has developed an animal model and cellular system for the further study of the interrelationship among cancer, aging and differentiation. Using this system, we have identified a progenitor-like cell which is characterized by a lack of contact- or density-dependent growth control in culture. The loss or absence of these CS^- cells is correlated with (i) the reduction in *in vitro* proliferative capacity with serial passage (cellular senescence), (ii) the reduced in vitro proliferative capacity of cells from older tissue, (iii) a reduced response to growth promoting factors (tumor promoters and growth factors), and (iv) a reduced susceptibility to carcinogen-induced in vitro neoplastic transformation.

Additional studies at the cellular level both in vitro and in vivo are required to test the hypothesis shown in Figure 1. Specifically, it is necessary to (i) isolate pure populations of the progenitor-like cells identified in Syrian hamster cell cultures, (ii) to characterize further the role of these cells in senescence and to determine the relationship between mesenchymal cell senescence and mesenchymal cell differentiation, and (iii) to identify the in vivo cellular equivalent of the progenitor-like cells observed in vitro and to quantitate any age-related changes in these in vivo progenitor cells. These cellular studies as well as investigations of the molecular basis of the cellular event are underway using the Syrian hamster system which we believe is particularly well suited for these studies.

REFERENCES

1. DeVita, V.T., Hellman, S., and Rosenberg, S.A., editors: 1982, "Cancer - Principles and Practive of Oncology", J.B. Lippicott Comp., Philadelphia, pp. 3-32.
2. Doll, R.: 1968, "Cancer and Aging", Engel, A. and Larsson, T., eds., Nordiska Bokhandelns Forlag, Stockholm, pp. 15-36.
3. Burch, P.R.J.: 1976, "The Biology of Cancer: A New Approach", University Park Press, Baltimore.
4. Walton, J.: 1982, Mech. Ageing and Devel. 19, pp. 217-244.
5. Homberger, F., editor: 1972, "Pathology of the Syrian Hamster", Prog. Exptl. Tumor Res., Volume 16.
6. Heidelberger, C., Freeman, A.E., Pienta, R.J., Sivak, A., Bertram, J.S., Casto, B.C., Dunkel, V.C., Francis, M.W., Kakunaga, T., Little, J.B., and Schectman, L.M.: 1983, Mutation Res., 114, pp. 283-385.
7. National Research Council (U.S.), Committee on Animal Models for Research on Aging: 1981, "Mammalian Models for Research on Aging", National Academy Press, Washington, D.C., pp. 146-154.
8. Blaha, G.C. and Leavitt, W.W.: 1978, J. Gerontol., 33, pp. 810-814.
9. Rahima, A.: 1981, Exp. Gerontol., 16, pp. 343-346.
10. Lamperti, A., and Blaha, G.: 1980, J. Gerontol., 35, pp. 335-338.
11. Parkening, T.A., Collins, T.J., Lau, I.F., and Saksena, S.K.: 1982, J. Repro. Fert., 64, pp. 37-46.

12. Reiter, R.J., Vriend, J., Brainard, G.C., Matthews, S.A., and Craft, C.M.: 1982, Exptl. Aging Res., 8, pp. 27-30.
13. Buschman, M.T., Geoffroy, J.S., and LaVelle, A.: 1981. Neurobiol. of Aging, 2, pp. 27-32.
14. Todaro, G.J. and Green, H.: 1963, J. Cell Biol., 17, pp. 299-313.
15. Freeman, A.E. and Igel, H.I.: 1975, In Vitro, 11, pp. 107-116.
16. Sandford, K.K. and Evans, V.J.: 1982, J. Natl. Cancer Inst., 68, pp. 895-913.
17. Kramer, P.M., Travis, G.L., Ray, F.A., and Cram, L.S.: 1983, Cancer Res., 43, pp. 4822-4827.
18. Hayflick, L. and Moorhead, P.S.: 1961, Exptl. Cell Res., 25, pp. 585-621.
19. Hayflick, L.: 1965, Exptl. Cell Res., 37, pp. 614-636.
20. Bruce, S.A., Gyi, K.K., Nakano, S., Ueo, H., Zajac-Kaye, M., and Ts'o, P.O.P.: 1984, "Biochemical Basis of Chemical Carcinogenesis", Greim, H., Jung, R., Kramer, M., Marquardt, H., and Oesch, F., editors, Raven Press, New York, pp. 159-174.
21. Boyer, C.C.: 1968, "The Golden Hamster", Hoffman, R.A., Robinson, P.F., and Magalhaes, H., editors, Iowa State University Press, Ames, pp. 73-89.
22. Bruce, S.A., Deamond, S.F., and Ts'o, P.O.P.: 1985, Mech. Ageing and Develop., in press.
23. Casto, B.C.: 1973, Cancer Res., 33, pp. 402-407.
24. Martin, G.M., Sprague, C.A. and Epstein, C.J.: 1970, Lab. Invest., 23, pp. 86-92.
25. Schneider, E.L. and Mitsui, Y.: 1976, Proc. Natl. Acad. Sci. USA, 73, pp. 3584-3588.
26. Raes, M. and Remacle, J.: 1983, Exptl. Gerontol., 18, pp. 223-240.
27. Raes, M., Geuens, G., de Brabander, M., and Remacle, J.: 1983. Exptl. Gerontol., 18, pp. 241-254.
28. Martin, G.M., Sprague, C.A., Norwood, T.H., and Pendergrass, W.R.: 1974, Am. J. Path., 74, pp. 137-154.
29. Bell, E., Marek, L.F., Levinston, D.S., Merrill, C., Sher, S., Young, I.T., and Eden, M.: 1978, Science, 202, pp. 1158-1163.
30. Ueo, H., Bruce, S.A., Gyi, K.K., Nakano, S., and Ts'o, P.O.P.: 1982, J. Cell Biol., 95, pp. 58a.
31. Diamond, L., O'Brien, T.G., and Rovera, G.: 1977, Nature, 269, pp. 247-248.
32. Cohen, R., Pacifici, M., Rubinstein, N., Biehl, J., and Holtzer, H.: 1977, 266, pp. 538-540.
33. Bruce, S.A., Deamond, S.F., Ueo, H., and Ts'o, P.O.P.: 1983, J. Cell Biol., 97, pp. 346a.
34. Okeda, T., Ueo, H., Bruce, S.A., Bury, M.A., and Ts'o, P.O.P.: 1984, J. Cell Biol., 99, pp. 337a.

35. Barrett, J.C. and Ts'o, P.O.P.: 1978, Proc. Natl. Acad. Sci. USA, 75, pp. 3761-3765.
36. Nakano, S. and Ts'o, P.O.P.: 1981, Proc. Natl. Acad. Sci. USA, 78, pp. 4995-4999.
37. Ueo, H., Bruce, S.A., Nakano, S., and Ts'o, P.O.P.: 1985, J. Cell. Physiol., in press.
38. Nakano, S., Ueo, H., Bruce, S.A., and Ts'o, P.O.P.: 1985, Proc. Natl. Acad. Sci. USA, in press.
39. Silverberg, E.: 1983, CA (Amer. Cancer Society), 33, pp. 9-25.
40. Sachs, L.: 1982, Cancer Surveys, Volume 1.

INDEX